SELF-ORGANIZED ORGANIC SEMICONDUCTORS

SELF-ORGANIZED ORGANIC SEMICONDUCTORS

From Materials to Device Applications

Edited by

Quan Li
Liquid Crystal Institute
Kent State University
Kent, Ohio

A JOHN WILEY & SONS, INC., PUBLICATION

Library of Congress Cataloging-in-Publication Data:

Li, Quan, 1965-
   Self-Organized organic semiconductors : from materials to device applications / edited by Quan Li.
       p. cm.
     Includes index.
     ISBN 978-0-470-55973-4 (hardback)
  1. Organic semiconductors. 2. Self-assembly (Chemistry) 3. Self-Organizing systems.
   TK7871.99.O74S45 2011
   621.3815′2–dc22

                                                    2010036841

Printed in Singapore

10 9 8 7 6 5 4 3 2 1

CONTENTS

9   Selective Molecular Assembly for Bottom-Up Fabrication of Organic Thin-Film Transistors   267

Takeo Minari, Masataka Kano, and Kazuhito Tsukagoshi

Organic semiconductors are attracting tremendous attention because of the promise of low cost and the possibility of roll-to-roll processing at ambient temperature and pressure. Among all organic semiconductors, self-organized organic semiconductors such as large π-conjugated liquid crystals and conjugated block copolymers undoubtedly represent a most exciting material today. The unique self-organized feature offers a brand new scientific frontier that holds immense opportunities as well as challenges in fundamental science that is opening the door for numerous applications such as organic photovoltaics, organic light-emitting diodes and organic field-effect transistors.

This book does not attempt to cover the whole field of self-organized organic semiconductors as this is extremely difficult to cover within a single book. Instead, the book focuses on the most fascinating topics in this field. Here self-organized organic semiconductors including crystal engineering organic semiconductors, conjugated block copolymers and cooligomers, charge transport and its modeling in liquid crystals, self-organized discotic liquid crystals, self-organized smectic liquid crystals, self-assembling of carbon nanotubes, and self-organized fullerene-based organic semiconductors are presented. The self-organized semiconducting materials, characterizations, and principles of devices are described. The applications of high-efficiency organic solar cells using self-organized materials and selective molecular assembly for bottom-up fabrication of organic thin-film transistors are also presented.

This book provides up-to-date and accessible coverage of self-organized semiconductors for graduate students and researchers in organic chemistry, polymer science, liquid crystals, materials science, material engineering, electrical engineering, chemical engineering, optics, optic-electronics, nanotechnology, and semiconductors. It can be used as a database and a reference by readers in both academia and industry. I sincerely hope that all those involved in research and education in this field will find the book to be useful.

Finally, I would like to express my gratitude to Jonathan Rose at John Wiley & Sons, Inc. for inviting us to bring this exciting field of research to a wider audience, and to all our distinguished contributors for their efforts. I am indebted to my wife Changshu and our two boys Daniel and Songqiao for their great support and encouragement.

Kent, Ohio   QUAN LI
June 18, 2010

Liming Dai, Department of Chemical Engineering, Case School of Engineering, Case Western Reserve University, Cleveland, Ohio

Jun-ichi Hanna, Imaging Science and Engineering Laboratory, Tokyo Institute of Technology, Yokohama, Japan

Li-Mei Jin, Liquid Crystal Institute, Kent State University, Kent, Ohio

Masataka Kano, Dai Nippon Printing Co., Ltd., Kashiwa, Chiba, Japan

Paul A. Lane, U.S. Naval Research Laboratory, Washington DC

Quan Li, Liquid Crystal Institute, Kent State University, Kent, Ohio

Yongye Liang, Department of Chemistry and the James Franck Institute, The University of Chicago, Chicago, Illinois

Leonard R. MacGillivray, Department of Chemistry, University of Iowa, Iowa City, Iowa

Ji Ma, Liquid Crystal Institute, Kent State University, Kent, Ohio

Manoj Mathew, Liquid Crystal Institute, Kent State University, Kent, Ohio

Takeo Minari, MANA, NIMS, Tsukuba, Ibaraki, Japan; and RIKEN, Wako, Saitama, Japan

Akira Ohno, Imaging Science and Engineering Laboratory, Tokyo Institute of Technology, Yokohama, Japan

Anatoliy N. Sokolov, Department of Chemistry, University of Iowa, Iowa City, Iowa

Joseph C. Sumrak, Department of Chemistry, University of Iowa, Iowa City, Iowa

Kazuhito Tsukagoshi, MANA, NIMS, Tsukuba, Ibaraki, Japan; AIST, Tsukuba, Ibaraki, Japan; and CREST, JST, Kawaguchi, Saitama, Japan

Luping Yu, Department of Chemistry and the James Franck Institute, The University of Chicago, Chicago, Illinois

CHAPTER 1

Crystal Engineering Organic Semiconductors

JOSEPH C. SUMRAK, ANATOLIY N. SOKOLOV, and LEONARD R. MACGILLIVRAY

Department of Chemistry, University of Iowa, Iowa City, Iowa

1.1. INTRODUCTION

Organic semiconductors are of great interest owing to the promise of low-cost flexible electronics (e.g., RFID tags, displays, e-paper) [1–3]. A variety of conjugated organic polymers and oligomers have been synthesized and studied as semiconductors [4]. Research has demonstrated semiconductors based on oligoacenes or oligothiophenes to be some of the most promising candidates for organic electronics. Pentacene has been one of the most widely studied organic semiconductors and has set a benchmark with room temperature mobilities as high as 35 $cm^2V^{-1}s^{-1}$ for ultrapure single crystals [5]. Oligomers, compared to polymer counterparts, offer samples of high purity and well-defined structure. While both polymer- and oligomer-based materials have been extensively studied for electronics application, the materials show varying mechanisms for charge transport. In oligomers charge-hopping between overlapping wavefunctions of nearest neighboring π-faces dominates the transport mechanism, while intrachain transport is observed in polymeric materials. Thus, the packing of small molecules in the solid state plays a key role in the charge transport properties [6]. In this chapter, we discuss the influence of solid-state packing and the effects of recent attempts to establish control over the placement of molecules on the electronic properties of organic semiconductor solids.

A prerequisite for a molecule to function as an organic semiconductor is the presence of an extended π-conjugated surface. The extended packing of these molecules within single crystals or thin films establishes a degree of overlap

Self-Organized Organic Semiconductors: From Materials to Device Applications, First Edition.
Edited by Quan Li.
© 2011 John Wiley & Sons, Inc. Published 2011 by John Wiley & Sons, Inc.

between neighboring π-faces, which is characterized as the transfer integral. The extent of the overlap plays a great role in the increase of the ease of charge transport within a solid, or charge-carrier mobility. Indeed, studies have shown that cofacial stacking can lead to higher mobilities owing to increased orbital-orbital overlap between neighboring molecules [6, 7]. However, the π-surfaces of most commonly used organic semiconductors (i.e., pentacene and oligothiophene) tend to crystallize in a herringbone, or edge-to-face, motif. Edge-to-face packing is nonideal to achieve maximum performance of an organic semiconductor owing to poor orbital overlap. Thus it is of great value to establish control of π-orbital overlap within semiconductor solids.

Crystal engineering is the use of intermolecular interactions to assemble molecules into a specific solid-state arrangement to achieve desired physical and chemical properties [8]. Examples of such intermolecular forces include lipophilic, dipolar, and quadrupole interactions, as well as hydrogen bonding. Control of dimensionality in the solid state with noncovalent bonds has been realized through the formation of zero-dimensional (0-D), 1-D, 2-D, and 3-D assemblies [8]. Recently, great interest has developed in the utilization of crystal engineering as a bottom-up approach to achieve increased overlap of molecular orbitals between neighboring semiconductor molecules. Improvement in the control of intermolecular interactions could also be used to stabilize the lattice of the transport media, resulting in increased maximum charge-carrier mobilities [9].

With these ideas in mind, this chapter discusses crystal engineering strategies in the context of semiconductor solids. It is first important to understand the nature of the structural problem that essentially plagues oligomer-based semiconductors (i.e., crystal packing). From there, strategies are described that utilize a range of interactions from steric crowding to circumvent edge-to-face packing to attractive forces (e.g., lipophilic) to enforce face-to-face geometries. We also describe a modular approach developed in our laboratory that achieves face-to-face stacking through hydrogen bonding in the form of molecular cocrystals. It should be noted that while the strategies described herein may be applicable for thin-film devices, the structure of a thin film will not necessarily correlate to that of a single crystal [10].

1.2. PACKING AND MOBILITY OF ORGANIC SEMICONDUCTORS

Organic semiconductors consist of one or more classes of molecular species that possess an aromatic structure (e.g., acenes, thiophenes) (Fig. 1-1). These materials typically begin to demonstrate field effect transistor (FET) mobilities upon reaching four units of conjugation. Molecules based on the aromatic moieties can be either covalently linked through single bonds (e.g., α-linked, β-linked thiophenes) or fused together (e.g., pentacene). Initial research on transport in organic semiconductor solids focused on the unsubstituted derivatives of the aromatic molecules. Indeed, knowledge of the packing and structure of unsubstituted organic semiconductors is generally required to draw comparisons to substituted products.

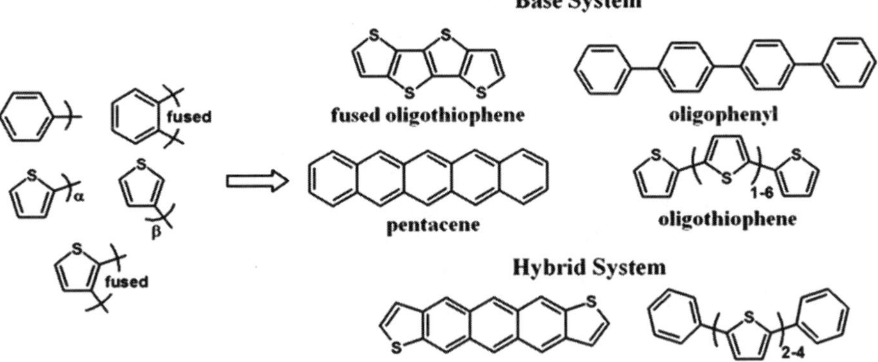

Figure 1-1 Base and hybrid organic semiconductors.

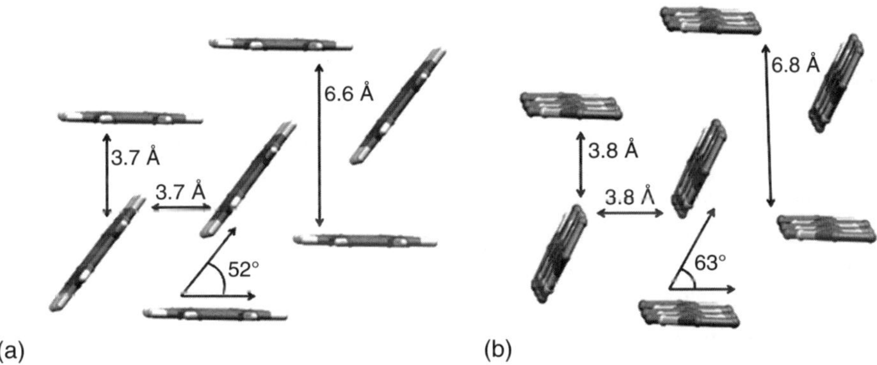

(a)   (b)

Figure 1-2 Crystal packing: a) pentacene and b) sexithiophene.

Of the oligoacenes, tetracene and pentacene have been the main focuses of research. Pure tetracene and pentacene pack in a similar manner (Fig. 1-2a) [11]. Both molecules exhibit herringbone packing that is directed by C–H$\cdots\pi$ interactions, with an angle approximately $52°$ between the planar surfaces of nearest-neighbor molecules. The shortest carbon-carbon (C–C) distance between pentacene and tetracene molecules in the solid state is approximately 3.7 Å, with separations on the order of 6.6 Å being between columns. Reported mobilities for pentacene are as high as 35 $cm^2V^{-1}s^{-1}$ for extremely pure single crystals [5]. Without rigorous purification, mobilities of approximately 3 $cm^2V^{-1}s^{-1}$ have been reported [12]. The highest reported mobility for a single crystal of unsubstituted tetracene is 1.3 $cm^2V^{-1}s^{-1}$ [13].

Oligothiophenes are of great interest owing to ease of synthesis and modifications. Thus far, the solid-state packings for all unsubstituted oligothiophenes up to octithiophene have been determined via single-crystal X-ray diffraction [14–17]. As with the oligoacene counterparts, oligothiophenes exhibit herringbone packing

Table 1-1   Mobilities of [n]-Oligothiophenes.

Number of Thiopheners	4	5	6	7	8
Mobility ($cm^2V^{-1}s^{-1}$)	0.006	0.08	0.075	0.17	0.33

in the solid state (Fig. 1-2b). The angle between planar surfaces of oligothiophenes is typically $63°$, which is slightly higher than the oligoacenes. The
shortest distance between two thiophene rings is on the order of 3.8 Å. A distance
of approximately 6.8 Å separates thiophenes in the column of the herringbone
structure. As expected, an increase in the number of thiophene rings in an oligothiophene tends to lead to an increase in mobility (Table 1-1).

1.3. OVERCOMING THE PACKING PROBLEM

The solid-state structures of unsubstituted organic semiconductors are not ideally
suited to achieve high mobilities. To improve mobilities, the structures must be
designed to produce more efficient π-π overlap. To this end, attempts to control
the arrangements of organic semiconductors have typically involved modifying
the intermolecular forces that govern the solid-state packing. Changes to packing
have been accomplished through the introduction, or modification, of substituents
along the periphery of semiconductor molecules. Steric interactions using bulky
groups, for example, have been used to destabilize edge-to-face $C-H \cdots \pi$ forces.
The introduction of bulky groups to obtain reduced π-π separations may appear
counterintuitive; however, the approach has been reliable. Electronic interactions
can also be used to compete with $C-H \cdots \pi$ interactions. Quadrupole and dipolar
interactions, as well as hydrogen bonds, have all been used to improve π-stacking.

1.3.1. Steric Interactions to Prevent $C-H \cdots \pi$ Interactions

One method to improve packing has been the use of bulky side groups to destabilize or eliminate close $C-H \cdots \pi$ contacts. $C-H \cdots \pi$ forces are based on electrostatic interactions between the δ^+ charges of H-atoms located on the edges of
the rings and the δ^- charges on the π-face of the internal ring C-atoms. With the
driving force for edge-to-face packing eliminated, there is an increased likelihood
for face-to-face stacking. The approach has been developed by Anthony et al.,
where a pentacene functionalized with triisopropylsilylethynyl (TIPS) groups not
only prevented herringbone packing but improved solubility and stability [18, 19].
With a reduction in $C-H \cdots \pi$ interactions achieved, the TIPS-functionalized pentacene packed in a brick motif based on cofacial columns with a 3.5-Å separation
between nearest-neighbor acenes (Fig. 1-3). Whereas such changes to packing
led to a significant improvement in π-orbital overlap compared to unsubstituted
pentacene, control of a slip-stack arrangement of the pentacenes on a long molecular axis remained difficult to achieve [20]. Mobilities as high as 1.8 $cm^2V^{-1}s^{-1}$
were obtained for TIPS-functionalized pentacenes prepared via solution process

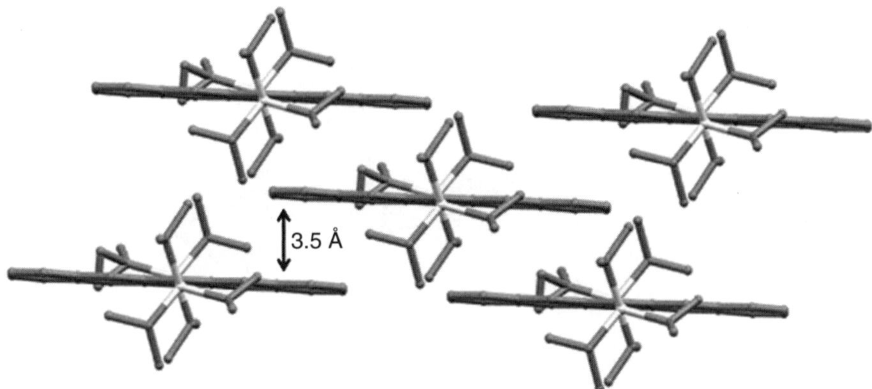

Figure 1-3 Slip-stacked packing of triisopropylsilylethynyl (TIPS) functionalized pentacene.

methods [21]. Anthony et al. have also successfully extended the method to hexacene and heptacene [22]. The method has been highly successful in both single crystals and thin-film devices, which demonstrates the robustness of steric interactions for potential applications in organic electronics.

Along with TIPS groups, phenyl groups have been used to reduce C–H$\cdots\pi$ interactions. Ultrapure single crystals of 5,6,11,12-tetraphenyltetracene, or rubrene, have afforded a hole mobility of 20 cm^{2}V^{-1}s^{-1} [23]. The mobility is significantly higher than that for single crystals of the parent tetracene [24]. The increase in mobility can be attributed to the increase in cofacial π-stacked interactions along the a-axis in single crystals of the molecule [25]. Mobilities of rubrene in thin films, however, have underperformed single crystals owing to difficulties in achieving crystalline films of the compound.

Nuckolls et al. have extended phenyl substitution to pentacene with a series of cruciform-based π-systems; specifically, with 6-phenyl-, 6,13-diphenyl-, 6,13-dithienyl-, 5,7,12,14-tetraphenyl-, 1,4,6,8,11,13-hexaphenyl-, and 1,2,3,4,6,8,9,10,11,13-decaphenylpentacene [26]. The phenyl substituents circumvented herringbone packing; however, the packing motifs varied depending upon number of phenyl substitutents. A single phenyl substitutent resulted in both edge-to-face and face-to-face packing in the solid. Hexaphenyl- and decaphenylpentacene packed in layered structures directed by edge-to-face C–H$\cdots\pi$ interactions between the phenyl substituents with the pentacene backbones being separated by 5 Å. The resulting solids did not display measurable mobilities as thin films (Fig. 1-4a). The diphenyl-substituted pentacene packed cofacially; however, the long axis of nearest-neighbor acenes were oriented orthogonally and resulted in cagelike superstructures. The solid exhibited a mobility of 8 × 10^{-5} cm^{2}V^{-1}s^{-1} (Fig. 1-4b). A mobility of 0.1 cm^{2}V^{-1}s^{-1} was obtained for the dithienyl derivative, which exhibited the greatest π-π overlap (Fig. 1-4c). Each of these methods circumvented C–H$\cdots\pi$ interactions to vary the packing

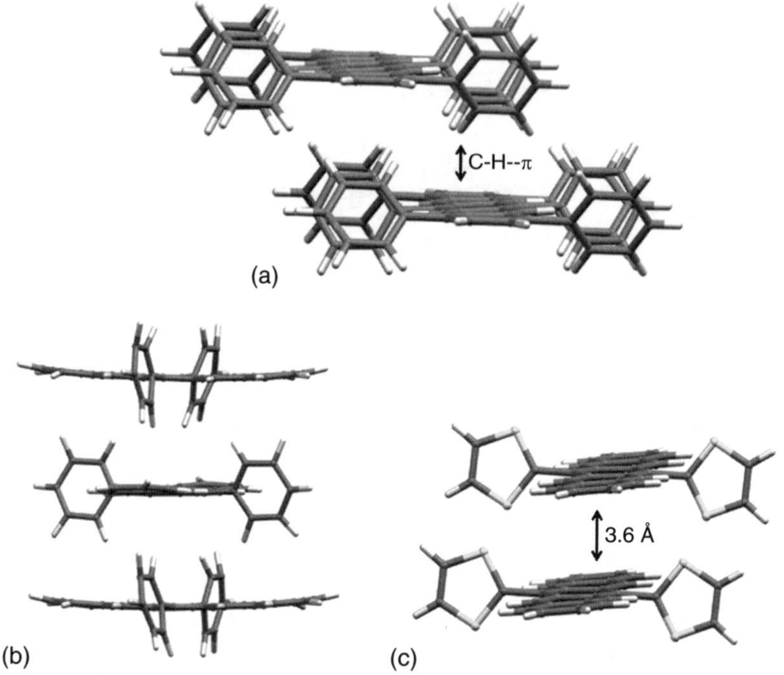

Figure 1-4 X-ray structures of layered packings: a) hexylphenyl-, b) diphenyl-, c) dithienyl-pentacene.

motifs. The lack of mobility in hexaphenyl- and decaphenyl-substituted pentacene demonstrated the importance of π-π overlap to improve mobility.

1.3.2. Alkyl-Alkyl Interactions

Alkyl chains are known to promote layered structures in the solid state through lipophilic interactions. Garnier et al. synthesized 2,5-dihexylsexithiophene (DH6T) to incorporate the self-assembly of alkyl chains into thin films of the semiconductor molecule [27]. Although X-ray diffraction (XRD) revealed the thin film to exhibit herringbone packing similar to unsubstituted sexithiophene (6T), mobility measurements showed a 40-fold increase compared to 6T when both were deposited at room temperature (6T = 2×10^{-3} cm^{2}V^{-1}s^{-1}, DH6T = 8×10^{-2} cm^{2}V^{-1}s^{-1}) [28]. The increase in thin-film mobility in α-alkyl-substituted oliogthiophenes was attributed to an increase in long-range order owing to a change in the growth mechanism of the thin film from a 3-D island to a more 2-D layered structure (Fig. 1-5) [29]. Halik et al. compared diethyl, dihexyl, and didecyl α, ω-substituted sexithiophenes to study the impact of alkyl substituent length on mobility. Diethyl- and dihexyl-substituted sexithiophene were determined to exhibit superior thin-film mobilities compared to sexithiophene, with the respective mobilities being

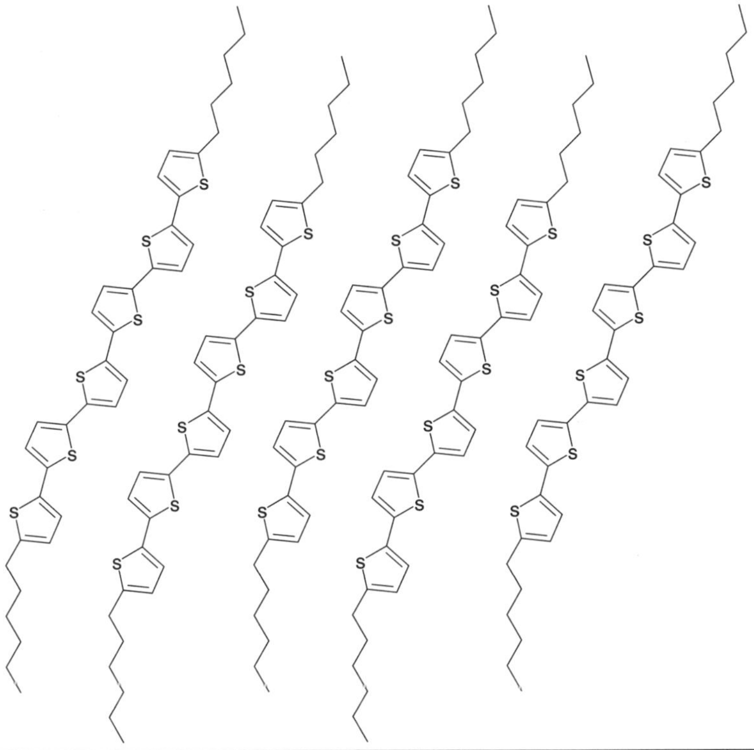

Figure 1-5 Schematic representation of DH6T thin film.

1.1 and 1.0 $cm^2V^{-1}s^{-1}$. α,ω-Didecylsexithiophene, however, showed little improvement versus sexithiophene, with mobilities of 0.1 and 0.07 $cm^2V^{-1}s^{-1}$, respectively, suggesting an optimum alkyl substitution length [30].

More recent studies on alkyl-substituted organic semiconductors have involved a hybrid phenyl-thiophene system. Marks et al. demonstrated entirely solution-processed devices composed of 5,5′-bis(4-*n*-hexylphenyl)-2,2′-bithiophene (dHPTTP) that exhibited higher than expected mobilities (0.07 $cm^2V^{-1}s^{-1}$) for a system with a relatively small conjugated π-system [31]. Single-crystal studies on dHPTTP revealed the molecule to exhibit herringbone packing similar to most oligothiophenes (Fig. 1-6). The improvement in mobility was attributed to a decrease in traps in the thin film owing to alkyl-alkyl interactions similar to DH6T [32]. In related work, Bao et al. have synthesized a series of alkyl- and alkoxy-substituted bisphenylbithiophenes (PTTP) and bisphenylterthiophenes (PTTTP) to study side chain effects on mobility. It was discovered that branching in the side chain leads to a decrease in mobility. The decrease was attributed to increased steric interactions to an extent that prevented close packing and efficient π-π overlap. The growth modes of the branched substituents were also more 3-D, as opposed to the linear

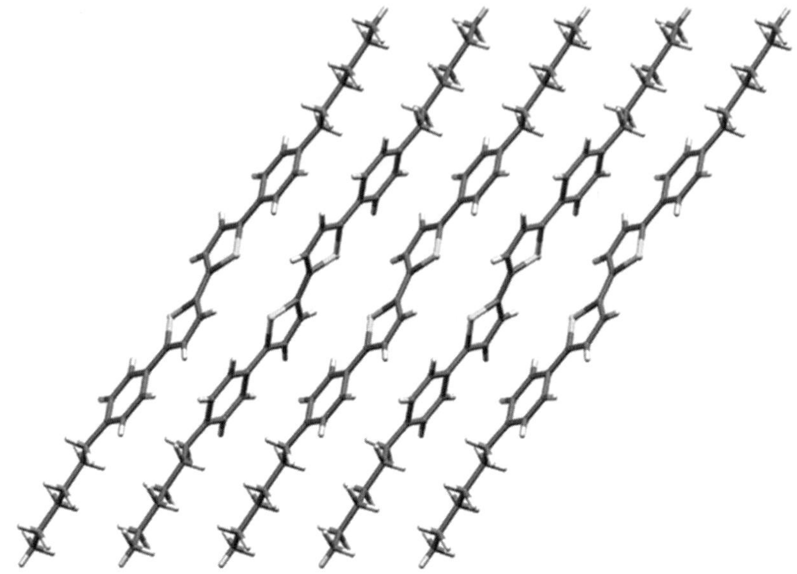

Figure 1-6 Crystal structure of dHPTTP.

chains, further contributing in the lowering of mobilities and making it difficult to decouple steric from growth mode effects. When the semiconductor core was changed from PTTP to PTTTP, the mobility peaked at a shorter side chain length, suggesting an optimal molecular dimension for charge transport [33]. The changes in alkyl lengths played an important role in altering the molecular packing, and thus can be considered a tool to gauge finer effects of packing on charge transport.

1.3.3. Dipolar Interactions

The introduction of dipolar interactions within an organic semiconductor molecule provides a means to enforce face-to-face π-π stacking arrangements. The arrangements are based on the stacked molecules adopting either an antiparallel or a parallel geometry. In this context, Kobayashi et al. have introduced chalcogens (O, S, Te) at the 9,10-position of anthracene and alkylthio groups at the 6,13-position of pentacene to induce π-π stacking through chalcogen-chalcogen interactions [34]. 9,10-Dimethoxyanthracene was determined to pack into a herringbone motif with a lack of heteroatom interactions. However, 9,10-bis(methylthio)-anthracene crystallized into 2-D sheets, directed by S$\cdots$S interactions. Anthracene rings of adjacent stacked sheets formed 1-D π-stacked columns where the S$\cdots$S interactions were preserved. The same packing motif was observed in the case of 6,13-bis(methylthio)pentacene (Fig. 1-7) [35]. 9,10-bis(methyltelluro)anthracene also formed sheets through Te$\cdots$Te interactions [34]. Mobility measurements were not reported for the compounds.

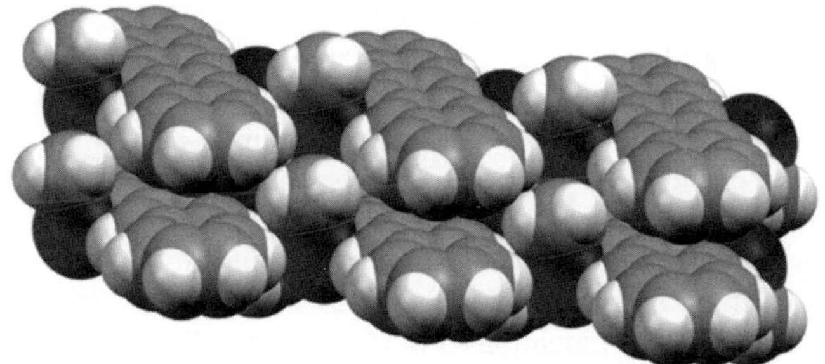

Figure 1-7   Crystal structure of 6,13-bis(methylthio)pentacene.

Halogen-halogen interactions have been investigated by Bao and colleagues to increase π-π stacking in crystals of 5-chlorotetracene (CT), 5,11-dichlorotetracene (DCT), and 5-bromotetracene (BT) [24]. CT and BT exhibited isostructural herringbone arrangements, while DCT adopted a face-to-face slipped π-stacking motif with an intermolecular distance of 3.48 Å (Fig. 1-8). Crystals grown by vapor growth methods for BT exhibited a mobility of 0.3 cm^{2}V^{-1}s^{-1}. Crystals of DCT grown by the same method exhibited a mobility of 1.6 cm^{2}V^{-1}s^{-1}, which is the highest reported for tetracene. The increase in mobility in the case of DCT relative to tetracene was attributed to the enhanced π-orbital overlap in the mode of packing.

1.3.4. Quadrupole Interactions

Molecular complexes between electron-rich and electron-poor aromatic rings have been known since 1960 [36]. Self-assembly via quadrupole interactions has been applied in supramolecular construction since the late 1990s [37, 38]. In 2003, Marks and colleagues reported organic semiconductors that incorporate quadrupole interactions. Two fluoroarene units were incorporated with four thiophene units to make three symmetrical molecules, specifically, 2,5′′′-bis(2,3,4,5,6-pentafluorophenyl)-quaterthiophene,   5,5′-bis[2,3,5,6-tetrafluoro-4-

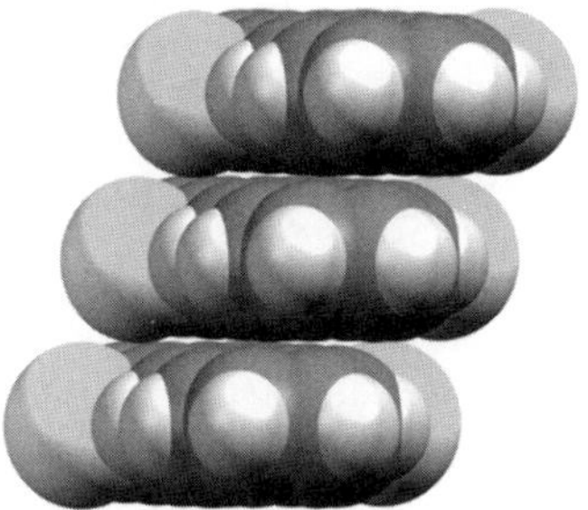

Figure 1-8   Packing of DCT.

(2-thienyl)phenyl]-bithiophene, and 5,5″-(2,2′,3,3′,5,5′,6,6′-octafluoro[1,1′-biphenyl]-4,4′-diyl)bisbithiophene [39]. All three molecules exhibited cofacial packing of the electron-rich thiophene units with the electron-deficient fluoroarene units. The compounds in which the fluoroarene units were spaced by a quaterthiophene or bithiophene space displayed cofacial π-π stacking distances of 3.20 Å and 3.37 Å, respectively. The distances were remarkably short for thiophene-based oligomers (Fig. 1-9). The compound in which the fluoroarene units were adjacent to each other exhibited a torsion angle of approximately 54°, which disrupted the π-conjugation (Fig. 1-9c). Mobilities of 0.08, 0.01, and 4×10^{-5} $cm^2V^{-1}s^{-1}$ were reported for the fluoroarenes with the quaterthiophene spacer, bithiophene spacer, and no spacer, respectively, which generally compare to that of α-6T (0.03 $cm^2V^{-1}s^{-1}$) [39].

Anthony et al. have integrated quadrupole interactions into TIPS-based pentacenes to improve packing [40]. In particular, changing one or both of the end benzene rings in pentacene to TIPS-tetrafluoropentacene or TIPS-octafluoropentacene led to increased face-to-face π-interactions. Both compounds adopted similar 2-D π-stacked arrangements similar to TIPS pentacene. The compounds were stable in solution and the solid state when exposed to air and light. The intermolecular π-stacking distances decreased with an increase in the number of fluoro groups, with the average intermolecular distances being reduced from

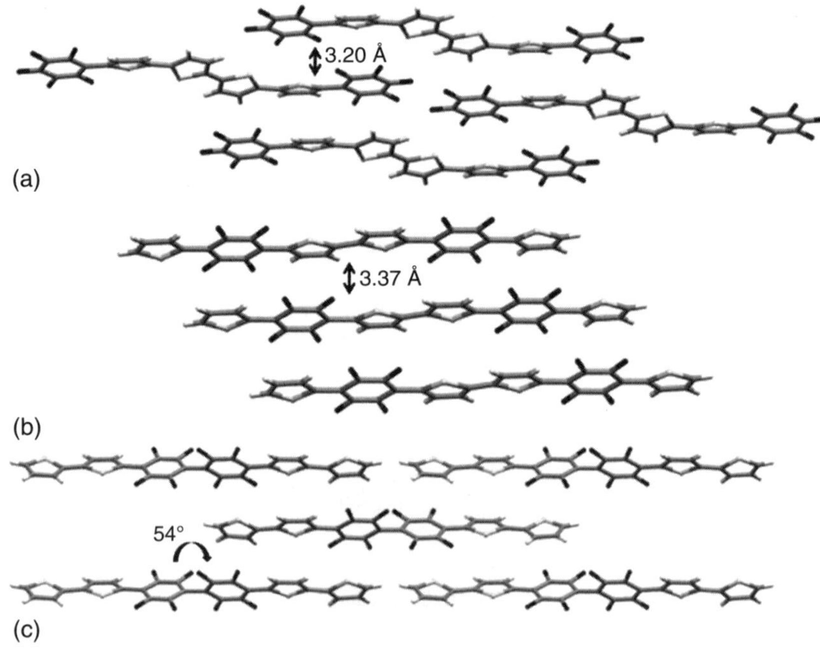

Figure 1-9 Packing: a) 2,5‴-bis(2,3,4,5,6-pentafluorophenyl)-quaterthiophene, b) 5,5′-bis[2,3,5,6-tetrafluoro-4-(2-thienyl)phenyl]-bithiophene, c) 5,5″-(2,2′,3,36′,5,5′,6,6′-octafluoro[1,1′-biphenyl]-4,4′-diyl)bisbithiophene.

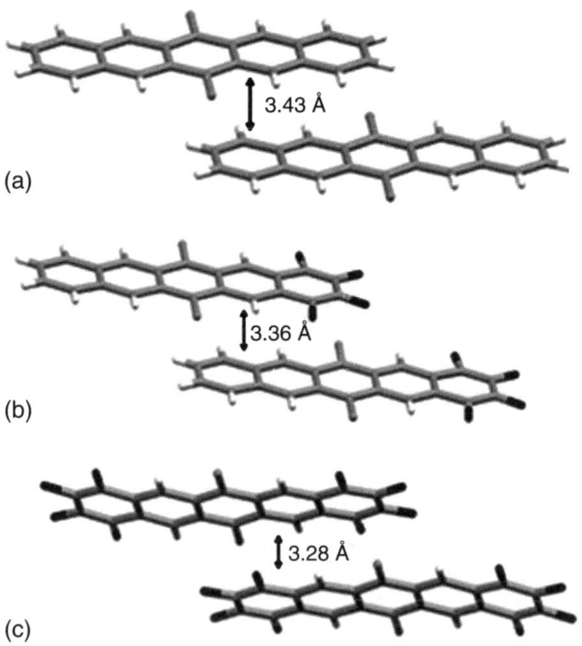

Figure 1-10 Stacking of a) TIPS-pentacene, b) TIPS-tetrafluoropentacene, c) TIPS-octafluoropentacene (TIPS groups removed for clarity).

3.43 Å in the nonfluorinated compound to 3.36 Å in tetrafluoro and 3.28 Å in octaflouro compound (Fig. 1-10). However, in all cases the slip-stacked arrangement along the long axis of the pentacene molecules was prevalent. Nevertheless, the reduction in the intermolecular spacing was accompanied by an increase in mobility from 0.001 to 0.014 to 0.045 $cm^2V^{-1}s^{-1}$ for TIPS-pentacene, TIPS-tetrafluoropentacne, and TIPS-octafluoropentacene when prepared under the same evaporation conditions, respectively. A similar introduction of perfluoroarene groups to improve stacking has been demonstrated by Swager et al. involving tetracene derivatives [41] and Watson et al. using polycyclic aromatics [42].

Nuckolls et al. have used quinone moieties to achieve a quadrupole interaction that improved stacking. The syntheses, packings, and mobilities of three acene-quinone compounds, namely, 2,3-dimethyl-1,4-hexacene-quinone, 2,3-dimethyl-1,4-pentacene-quinone, and 5,6-hexacene-quinone were described. Single crystals of the pentacene-quinone revealed a head-to-tail arrangement with an intermolecular distance of 3.25 Å between π-surfaces and a mobility of 2×10^{-3} $cm^2V^{-1}s^{-1}$ was obtained for thin films based on pentacene-quinone. Single crystals for dimethyl hexacene-quinone were determined to be unsuitable for X-ray diffraction; however, a mobility of 5.2×10^{-2} $cm^2V^{-1}s^{-1}$ with an on/off ratio of greater than 10^6 was obtained for thin films. Experimental data suggested that dimethyl hexacene-quinone adopted the same head-to-tail arrangement as the pentacene-quinone (Fig. 1-11) [26]. Chang et al. have also recently

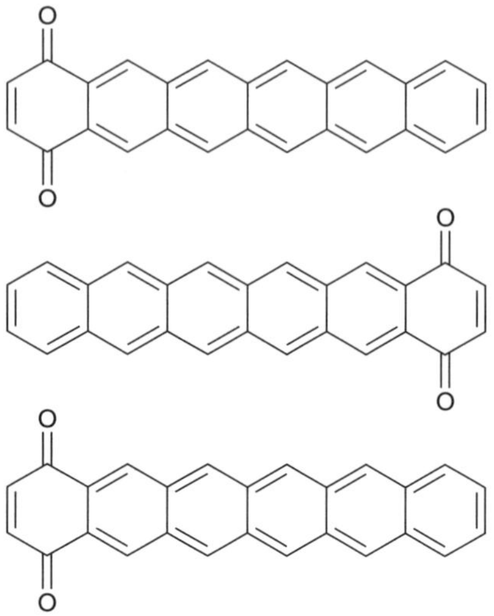

Figure 1-11   Head-to-tail orientation of 5,6-hexacene-quinone.

described a donor-acceptor interaction integrated into a thiophene. In particular, the solid-state structure of 2,5-di(pyrimidin-5-yl)thieno[3,2-*b*]thiophene revealed π-stacking achieved via the stacked electron-rich S-atom of the thiophene and the electron-poor C-atoms of the pyrimidine [43]. The use of such quadrupolar interactions is promising to control π-π arrangements; however, low mobilities may be attributed to the electronegative substituents (i.e., O-atom) along the aromatic core. The direct placement may cause a trapping effect, effectively lowering the observed performance.

1.3.5. Hydrogen Bonding

The first study that incorporated hydrogen bonds as a means to affect the self-assembly of an organic semiconductor in the solid state was reported by Barbarella et al. in 1996. 2-Hydroxyethyl groups were added the to the backbones of 2T, 4T, and 6T in the forms of 3,3′-bis(2-hydroxyethyl)-2,2′-bithiophene, 3,3′,4″,3‴-tetrakis(2-hydroxyethyl)-2,2′ : 5′,2″ : 5″,2‴-quaterthiophene, and 3,3, 4″,3‴,4⁗,3‴‴‴-hexakis(2-hydroxyethyl)-2,2′ : 5′,2″ : 5″,2‴ : 5‴,2⁗ : 5⁗,2‴‴‴-sexi-thiophene. Although a crystal structure of the bithiophene revealed intermolecular O−H···O hydrogen bonds, a large twist angle of approximately $67°$ between thiophene units of the backbone effectively disrupted the conjugation of the molecule (Fig. 1-12) [44] and circumvented the formation of face-to-face stacking. The bithiophenes, however, packed into a columnar arrangement. Crystal structures were not obtained for the quaterthiophene or sexithiophene derivative. UV adsorption studies involving the longer derivatives suggested each to exhibit a largely twisted conformation, a nonideal conformation for achieving high mobility, in solution similar to the bithiophene derivative.

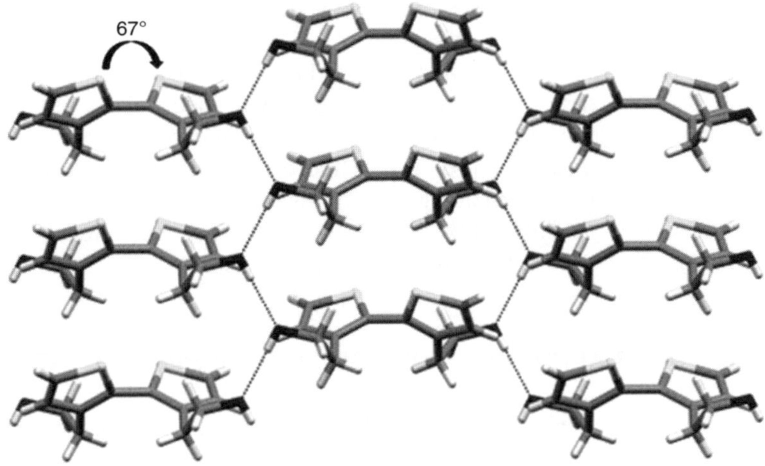

Figure 1-12 X-ray structure of hydrogen-bonded network of 3,3′-bis(2-hydroxyethyl)-2,2′-bithiophene.

Feringa et al. have synthesized a series of bisurea compounds with a thiophene spacer in an attempt to form infinite 1-D π-stacked columns. Specifically, 2,5-di(4-(3-dodecyl-ureido)butyl)thiophene and 5,5′-di(4-(3-dodecylureido)butyl)-2,2′-bithiophene were described. Organogels formed upon cooling of each compound to room temperature in tetralin and 1,2-dichloroethane. Characterization of the gels by infrared spectroscopy, scanning electron microscopy, and powder X-ray diffraction revealed the formation of lamellar fibers formed through the hydrogen-bonded urea groups. It was suggested that the lamellar arrangement of the bithiophene derivative corresponds to cofacially stacked bithiophene moieties with an intermolecular distance on the order of 3.0 Å (Fig. 1-13). Conductive properties of the bisurea compounds

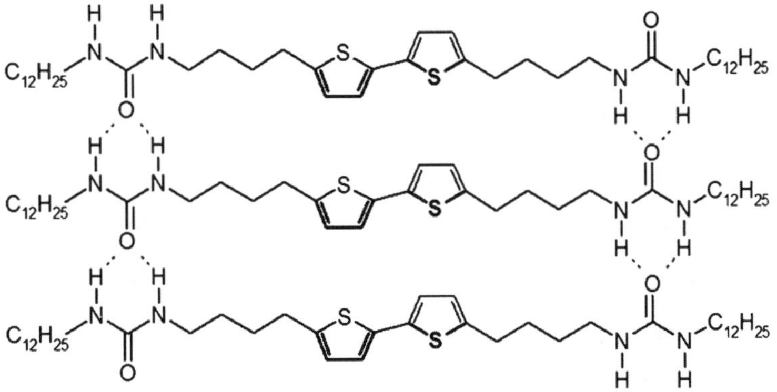

Figure 1-13 Schematic representation of hydrogen-bonding of 5,5′-di(4-(3-dodecylureido) butyl)-2,2′-bithiophene.

were studied via pulse-radiolysis time-resolved microwave conductivity. The bithiophene bisurea displayed a mobility of $(5 \pm 0.2) \times 10^{-3}$ cm^{2}V^{-1}s^{-1}, which is higher than unsubstituted quarterthiophene $(1 \times 10^{-3}$ cm^{2}V^{-1}s$^{-1})$ when studied by the same technique [45]. The growth mechanisms of the fibers from solution and electronic properties were further probed by Rep et al. and DeSchryver et al., respectively [46, 47]. While the mobilities have not resulted in devices, the finding is an encouraging step toward the use of hydrogen bonds to direct molecular self-assembly where electronegative substituents do not appear to have an adverse effect on transport properties.

1.4. A MODULAR APPROACH TOWARD ENGINEERING π-STACKING

The approaches discussed above achieve face-to-face π-stacking by introducing or modifying a substituent on the backbone of a semiconductor molecule. The methods share a common theme of utilizing interactions provided by covalently attached substituents to direct molecular packing. Recently, we have described how face-to-face stacking of semiconductor molecules can be achieved in two-component solids, or cocrystals. The method involves cocrystallizing a semiconductor molecule with a second molecule, termed the semiconductor cocrystal former (SCCF), which is designed to enforce, via intermolecular bonds, the stacking of the semiconductor in a face-to-face arrangement [48]. The key to the cocrystal approach is that the SCCF effectively decouples the organization of the semiconductor within the crystal from the effects of long-range crystal packing. More specifically, the bifunctional SCCF utilizes hydrogen bonds to enforce stacking of acenes and thiophenes functionalized with appropriate molecular recognition sites (Fig. 1-14). Using a series of 1,3-dihydroxybenzenes, or resorcinols, as SCCFs, the approach has been used to successfully enforce π-π stacking between an anthracene and a thiophene within cocrystals of 2(5-iodoresorcinol)·2(9,10-bis(4-pyridylethynyl)anthracene) and 2(5-methylresorcinol)·2(2,5-bis(4-pyridylethynyl)thiophene), respectively. The acene and thiophene were forced into face-to-face stacks with separation

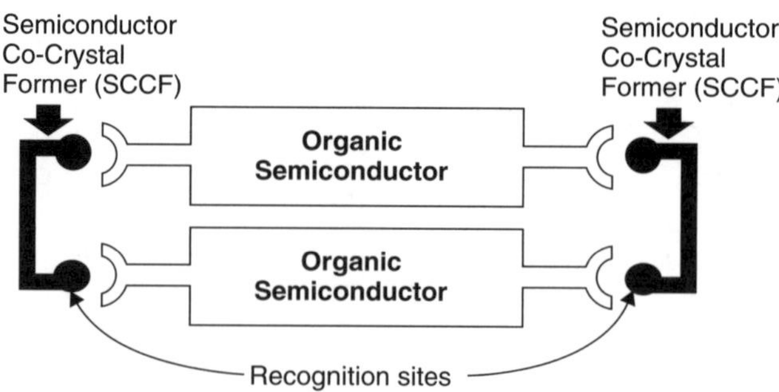

Figure 1-14 Scheme of cocrystallization of organic semiconductor and SCCF.

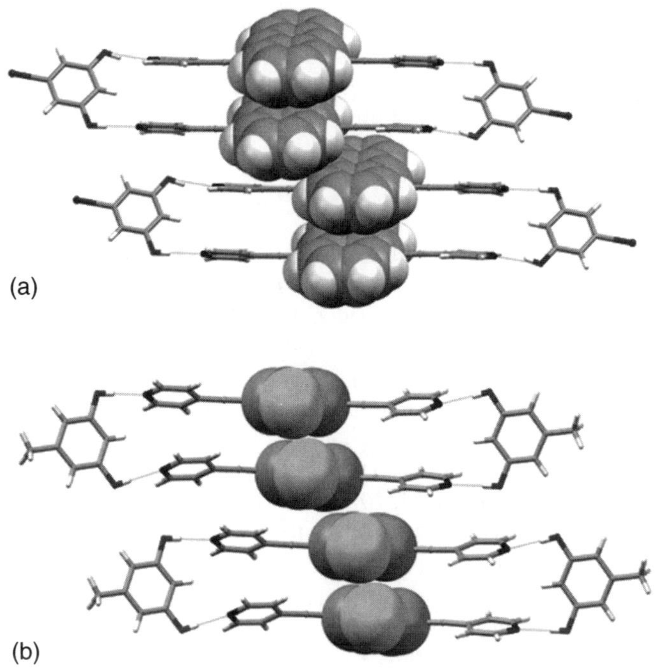

Figure 1-15 X-ray structures: a) 2(5-iodoresorcinol)·2(9,10 bis(4-pyridylethynyl) anthracene), b) 2(5-methylresorcinol)·2(2,5-bis(4-pyridylethynyl)thiophene). (A full color version of this figure appears in the color plate section.)

distances of 3.44 Å and 3.56 Å, respectively (Fig. 1-15). The hydrogen-bonded assemblies were shown to self-assemble into face-to-face arrangements, which enabled the formation of extended π-stacked structures. We are working to expand the components of the modular approach to semiconductors of increasing complexity and determine the mobilities of the resulting solids.

1.5. CONCLUSION

In this chapter, we have shown that crystal engineering can be used to improve the π-overlap of organic semiconductor molecules in the solid state. This goal is accomplished by several approaches, from disrupting C–H···π interactions that dictate edge-to-face arrangements, to enhancing intermolecular forces that overcome edge-to-face interactions, supporting face-to-face stacking. Face-to-face stacking typically results in higher mobilities in solids. Steric interactions have been shown to disfavor herringbone stacking, while attractive forces based on lipophilic, dipolar, quadrupolar, and hydrogen bonding interactions have overcome effects of long-range packing, resulting in face-to-face stacking arrangements. Whereas virtually all approaches have relied on covalent modification of a semiconductor substituent, a new approach describes a method that employs a second component, in the form of a cocrystal, to direct face-to-face stacking. The approaches to date have been highly encouraging, and we anticipate that crystal engineering will continue to provide steps that lead to rational design of semiconductor solids with increasing mobilities and novel applications.

REFERENCES

1. C. J. Drury, C. M. J. Mutsaers, C. M. Hart, M. Matters, and D. M. de Leeuw. Low-cost all-polymer integrated circuits. *Appl. Phys. Lett.* **1998**, *73*, 108–110.

2. C. Reese, M. Roberts, M.-m. Ling, and Z. Bao. Organic thin film transistors. *Mater. Today* **2004**, *7*, 20–27.

3. H. E. Katz, Z. Bao, and S. L. Gilat. Synthetic chemistry for ultrapure, processable, and high-mobility organic transistor semiconductors. *Acc. Chem. Res.* **2001**, *34*, 359–369.

4. A. R. Murphy and J. M. J. Frechet. Organic semiconducting oligomers for use in thin film transistors. *Chem. Rev.* **2007**, *107*, 1066–1096.

5. O. D. Jurchescu, J. Baas, and T. T. M. Palstra. Effect of impurities on the mobility of single crystal pentacene. *Appl. Phys. Lett.* **2004**, *84*, 3061–3063.

6. J. Cornil, D. Beljonne, J. P. Calbert, and J. L. Bredas. Interchain interactions in organic pi-conjugated materials: Impact on electronic structure, optical response, and charge transport. *Adv. Mater.* **2001**, *13*, 1053–1067.

7. G. R. Hutchison, M. A. Ratner, and T. J. Marks. Intermolecular charge transfer between heterocyclic oligomers. Effects of heteroatom and molecular packing on hopping transport in organic semiconductors. *J. Am. Chem. Soc.* **2005**, *127*, 16866–16881.

8. G. R. Desiraju. Supramolecular synthons in crystal engeniring—a new organic-synthesis. *Angew. Chem. Int. Ed. Engl.* **1995**, *34*, 2311–2327.

9. C. D. Dimitrakopoulos and P. R. L. Malenfant. Organic thin film transistors for large area electronics. *Adv. Mater.* **2002**, *14*, 99–117.

10. S. H. Liu, W. C. M. Wang, A. L. Briseno, S. C. E. Mannsfeld, and Z. N. Bao. Controlled deposition of crystalline organic semiconductors for field-effect-transistor applications. *Adv. Mater.* **2009**, *21*, 1217–1232.

11. D. Holmes, S. Kumaraswamy, A. J. Matzger, and K. P. C. Vollhardt. On the nature of nonplanarity in the N phenylenes. *Chemistry* **1999**, *5*, 3399–3412.

12. T. W. Kelley, L. D. Boardman, T. D. Dunbar, D. V. Muyres, M. J. Pellerite, and T. Y. P. Smith. High-performance OTFTs using surface-modified alumina dielectrics. *J. Phys. Chem. B* **2003**, *107*, 5877–5881.

13. C. Goidmann, S. Haas, C. Krellner, K. P. Pernstich, D. J. Gundlach, and B. Batlogg. Hole mobility in organic single crystals measured by a "flip-crystal" field-effect technique. *J. Appl. Phys.* **2004**, *96*, 2080–2086.

14. R. Azumi, M. Goto, K. Honda, and M. Matsumoto. Conformation and packing of odd-numbered alpha-oligothiophenes in single crystals. *Bull. Chem. Soc. Jpn.* **2003**, *76*, 1561–1567.

15. D. Fichou, B. Bachet, F. Demanze, I. Billy, G. Horowitz, and F. Garnier. Growth and structural characterization of the Quasi-2D single crystal of alpha-octithiophene. *Adv. Mater.* **1996**, *8*, 500–504.

16. T. Siegrist, C. Kloc, R. A. Laudise, H. E. Katz, and R. C. Haddon. Crystal growth, structure, and electronic band structure of alpha-4T polymorphs. *Adv. Mater.* **1998**, *10*, 379–382.

17. G. Horowitz, B. Bachet, A. Yassar, P. Lang, F. Demanze, J.-L. Fave, and F. Garnier. Growth and characterization of sexithiophene single crystals. *Chem. Mater.* **2002**, *7*, 1337–1341.

18. J. E. Anthony, J. S. Brooks, D. L. Eaton, and S. R. Parkin, Functionalized pentacene: Improved electronic properties from control of solid-state order. *J. Am. Chem. Soc.* **2001**, *123*, 9482–9483.

19. J. E. Anthony. Functionalized acenes and heteroacenes for organic electronics. *Chem. Rev.* **2006**, *106*, 5028–5048.

20. J. E. Anthony, D. L. Eaton, and S. R. Parkin. A road map to stable, soluble, easily crystallized pentacene derivatives. *Org. Lett.* **2002**, *4*, 15–18.

21. S. K. Park, T. N. Jackson, J. E. Anthony, and D. A. Mourey. High mobility solution processed 6,13-bis(triisopropyl-silylethynyl) pentacene organic thin film transistors. *Appl. Phys. Lett.* **2007**, *91*, 063514–063517.

22. M. M. Payne, S. R. Parkin, and J. E. Anthony. Functionalized higher acenes: Hexacene and heptacene. *J. Am. Chem. Soc.* **2005**, *127*, 8028–8029.

23. V. Podzorov, E. Menard, A. Borissov, V. Kiryukhin, J. A. Rogers, and M. E. Gershenson. Intrinsic charge transport on the surface of organic semiconductors. *Phys. Rev. Lett.* **2004**, *93*, 086602–086606.

24. H. Moon, R. Zeis, E. J. Borkent, C. Besnard, A. J. Lovinger, T. Siegrist, C. Kloc, and Z. N. Bao. Synthesis, crystal structure, and transistor performance of tetracene derivatives. *J. Am. Chem. Soc.* **2004**, *126*, 15322–15323.

25. J. Cornil, J. P. Calbert, and J. L. Bredas. Electronic structure of the pentacene single crystal: Relation to transport properties. *J. Am. Chem. Soc.* **2001**, *123*, 1250–1251.

26. Q. Miao, X. L. Chi, S. X. Xiao, R. Zeis, M. Lefenfeld, T. Siegrist, M. L. Steigerwald, and C. Nuckolls. Organization of acenes with a cruciform assembly motif. *J. the Am. Chem. Soc.* **2006**, *128*, 1340–1345.

27. F. Garnier, F. Deloffre, G. Horowitz, and R. Hajlaoui. Structure effect on transport of charge-carriers in conjugated oligomers. *Synth. Metals* **1993**, *57*, 4747–4754.

28. F. Garnier, G. Horowitz, D. Fichou, and A. Yassar. Molecular order in organic-based field-effect transistors. *Synth. Metals* **1996**, *81*, 163–171.

29. M. Moret, M. Campione, A. Borghesi, L. Miozzo, A. Sassella, S. Trabattoni, B. Lotz, and A. Thierry. Structural characterisation of single crystals and thin films of alpha,omega-dihexylquaterthiophene. *J. Mater. Chem.* **2005**, *15*, 2444–2449.

30. M. Halik, H. Klauk, U. Zschieschang, G. Schmid, S. Ponomarenko, S. Kirchmeyer, and W. Weber. Relationship between molecular structure and electrical performance of oligothiophene organic thin film transistors. *Adv. Mater.* **2003**, *15*, 917–922.

31. M. Mushrush, A. Facchetti, M. Lefenfeld, H. E. Katz, and T. J. Marks. Easily processable phenylene-thiophene-based organic field-effect transistors and solution-fabricated nonvolatile transistor memory elements. *J. Am. Chem. Soc.* **2003**, *125*, 9414–9423.

32. J. C. Maunoury, J. R. Howse, and M. L. Turner. Melt-processing of conjugated liquid crystals: A simple route to fabricate OFETs. *Adv. Mater.* **2007**, *19*, 805–809.

33. A. Sung, M. M. Ling, M. L. Tang, Z. A. Bao, and J. Locklin. Correlating molecular structure to field-effect mobility: The investigation of side-chain functionality in phenylene—Thiophene oligomers and their application in field effect transistors. *Chem. Mater.* **2007**, *19*, 2342–2351.

34. K. Kobayashi, H. Masu, A. Shuto, and K. Yamaguchi. Control of face-to-face pi-pi stacked packing arrangement of anthracene rings via chalcogen-chalcogen interaction: 9,10-bis(methylchalcogeno)anthracenes. *Chem. Mater.* **2005**, *17*, 6666–6673.

35. K. Kobayashi, R. Shimaoka, M. Kawahata, M. Yamanaka, and K. Yamaguchi. Synthesis and cofacial pi-stacked packing arrangement of 6,13-bis(alkylthio)pentacene. *Org. Lett.* **2006**, *8*, 2385–2388.

36. C. R. Patrick and G. S. Prosser. Molecular complex of benzene and hexafluorobenzene. *Nature* **1960**, *187*, 1021.

37. W. J. Feast, P. W. Lovenich, H. Puschmann, and C. Taliani. Synthesis and structure of 4,4′-bis(2,3,4,5,6-pentafluorostyryl)stilbene, a self-assembling J aggregate based on aryl-fluoroaryl interactions. *Chem. Commun.* **2001**, 505–506.

38. G. W. Coates, A. R. Dunn, L. M. Henling, D. A. Dougherty, and R. H. Grubbs. Phenyl-perfluorophenyl stacking interactions: A new strategy for supermolecule construction. *Angew. Chem. Int. Ed. Engl.* **1997**, *36*, 248–251.

39. A. Facchetti, M. H. Yoon, C. L. Stern, H. E. Katz, and T. J. Marks. Building blocks for *n*-type organic electronics: Regiochemically modulated inversion of majority carrier sign in perfluoroarene-modified polythiophene semiconductors. *Angew. Chem. Int. Ed.* **2003**, *42*, 3900–3903.

40. C. R. Swartz, S. R. Parkin, J. E. Bullock, J. E. Anthony, A. C. Mayer, and G. G. Malliaras. Synthesis and characterization of electron-deficient pentacenes. *Org. Lett.* **2005**, *7*, 3163–3166.

41. Z. H. Chen, P. Muller, and T. M. Swager. Syntheses of soluble, pi-stacking tetracene derivatives. *Org. Lett.* **2006**, *8*, 273–276.

42. D. M. Cho, S. R. Parkin, and M. D. Watson. Partial fluorination overcomes herringbone crystal packing in small polycyclic aromatics. *Org. Lett.* **2005**, *7*, 1067–1068.

43. Y. C. Chang, Y. D. Chen, C. H. Chen, Y. S. Wen, J. T. Lin, H. Y. Chen, M. Y. Kuo, and I. Chao. Crystal engineering for pi-pi stacking via interaction between electron-rich and electron-deficient heteroaromatics. *J. Org. Chem.* **2008**, *73*, 4608–4614.

44. G. Barbarella, M. Zambianchi, A. Bongini, and L. Antolini. Polyhydroxyoligothiophenes. 2. Hydrogen-bonding-oriented solid state conformation of 3,3′-bis(2-hydroxyethyl)-2,2′-bithiophene and regioselective synthesis of the corresponding head-to-head/tail-to-tail quater- and sexithiophene. *J. Org. Chem.* **1996**, *61*, 4708–4715.

45. F. S. Schoonbeek, J. H. van Esch, B. Wegewijs, D. B. A. Rep, M. P. de Haas, T. M. Klapwijk, R. M. Kellogg, and B. L. Feringa. Efficient intermolecular charge transport in self-assembled fibers of mono- and bithiophene bisurea compounds. *Angew. Chem. Inter. Ed.* **1999**, *38*, 1393–1397.

46. D. B. A. Rep, R. Roelfsema, J. H. van Esch, F. S. Schoonbeek, R. M. Kellogg, B. L. Feringa, T. T. M. Palstra, and T. M. Klapwijk. Self-assembly of low-dimensional arrays of thiophene oligomers from solution on solid substrates. *Adv. Mater.* **2000**, *12*, 563–566.

47. A. Gesquiere, S. De Feyter, F. C. De Schryver, F. Schoonbeek, J. van Esch, R. M. Kellogg, and B. L. Feringa. Supramolecular pi-stacked assemblies of bis(urea)-substituted thiophene derivatives and their electronic properties probed with scanning tunneling microscopy and scanning tunneling spectroscopy. *Nano Lett.* **2001**, *1*, 201–206.

48. A. N. Sokolov, T. Friscic, and L. R. MacGillivray. Enforced face-to-face stacking of organic semiconductor building blocks within hydrogen-bonded molecular cocrystals. *J. Am. Chem. Soc.* **2006**, *128*, 2806–2807.

CHAPTER 2

Conjugated Block Copolymers and Cooligomers

YONGYE LIANG and LUPING YU

Department of Chemistry and the James Franck Institute, The University of Chicago, Chicago, Illinois

2.1. INTRODUCTION

Since the discovery of polyacetylene as conducting polymers, organic materials bearing conjugated systems have been shown to exhibit interesting optical and electronic properties [1]. These novel materials have found a variety of applications and show comparable or even superior performance to inorganic counterparts, such as field-effect transistors (FETs) [2], light-emitting diodes (LEDs) [3], solar cells [4], memory devices [5], and sensors [6]. Extensive research effort is devoted to the understanding of structure-property relationships through investigation of synthesis, physical properties, and new types of devices from these novel materials. These studies have shown that the properties of these conjugated materials are determined not only by the chemical structures but also by the aggregated structures and supramolecular organizations [7]. However, controlling the assembly process for these conjugated materials to obtain the desired aggregation and microstructures is still a challenge.

Block copolymers consisting of flexible chains have been studied extensively for decades [8], and they offer a rich opportunity for developing simple and reliable ways to control the formation of nanoscale structures [9]. The correlation between chemical structure, block length, and molecular weight distribution and the structure of the resulting molecular assemblies and nanostructure formation has been well investigated [10]. The conjugated molecules have rigid backbones, and incorporation of such rodlike segments into heterogeneous block copolymers creates a new class of self-assembling materials in which the conjugated blocks

Self-Organized Organic Semiconductors: From Materials to Device Applications, First Edition.
Edited by Quan Li.
© 2011 John Wiley & Sons, Inc. Published 2011 by John Wiley & Sons, Inc.

can be organized into supramolecular nanostructures with various morphologies such as classical lamella, cylindrical, and spherical domains [11]. Therefore, studies on copolymers/cooligomers with conjugated blocks are very interesting and have received a wide range of attention [12]. The difference between the chain rigidity of the stiff rodlike conjugated block and the flexibility of the coillike block greatly affects the details of molecular packing and thus the nature of thermodynamically stable supramolecular structures. In systems with conjugated blocks, some interactions from the extended π-system, like the π-π interaction, offer strong driving forces for assembly and phase separation. The combination of these effects leads to rich self-assembly behaviors and novel structures, which can result in interesting opto-electronic properties in the systems.

In this chapter, we summarize the research results from our group and other research groups on the synthesis and supramolecular assemblies of conjugated block systems, which contain copolymers/cooligomers with rod and coil blocks, and systems with all-rod blocks. Interesting features of these studies are illustrated.

2.2. CONJUGATED COPOLYMERS/COOLIGOMERS CONTAINING COIL AND ROD BLOCKS

2.2.1. Polymeric Conjugated Rod Systems

Synthesis of rod-coil copolymers with polymeric conjugated rod systems can be traced back to an early stage of conjugated polymer research [13]. The incorporation of a flexible coil segment into a conjugated polymer block was initially performed to make the conjugated polymer block soluble in solvents [14]. Several systematic methods were developed to synthesize rod-coil copolymers with a conjugated polymer rod.

One method is to prepare diblock coil copolymer precursors containing one block that can be converted into a conjugated rod block. For example, poly(phenyl vinyl sulfoxide)-polystyrene block copolymers precursor was synthesized via living polymerization; subsequent thermal elimination of phenylsulfenic acid resulted in polyacetylene-polystyrene diblock or triblock copolymers [15]. Similar precursor approaches were also used to synthesize block copolymers with poly(*para*-phenylene) (PPP) or polythiophene (PT) as the conjugated rod block [16]. To synthesize PPP-PS block copolymers, polystyrene-poly(1,3-cyclohexadiene) was synthesized first by sequential polymerization of styrene and then cyclohexadiene. Then aromatization with *p*-chloranil yielded the poly(*para*-phenylene) block (Fig. 2-1). Interestingly, the film formed by such a copolymer from evaporating solutions in carbon disulfide under a flow of moist gas presents a honeycomb morphology in which monodispersed pores arrange in a hexagonal array [17]. There is an obvious disadvantage of such a precursor method. Defects in the molecules that break the conjugation are very difficult to avoid during the elimination reaction or

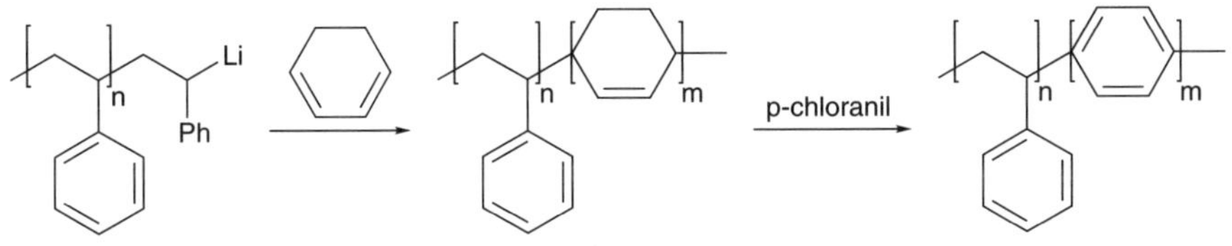

Figure 2-1   Synthesis of PPP-PS block copolymers via precursor method.

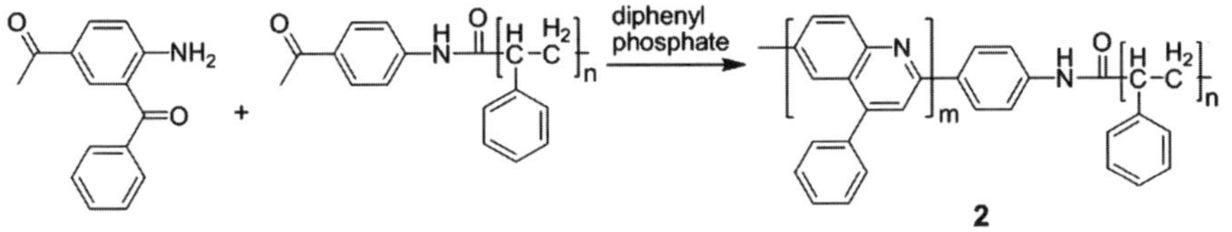

Figure 2-2   Synthesis of PPQ-b-PS.

aromatization of the precursor polymer, which adversely affects the properties of the conjugated block.

Similarly, conjugated rod-coil copolymers can also be prepared from coil polymer blocks end-capped with a functional group for the condensation polymerization of conjugated blocks. This method requires 100% conversion to avoid the presence of homopolymer, and it is better if the end-cap is much more reactive than the other reactive group to reduce polydispersity. Jenekhe et al. reported the synthesis of poly(phenylquinoline)-b-polystyrene (PPQ-b-PS) from an end-cap on a functionalized polystyrene block [18]. The copolymer was prepared by condensation reaction of polystyrene with a ketone methylene-terminated group and 5-acetyl-2-aminobenzophenone with diphenyl phosphate. The polymerization degree of conjugated segments is controlled by the stoichiometry between the two reactants (Fig. 2-2). Such copolymers could self-assemble into a variety of supramolecular structures, like hollow spheres, lamellar, and hollow cylinders.

Grafting of the end-functionalized conjugated rod block with a coil block is another way to synthesize conjugated rod-coil copolymers. For example, our group reported oligothiophene (OT)-based rod-coil copolymers in 1996 [19]. A two-stage approach was used to synthesize the diblock copolymers. First, an aldehyde functionality was introduced into the OTs via the formylation of lithiliated OTs. Then it was used to couple with living anionic polystyrene species, and the diblock copolymers were generated (Fig. 2-3). Three living polystyrenes with different chain lengths were used to generate three diblock copolymers. All of these have glass transition temperatures similar to the corresponding polystyrene homopolymers and no phase separation was observed, which is due to the small weight fraction and small volume fraction of the oligothiophene blocks compared to that of polystyrene.

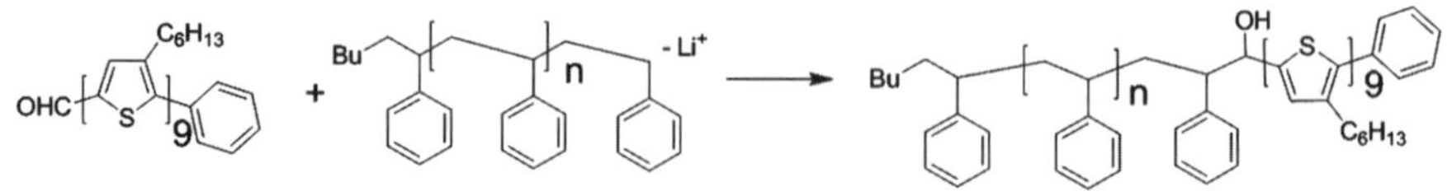

Figure 2-3   Synthesis of OT containing rod-coil copolymers.

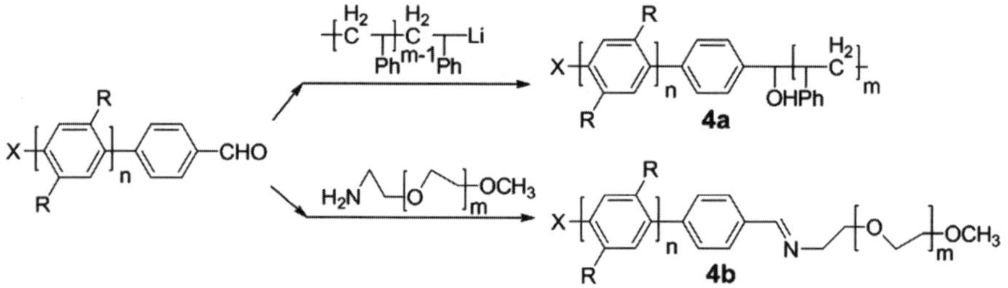

Figure 2-4   Synthesis of PPP-b-PS and PPP-b-PEO by grafting method.

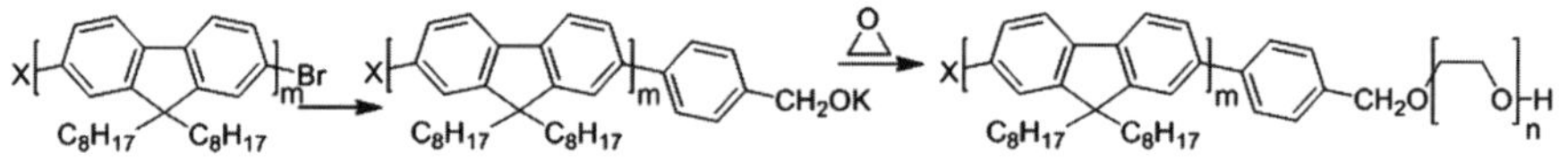

Figure 2-5   Synthesis of polyfluorene-polyethylene oxide block copolymers.

This approach was also adopted by Mullen et al., who prepared the aldehyde-functionalized polyphenylene, followed by reaction with lithium-activated polystyrene, resulting in PPP-b-PS copolymers [20]. Another reaction with amine-functionalized poly(ethylene oxide) yielded the PPP-b-PEO rod-coil copolymers (Fig. 2-4).

Another method is to prepare conjugated rod polymer as the microinitiator for living polymerization of the coil block. Mullen et al. synthesized a polyfluorene microinitiator by coupling of the benzyl alcohol group to polyfluorene and further conversion to potassium alcoholate. The following anionic polymerization of ethylene oxide yielded the corresponding conjugated rod-coil block copolymer [21] (Fig. 2-5). Hadziioannou et al. also used a similar approach to prepare poly(phenylenevinylene) (PPV)-PS block copolymers with C_{60} as the pendant group on the PS block [22]. Such copolymers were synthesized from end-functionalized PPV with nitroxide initiator. The radical polymerization yielded the poly(styrene-chloromethylstyrene) block, and the chloromethyl groups were further transformed into C_{60} by atom transfer radical addition, leading to the final copolymer. This system showed enhanced photovoltaic response compared to the blend systems, but its power conversion efficiency is low [23]. Poly-thiophenes with functionalized initiators at end groups have been developed by

Osaka and McCullough by in situ or postpolymerization methods for the synthesis of polythiophene rod-coil copolymers [24]. The poly(3-hexylthiophene)-block-poly(methylacrylate) (P3HT-b-PMA) copolymers have demonstrated high FET mobility approaching to that of P3HT, which showed that the ordering of P3HT segment is preserved in the copolymer system [25].

Generally, the conjugated polymeric rod segments are synthesized by polycondensation reactions, which leads to broad polydispersity. This can cause problems for rod-coil copolymers with polymeric conjugated blocks as it may not only make the purification of the final copolymer difficult but also affect the phase separation behavior and properties of these rod-coil copolymers.

2.2.2. Oligomeric Conjugated Rod Systems

Because of the high stiffness of conjugated systems, microphase separation of the rod and coil blocks into ordered periodic structures can occur even when the conjugated rod segment has a relatively low molecular weight, such as in oligomers. Conjugated rod-coil block systems containing well-defined, monodispersed oligomers can provide more defined structure-property relationships than the polymeric systems. A large number of rod-coil block systems with varied oligomer rods have been widely studied.

2.2.2.1. Oligo (phenylenevinylene)s as Conjugated Backbones.   Oligo-(phenylenevinylene) (OPV) consists of alternating vinylene and phenylene groups, which can offer a rigid backbone for microphase separation with flexible coil blocks. Meanwhile, OPVs exhibit very good thermal stability [26] and interesting photoelectronic properties due to the high coplanarity of the conjugated backbones [27].

One of the first OPV rod-coil copolymer systems was the OPV-polyisoprene system [28]. The selection of polyisoprene as a coil block is mainly due to its large structural difference with OPV, which can make microphase separation possible. An orthogonal approach was developed to synthesize the OPV block with a defined molecular length, which utilizes two types of reactions to grow the conjugated chain stepwise without the need for protecting group chemistry [29]. The end-functionalized OPVs were further coupled with living polyisoprene species to form diblock copolymers (Fig. 2-6). A series of copolymers were synthesized with different molecular weights in polyisoprene blocks. TEM images of these copolymer films showed obvious nanophase separation, and alternating strips of OPV and polyisoprene lamellar domains were observed. It was found that the phase separation process for these copolymers needs a high annealing temperature for a long time, which may cause degradation of the polymers.

To increase the tendency for phase separation of these rod-coil copolymers, the difference between rod and coil segments can be increased. The rod and coil blocks in the above copolymers are both hydrophobic. To enhance the driving force for the phase separation, the flexible block was replaced with a hydrophilic block. A series of copolymers consisting of hydrophobic OPV and

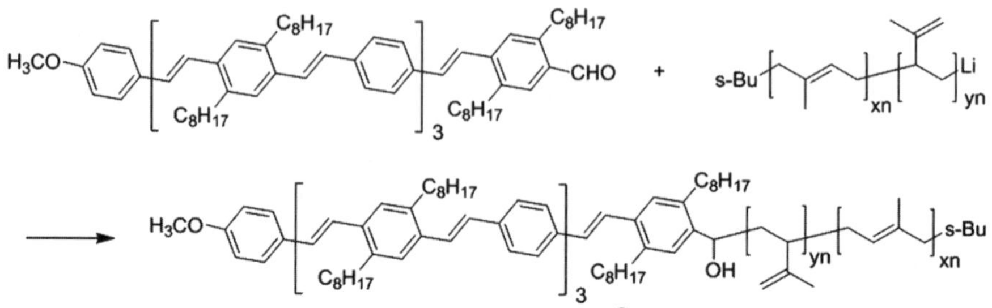

Figure 2-6   OPV-polyisoprene diblock copolymers via a two-stage approach.

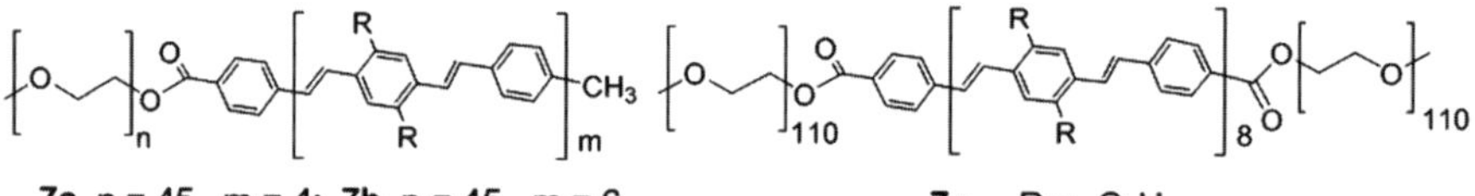

7a. n = 45, m = 4; **7b.** n = 45, m = 6
7c. n = 110, m = 4; **7d.** n = 110, m = 6

7e   R = -C_6H_{13}

Figure 2-7   Structures of OPV-PEG rod-coil amphiphilic block copolymers.

hydrophilic poly(ethylene glycol) (PEG) were synthesized [30] (Fig. 2-7). These copolymers are synthesized from the coupling reaction between carboxylic acid-functionalized OPV and PEG monoester. Each diblock copolymer consists of a perfectly monodispersed OPV block covalently bonded to a PEG block with a very low polydispersity. These copolymers exhibit remarkable self-assembling capability to form wormlike cylindrical micelles (Fig. 2-8a). TEM and SANS studies revealed that these OPV-PEG micelles have a cylindrical OPV core surrounded by a PEG corona. Cryo-TEM and SANS studies indicate that fibers were formed in dilute THF/H_2O solution. Since the PEG block can form complexes with cationic species, the nanofibers formed in solution can be further bundled when silver triflate solution is added (Fig. 2-8b).

To further investigate structural effects, another diblock copolymer system consisting of OPV covalently bonded to a coillike segment of poly(propyleneglycol) (OPV-PPG) was synthesized [31] (Fig. 2-9). Although the PPG block is less crystalline than the PEG block, it was found that the resulting copolymers still exhibit a strong driving force for phase separation. A schematic representation of self-assembly of **8c** is shown in Figure 2-10a. These copolymers could be aligned when the fiber solution was spun on the surface of mica (Fig. 2-10b). The average diameter of the fibers was about 17 nm, which corresponds to the SANS results of cylindrical micelles very well. These aggregates of fibers in solution were directly observed by fluorescence microscopy. The fibers had a strong yellow fluorescence, which comes from the aggregated OPV block (Fig. 2-10c).

OPV block copolymers connected to coil blocks with another functional group to facilitate phase separation were also studied by other groups. For

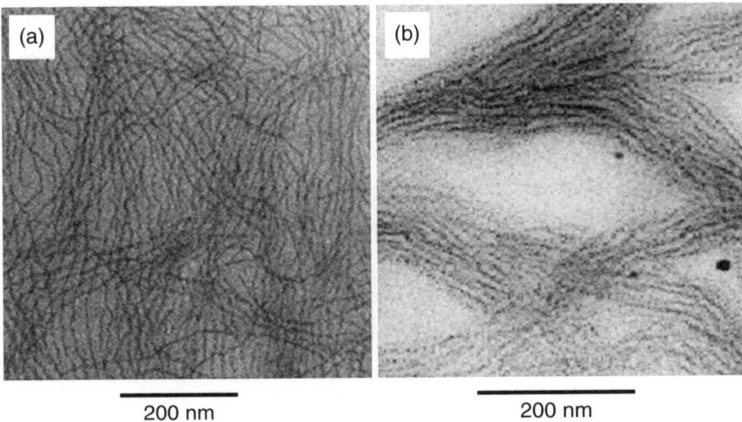

Figure 2-8 (a) TEM images of copolymer **7b** films on holey TEM grids. (b) TEM images of AgTf complexes of copolymer **7b**. Reproduced with permission from Ref. 30. Copyright 2000, American Chemical Society.

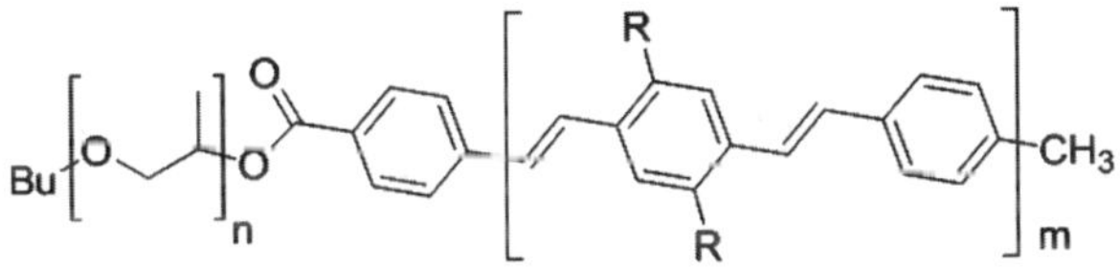

8a. n = 16, m = 6   **R** = n-C_6H_{13}
8b. n = 40, m = 6
8c. n = 70, m = 6

Figure 2-9   Structures of OPV-PPG copolymers.

example, OPV block co-oligomers with cholesterol segments were developed. The cholesterol segments can promote interesting self-assembly, and symmetrical or unsymmetrical functionalization can lead to different supramolecular organization [32] (Fig. 2-11). The OPV structures modified with symmetric cholesterol moieties at two terminals can aggregate through twisted packing by π-π interaction from the OPV backbones and van der Waals forces from the cholesterol. The twisted helical assembly leads to the formation of helical fibers. The monosubstituted molecules can dimerize via hydrogen bonding between the hydroxyl groups and aggregate through tilted and extended packing to form ribbonlike coiled nanostructures.

2.2.2.2. Oligothiophenes as Conjugated Backbones. Oligothiophene is another class of important π-conjugated materials, which has been extensively used in organic electronic devices [33]. The first diblock copolymer system was described in 1996 as shown above [19]. The change of alkyl-substituted thiophene to unsubstituted thiophene can increase the packing ability of

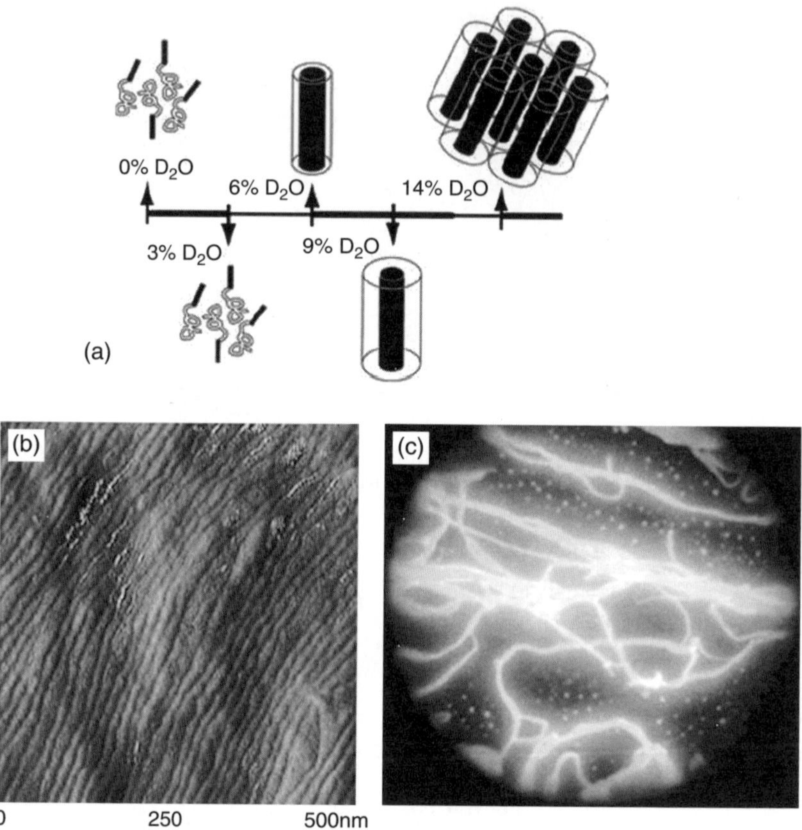

Figure 2-10   (a) Schematic representation of the self-assembly of copolymer **8c** in solution. (b) AFM image of **8c** on mica. (c) Fluorescence micrographs of copolymer **8c**. Reproduced with permission from Ref. 31. Copyright 2004, John Wiley & Sons, Inc. (A full color version of this figure appears in the color plate section.)

oligothiophene blocks, which increases the trend for phase separation. Some unsubstituted OT-b-PS copolymers have been reported to self-assemble into micellar [34] or microporous [35] structures. For example, copolymers **10** and **11** formed microporous structures on silicon wafers with hexagonal packing (Fig. 2-12). OTs with oligo(ethyleneglycol) chains [36] or cholesterol derivatives [37] as coil blocks have been studied by several groups; these copolymers/cooligomers can form interesting supramolecular structures, like fibrous networks, nanocapsules, and nanowires.

2.2.2.3. Oligo(p-phenylene)s as Conjugated Blocks.

Oligo(p-phenylene)s (OPPs) are chemically stable and rigid molecules. They have been used as the rigid components for assembling supramolecular structures [38]. Change of coil components can lead to a variety of supramolecular structures. For example,

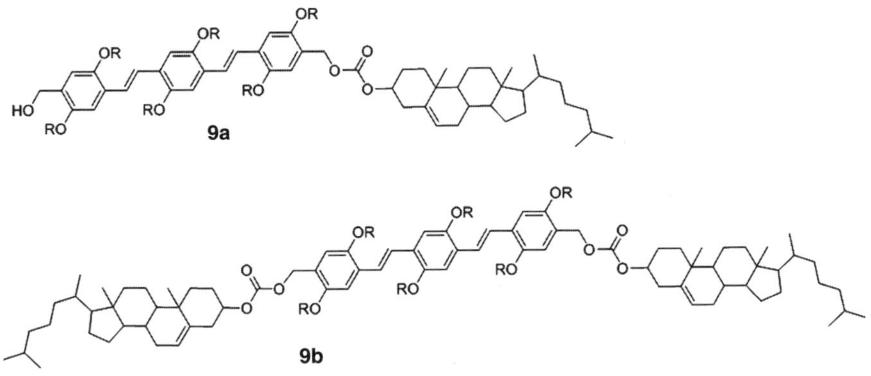

Figure 2-11   Molecular structures of OPVs with cholesterol moieties.

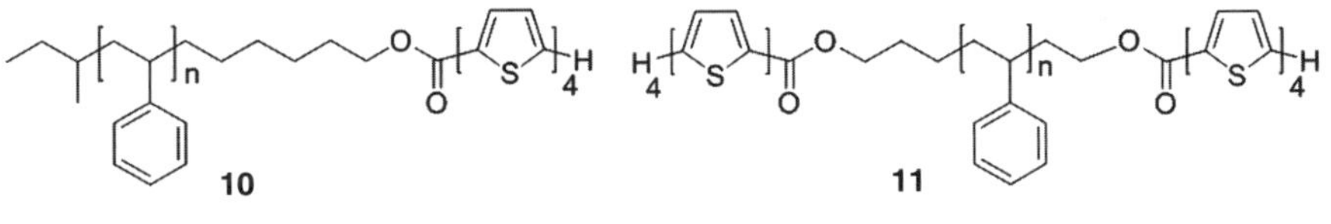

Figure 2-12   Structures of self-assembly OT-b-PS copolymers.

helical nanofibers are formed in aqueous solution when OPP is connected by aliphatic polyether dendrons at the terminals (**12a**) (Figure 2-13) [39]. The bulky dendritic wedges make the rod segments stack not in parallel, but with a certain rotation.

By changing the terminal dendrons to poly(ethylene oxide) in a macrocyclic ring compound (**12b**), similar conjugated backbones can assemble into different nanostructures, ranging from nanoribbons, and nanotubes to nanobarrels [40]. The formation of these structures was determined by the balance of the stacking forces from π-π interaction and the repulsive interactions among the adjacent flexible chains.

OPPs connected to carbohydrates by different linkages were synthesized, and different nanostructures in aqueous solution were obtained [41]. If a short PEG coil chain was used as a bridge, the molecules (**12c**) self-assembled to form nanocapsules, while smaller nanospheres were observed if longer PEG chains were used in **12d**. If two conjugated rods were connected to one coil chain, instead of one rod connected as in the first two cases, the molecules (**12e**) aggregate to form nanowires. It is interesting to find that different nanostructures show different binding activities to proteins, which illustrates that the tuning of the rod-coil structure can lead to different bioactivities by control of the assembling nanostructures.

2.2.2.4. Other Rod Systems.

In the past, various synthetic methods have been developed for π-conjugated systems, which offers approaches to modify

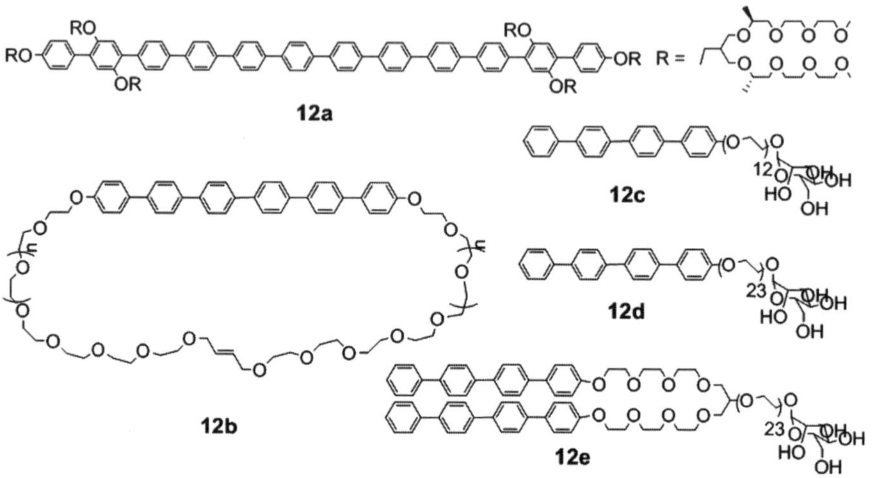

Figure 2-13   Structure of rod-coil molecules containing OPPs.

the conjugated rod structures. As a result, there are a large number of other conjugated rod systems used to construct the rod-coil copolymer/cooligomer, like oligofluorene [42], and oligo(phenylene ethynylene), to name a few [43]. Other conjugated systems will not be addressed here.

2.2.2.5. Donor-Acceptor Diblock Copolymers for Photovoltaic Applications.

The supramolecular assembly of the conjugated diblock copolymers offered an opportunity to develop new solar cell materials. This is the rationale for many research results published recently. Most of the conjugated polymers are well-known hole transport materials. If a proper electron acceptor block is introduced to the coil block, self-assembling will drive the two blocks into nanostructured domains, which provide large donor-acceptor interfaces and a continuous charge transport pathway. The donor blocks frequently used include PPV and P3HT blocks. Hadziioannou et al. synthesized PPV-C_{60} diblock copolymers via controlled free radical polymerization [22]. This approach was also extended to P3HT-C_{60} diblock copolymers [44]. Emrick et al. also utilized Grinard metathesis polymerization to prepare functional P3HT, which can be used to polymerize monomers with perylene diimide to obtain diblock copolymers [45]. Similar diblock copolymers are also reported by Thelakkat et al. [46]. The major problem with these materials is that their solar conversion efficiency is very low with a power conversion efficiency <1%, which is far inferior to the current PCE values in simple bulk heterojunction solar cells ($\sim$6–7.4%) [47]. The low efficiency is due to the unoptimized morphology and lack of the proper conjugated block that can efficiently harvest solar energy. In these materials, the bicontinuous network could not form properly, and charge-carrier transport in these diblock copolymers is poor. Further improvement of diblock copolymer solar

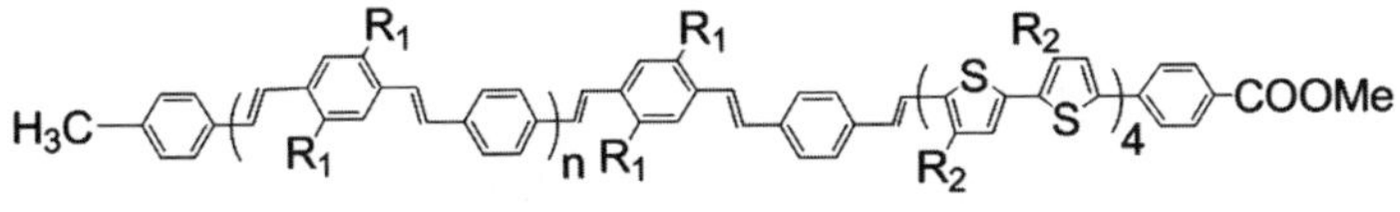

13a: n = 1, **13b:** n = 2, **13c:** n = 4.

Figure 2-14　Structures of OPV-OT conjugated diblock cooligomers.

cells requires more careful control of the design of new low-bandgap conjugated blocks and the phase separation and morphology.

2.3. CONJUGATED COPOLYMERS/COOLIGOMERS CONTAINING ALL-ROD BLOCKS

Although offering rich potential for self-assembly, the coil segments in block copolymers are usually not electronically active, which can block the unique electronic properties from conjugated rod segments. For example, the hole mobility of rod-coil block copolymers is usually lower than that of rod homopolymers. As a result, research on block copolymers/oligomers in which all the blocks are conjugated polymers/oligomers comes into focus. In these copolymers, all the blocks are rodlike. Structural studies on these would be interesting, as self-assembly of rod-rod copolymers is not well known. Meanwhile, as all the blocks are conjugated systems, interaction among them may lead to interesting photoelectronic properties, which can find application in electro-optic devices, molecular electronic components, and optical switches.

2.3.1. Oligomeric Rod Systems

The first conjugated diblock cooligomer system consisted of OPV and OT blocks [48] (Fig. 2-14). Each block was prepared in stepwise synthesis and finally connected together by the Heck coupling reaction. The self-assembly of these cooligomers is closely related to the structural composition. When the length of the OPV block increases with otherwise identical octothiophene block, the self-assembly of the molecules changes from an interwoven network to layered strip and to lamella (Fig. 2-15). Photophysical studies revealed that the excitation energy transfers from the higher-bandgap OPV block to the lower-bandgap OT block when the molecule is excited in the OPV absorption band.

Another conjugated block cooligomer contains CN-OPV in one block and OPV in another block (Fig. 2-16). This system shows high photoluminescent efficiency.

The combination of donor blocks and acceptor blocks in a conjugated oligomer can lead to *p-n* junction-type behavior, providing the opportunity to observe rectification and photodiode effects in organic molecules. A series of molecular diodes with an interesting rectification effect based on donor-acceptor conjugated block cooligomers was developed by Yu's group [49] (Fig. 2-17). Thiol groups

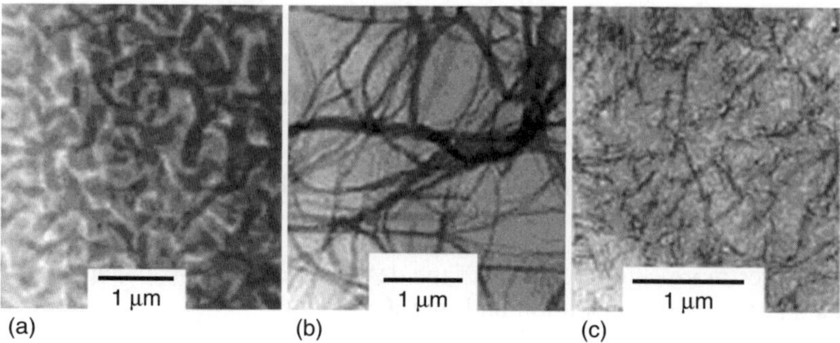

Figure 2-15 TEM images of the self-assembly by cooligomers **13a** (a), **13b** (b) and **13c** (c). Reproduced with permission from Ref. 48. Copyright 2002, Copyright 2004, John Wiley & Sons, Inc.

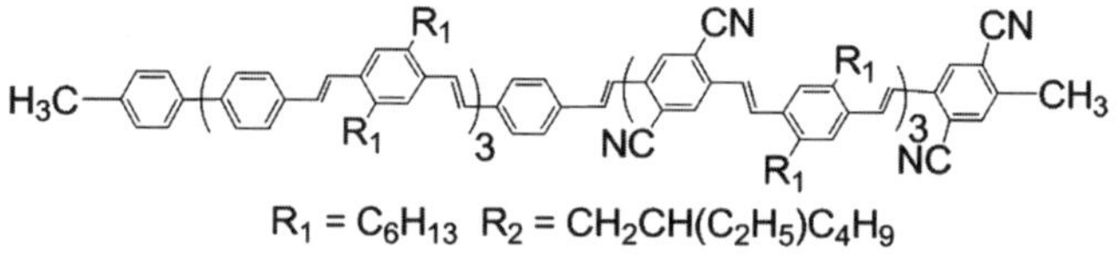

$$R_1 = C_6H_{13} \quad R_2 = CH_2CH(C_2H_5)C_4H_9$$

14

Figure 2-16 Structure of (CN-OPV)-OPV conjugated diblock cooliogomer.

are connected at the terminals of the molecules so that they can be immobilized on gold for the *I-V* measurement by STM.

The donor-acceptor cooligomers can also be used in photovoltaic applications, which requires a bicontinuous network with nanoscale phase separation for charge separation and transport [50]. Recently a series of cooligomers has been developed based on oligo(bithiophene-alt-fluorene) as a donor block and perylene diimide as an acceptor block [51] (Fig. 2-18). These cooligomers can form ordered films with alternating donor-acceptor lamellar structures, so charge carriers can transport efficiently in films. A power conversion efficiency of 1.5% has been achieved in **16c**-based single-component solar cells.

2.3.2. Polymeric Rod Systems

So far most of the conjugated polymers have been synthesized from polycondensation reactions, which usually generate a polydisperse product. As a result, synthesis of well-defined all-conjugated copolymers is still a challenge. One of the first all-conjugated copolymers was synthesized by Scherf et al. and contains polyaniline-*b*-polyfluorene-*b*-polyaniline triblock [52] (Fig. 2-19). The polyfluorene core was synthesized first by nickel-catalyzed homocoupling of the dibromofluorene. The obtained polyfluorene was end-capped by 4-bromoaniline. The aniline-functionalized polyfluorene was further polymerized with alkyl

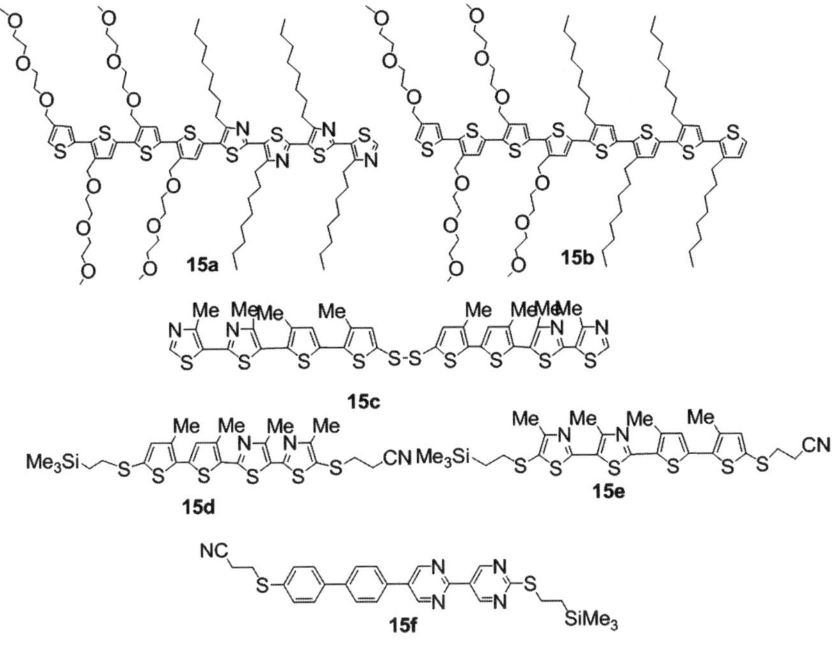

Figure 2-17 Structures for rectifying molecular diodes.

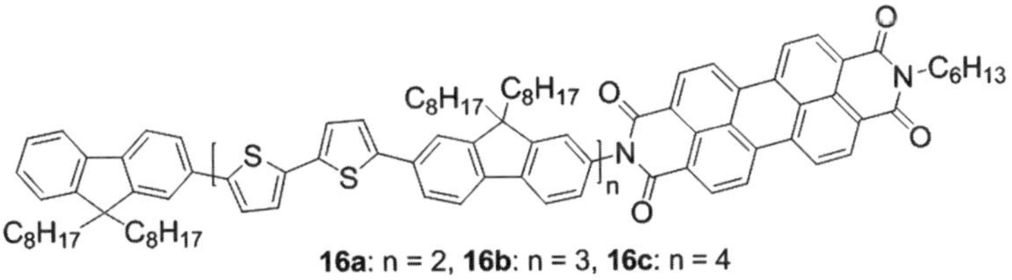

Figure 2-18 Structures of conjugated polymers for photovoltaic application.

aniline to yield the final triblock product. This method was also used to synthesize P3HT-*b*-PF-*b*-P3HT triblock copolymers [53]. Similarly, a series of donor-acceptor-donor triblock copolymers were synthesized by changing the polyfluorene core to an acceptor moiety, like cyan-substituted PPV, oligo(benzothiadiazole), and oligo(dicyanostillbene) [54]. The synthesis of these copolymers would lead to side products like homopolymer of the side block, and it is said that they can be removed by solvent extraction.

Synthesis of conjugated diblock copolymers by a "grafting from" approach was reported in 2007 in a PT-*b*-PF system [55] (Fig. 2-20). A monobromo-terminated polythiophene was first prepared. Then the Suzuki coupling reaction between such a polythiophene and 2-bromo-[9,9-bis(2-ethylhexyl)-fluorene]-y-pinacolato boronate, a bifunctional AB type monomer, led to the diblock copolymer. The

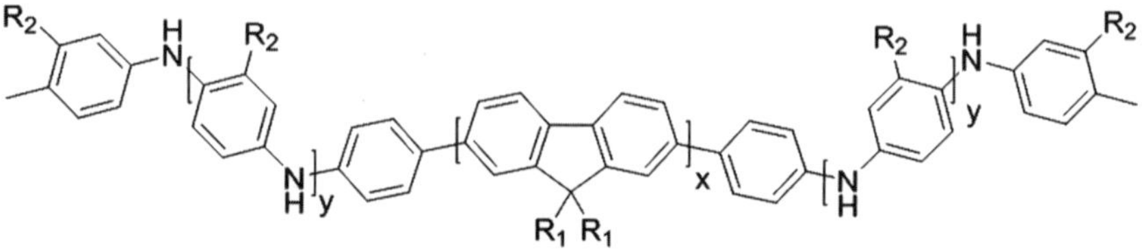

Figure 2-19 Structure of PAN-b-PF-b-PAN triblock copolymer.

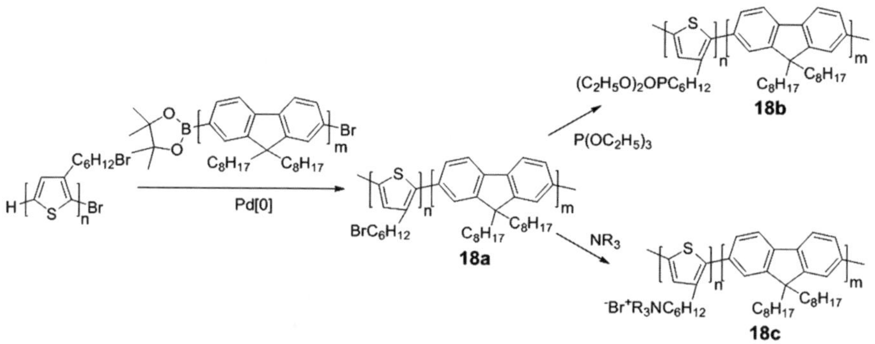

Figure 2-20 Synthesis of PT-b-PF diblock copolymers.

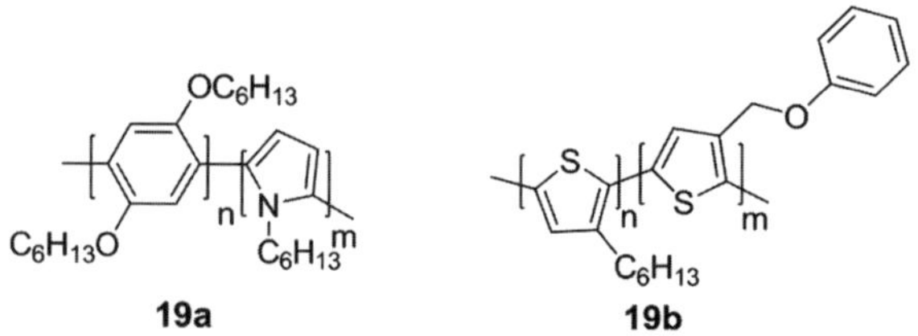

Figure 2-21 Structures of diblock copolymers from "grafting from" approach.

hydrophobic copolymer can be further changed to an amphiphilic copolymer by reacting with triethyl phosphite or trialkylamine [56]. The nonpolar copolymer **18b** can self-assemble into vesicles, while the polar copolymer **18c** forms larger, collapsed vesicles [57]. The assembly of **18c** shows a transition tendency from vesicular to lamellar morphology.

Several other conjugated diblock copolymers synthesized by a "grafting from" approach have been reported, showing rather narrow molecular weight distribution [58] (Fig. 2-21). The low polydispersity is due to the possible chain growth process from the step-by-step synthesis of two AB-type monomers. Copolymer **19b** shows lamellar or sheetlike supramolecular structures in the solid state.

2.4. CONCLUSION

Synthesis and characterization of block copolymers with conjugated rod blocks have been widely pursued with the intention of exploring the self-assembled electroactive nanostructures for unique applications. Incorporation of conjugated rods into block copolymers has been proven to be an effective way to manipulate the supramolecular assembly of the conjugated system in nanoscale dimensions. The self-assembly of block copolymers with rod and coil segments is affected by many factors, such as the crystallization of the rod segment, the stiffness asymmetry and relative volume fraction of the blocks, the interaction of functional groups in coil, and, of course, the chemical structure of these segments. Compared to the conjugated rod-coil system, studies on all-rod conjugated block systems are still in an early state. However, the last several years have witnessed fast development of synthetic approaches and supramolecular assembly of these materials. Interesting properties in molecular electronics, solar energy harvesting, and other applications are emerging. However, there are still great challenges in both systems, especially in manipulating the alignment and assembly of the conjugated rod systems in a controlled way to generate unique optical and electronic properties. With collective and continuous effort, it is believed that conjugated block copolymers will become an important class of nanostructured, functionalized materials with interesting applications.

ACKNOWLEDGMENTS

The preparation of this chapter benefited from support by NSF and NSF MRSEC at the University of Chicago.

REFERENCES

1. (a) N. C. Greenham, S. C. Moratti, D. D. C. Bradley, R. H. Friend, and A. B. Holmes. *Nature* **1993**, *365*, 628. (b) H. S. Nalwa, Ed. *Handbook of Organic Conductive Molecules and Polymers*, Marcel Dekker, New York, **1997**, Vol. 1–4. (c) T. A. Skotheim, R. L. Elsenbaumer, and J. R. Reynolds. *Handbook of Conducting Polymers*, 2nd ed., Marcel Dekker, New York, **1998**. (d) K. Müllen and G. Wegner. *Electronic Materials: The Oligomer Approach*, Wiley-VCH, Weinheim, **1998**. (e) A. J. Heeger. *Rev. Mod. Phys.* **2001**, *73*, 681.

2. H. Yan, Z. Chen, Y. Zheng, C. Newman, J. R. Quinn, F. Dotz, M. Kastler, and A. Facchetti. *Nature* **2009**, *457*, 679.

3. S. J. Su, E. Gonmori, H. Sasabe, and J. Kido. *Adv. Mater.* **2009**, *21*, 361.

4. Y. Y. Liang, D. Q. Feng, Y. Wu, S. T. Tsai, G. Li, C. Ray, and L. P. Yu. *J. Am. Chem. Soc.* **2009**, *131*, 7792.

5. J. C. Scott and L. D. Bozano. *Adv. Mater.* **2007**, *19*, 1452.

6. S. W. Thomas, G. D. Joly, and T. M. Swager. *Chem. Rev.* **2007**, *107*, 1339.

7. F. J. M. Hoeben, P. Jonkheijm, E. W. Meijer, and A. Schenning. *Chem. Rev.* **2005**, *105*, 1491.

8. (a) S. Aggarwal, Ed. *Block Copolymers*, Plenum Press, New York, **1970**. (b) F. S. Bates and G. H. Fredrickson. *Annu. Rev. Phys. Chem.* **1990**, *41*, 525.

9. (a) E. Helfand and Z. R. Wasserman. *Macromolecules* **1980**, *13*, 994. (b) T. Hashimoto, M. Shibayama, and H. Kawai. *Macromolecules* **1983**, *16*, 1093.

10. (a) F. S. Bates and G. H. Frederickson. *Phys. Today* **1999**, *52*, 32. (b) H. A. Klok and S. Lecommandoux. *Adv. Mater.* **2001**, *13*, 1217.

11. (a) Y. Y. Liang, H. B. Wang, S. W. Yuan, Y. G. Lee, L. Gan, and L. P. Yu. *J. Mater. Chem.* **2007**, *17*, 2183. (b) B. D. Olsen and R. A. Segalman. *Mater. Sci. Eng. R* **2008**, *62*, 37.

12. A. C. Grimsdale and K. Mullen. *Angew. Chem. Int. Ed.* **2005**, *44*, 5592.

13. (a) F. S. Bates and G. L. Baker. *Macromolecules* **1983**, *16*, 704. (b) M. Aldissi. *J. Chem. Soc. Chem. Commun.* **1984**, 1347. (c) S. P. Armes, B. Vincent, and J. W. White. *J. Chem. Soc. Chem. Commun.* **1986**, 1525. (d) S. A. Krouse and R. R. Schrock. *Macromolecules* **1988**, *21*, 1888. (e) J. A. Stowell, A. J. Amass, M. S. Beevers, and T. R. Farren. *Polymer* **1989**, *30*, 195. (f) R. S. Saunders, R. E. Cohen, and R. R. Schrock. *Macromolecules* **1991**, *24*, 5599.

14. (a) F. S. Bates and G. L. Baker. *Macromolecules* **1983**, *16*, 704. (b) M. E. Galvin and G. E. Wnek. *Mol. Cryst. Liq. Cryst.* **1985**, *117*, 33. (c) M. Aldissi. *J. Chem. Soc. Chem. Commun.* **1984**, 1347.

15. (a) R. S. Kanga, T. E. Hogen-Esch, E. Randrianalimanana, A. Soum, and M. Fontanille. *Macromolecules* **1990**, *23*, 4235. (b) R. S. Kanga, T. E. Hogen-Esch, E. Randrianalimanana, A. Soum, M., Fontanille. *Macromolecules* **1990**, *23*, 4241.

16. (a) X. F. Zhong and B. Francois. *Macromol. Chem.* **1990**, *191*, 2735. (b) X. F. Zhong and B. Francois. *Macromol. Chem.* **1990**, *191*, 2743. (c) X. F. Zhong and B. Francois. *Macromol. Chem.* **1991**, *192*, 2277. (d) B. Francois, O. Pitois, and J. Francois. *Adv. Mater.* **1995**, *7*, 1041.

17. G. Widawski, M. Rawiso, and B. Francois. *Nature* **1994**, *369*, 387.

18. (a) S. A. Jenekhe and X. L. Chen. *Science* **1998**, *279*, 1903. (b) X. L. Chen and S. A. Jenekhe. *Langmuir* **1999**, *15*, 8007.

19. W. J. Li, T. Maddux, and L. P. Yu. *Macromolecules* **1996**, *29*, 7329.

20. D. Marsitzky, T. Brand, Y. Geerts, M. Klaper, and K. Mullen. *Macromol. Rapid Commun.* **1998**, *19*, 385.

21. D. Marsitzky, M. Klapper, and K. Mullen. *Macromolecules* **1999**, *32*, 8685.

22. U. Stalmach, B. de Boer, C. Videlot, P. F. Hutten, and G. Hadziioannou. *J. Am. Chem. Soc.* **2000**, *122*, 5464.

23. B. de Boer, U. Stalmach, P. F. Hutten, V. V. Krasnikov, and G. Hadziioannou. *Polymer* **2001**, *42*, 9097.

24. I. Osaka and R. D. McCullough. *Acc. Chem. Res* **2008**, *41*, 1202.

25. G. Sauve and R. D. McCullough. *Adv. Mater.* **2007**, *19*, 1822.

26. H. Meier. *Angew. Chem. Int. Ed.* **1992**, *31*, 1399.

27. (a) J. G. Kushmerick, D. B. Holt, S. K. Pollack, M. A. Ratner, J. C. Yang, T. L. Schull, J. Naciri, M. H. Moore, and R. Shashidhar. *J. Am. Chem. Soc.* **2002**, *124*, 10654. (b) T. Goodson, W. J. Li, A. Gharavi, and L. P. Yu. *Adv. Mater.* **1997**. *9*, 639.

28. W. J. Li, H. B. Wang, and L. P. Yu. *Macromolecules*. **1999**, *32*, 3034.

29. T. Maddux, W. J. Li, and L. P. Yu. *J. Am. Chem. Soc.* **1997**, *119*, 844.

30. H. B. Wang, H. Wang, V. S. Urban, P. Thiyagarajan, K. C. Littrell, and L. P. Yu. *J. Am. Chem. Soc.* **2000**, *122*, 6855.

31. H. B. Wang, W. You, P. Jiang, L. P. Yu, and H. H. Wang. *Chem. Eur. J*. **2004**, *10*, 986.

32. A. Ajayaghosh, C. Vijayakumar, R. Varghese, and S. J. George. *Angew. Chem. Int. Ed*. **2006**, *45*, 456.

33. (a) D. Fichou. *Handbook of Oligo- and Polythiophenes*, Wiley-VCH, Weinheim, **1999**. (b) A. Mishra, C. Q. Ma, and P. Bauerle. *Chem. Rev*. **2009**, *109*, 1141.

34. M. A. Hempenius, B. M. W. Langeveld-Voss, J. A. E. H. Haare, R. A. J. Janssen, S. S. Sheiko, J. P. Spatz, M. Moller, and E. W. J. Meijer. *J. Am. Chem. Soc*. **1998**, *120*, 2798.

35. T. Hayakawa and H. Yokoyama. *Langmuir* **2005**, *21*, 10288.

36. (a) A. F. M. Kilbinger, A. P. H. J. Schenning, F. Goldoni, W. J. Feast, and E. W. Meijer. *J. Am. Chem. Soc*. **2000**, *122*, 1820. (b) A. P. H. J. Schenning, A. F. M. Kilbinger, F. Biscarini, M. Cavallini, H. J. Cooper, P. J. Derrick, W. J. Feast, R. Lazzaroni, P. Leclere, L. A. McDonell, E. W. Meijer, and S. C. J. Meskers. *J. Am. Chem. Soc*. **2002**, *124*, 1269. (c) O. Henze, W. J. Feast, F. Gardebien, P. Jonkheijm, R. Lazzaroni, P. Leclere, E. W. Meijer, and A. P. J. Schenning. *J. Am. Chem. Soc*. **2006**, *128*, 5923.

37. S. I. Kawano, N. Fujita, and S. Shinkai. *Chem. Eur. J*. **2005**, *11*, 4735.

38. P. Kovacic and M. B. Jones. *Chem. Rev*. **1987**, *87*, 357.

39. J. Bae, J. H. Choi, Y. S. Yoo, N. K Oh, B. S. Kim, and M. Lee. *J. Am. Chem. Soc*. **2005**, *127*, 9668.

40. (a) W. Y. Yang, E. Lee, and M. Lee. *J. Am. Chem. Soc*. **2006**, *128*, 3484. (b) W. Y. Yang, J. H. Ahn, Y. S. Yoo, N. K. Oh, and M. Lee. *Nat. Mater*. **2005**, *4*, 399.

41. B. S. Kim, D. J. Hong, J. Bae, and M. Lee. *J. Am. Chem. Soc*. **2005**, *127*, 16333.

42. C. L. Chochus, P. K. Tsolakis, V. G. Gregoriou, and J. K. Kallitsis. *Macromolecules*, **2004**, *37*, 2502.

43. S. Rosselli, A. D. Ramminger, T. Wagner, B. Silier, S. Wiegand, W. Häußler, G. Lieser, V. Scheumann, and S. Höger. *Angew. Chem. Int. Ed*. **2001**, *40*, 3138.

44. J. U. Lee, A. Cirpan, T. Emrick, T. P. Russell, and W. H. Jo. *J. Mater. Chem*. **2009**, *19*, 1483.

45. Q. Hang, A. Cirpan, T. P. Russell, and T. Emrick. *Macromolecules* **2009**, *42*, 1079.

46. M. Sommer, A. S. Lang, and M. Thelakkat. *Angew. Chem. Int. Ed*. **2008**, *47*, 7901.

47. (a) Y. Y. Liang, Y. Wu, D. Q. Feng, S. T. Tsai, H. J. Son, G. Li, and L. P. Yu. *J. Am. Chem. Soc*. **2009**, *131*, 56. (b) Y. Y. Liang, Z. Xu, J. B. Xia, S. T. Tsai, Y. Wu, G. Li, C. Ray, and L. P. Yu. *Adv. Mater*. **2010**, *20*, E135.

48. H. B. Wang, M. K. Ng, L. M. Wang, L. P. Yu, B. H. Lin, M. Meron, and Y. N. Xiao. *Chem. Eur. J*. **2002**, *8*, 3246.

49. (a) M.-K. Ng and L. P. Yu. *Angew. Chem. Int. Ed*. **2002**, *41*, 3598. (b) M.-K. Ng, D. C. Lee, and L. P. Yu. *J. Am. Chem. Soc*. **2002**, *124*, 11862. (c) P. Jiang, M. Gustavo, W. You, and L. P. Yu. *Angew. Chem. Int. Ed*. **2004**, *43*, 4471. (d) M. Gustavo, P. Jiang, S. W. Yuan, Y. G. Lee, S. Arturo, W. You, and L. P. Yu. *J. Am. Chem. Soc*. **2005**, *127*, 10456.

50. (a) B. C. Thompson and J. M. J. Frechet. *Angew. Chem. Int. Ed*. **2008**, *47*, 58. (b) S. Gnes, H. Neugebauer, and N. S. Sariciftci. *Chem. Rev*. **2007**, *107*, 1324.

51. L. J. Bu, X. Y. Guo, B. Yu, Y. Qu, Z. Y. Xie, D. H. Yan, Y. H. Geng, and F. S. Wang. *J. Am. Chem. Soc*. **2009**, *131*, 13242.

52. C. Schmitt, H. G. Nothofer, A. Falcou, and U. Scherf. *Macromol. Rapid Commun*. **2001**, *22*, 624.

53. U. Asawapirom, R. Guntner, M. Forster, and U. Scherf. *Thin Solid Films* **2005**, *477*, 48.

54. G. Tu, H. Li, M. Forster, R. Heiderhoff, L. J. Balk, and U. Scherf. *Macromolecules* **2006**, *39*, 4327.

55. G. Tu, H. Li, M. Forster, R. Heiderhoff, L. J. Balk, R. Sigel, and U. Scherf. *SMALL* **2007**, *3*, 1001.

56. A. Gutacker, N. Koenen, U. Scherf, J. Pina, S. M. Fonseca, J. Seixas de Melo, A. J. M. Valente, and H. D. Burrows. *Macromol. Rapid Commun*. **2008**, *29*, F50.

57. N. Koenen, M. Forster, R. Ponnapati, U. Scherf, and R. Advincula. *Macromolecules* **2008**, *41*, 6169.

58. (a) A. Yokoyama, A. Kato, R. Miyakoshi, and T. Yokozawa. *Macromolecules* **2008**, *41*, 7271.   (b) K. Oshimidzu and M. Ueda. *Macromolecules* **2008**, *41*, 5289.

CHAPTER 3

Charge-Carrier Transport and Its Modeling in Liquid Crystals

JUN-ICHI HANNA and AKIRA OHNO

Imaging Science and Engineering Laboratory, Tokyo Institute of Technology, Yokohama, Japan

3.1. INTRODUCTION

After establishment of a new paradigm of the *semiconductor* in inorganic materials in the 1940s, Inokuchi and Akamatsu paid primary attention to the electrical properties of organic materials including violanthrone in 1950 [1]. This led to the concept of *the organic semiconductor*. Since then, the electrical properties of various organic compounds including polyacenes such as naphthalene and anthracene were investigated extensively [2], especially in single crystals. However, this research had remained within scientific interests and did not find any practical applications before organic polycrystalline thin films of extended π-conjugated aromatic molecules such as oligothiophenes and pentacene were found to be available for field-effect transistors in the 1990s [3].

On the other hand, the electrical properties in organic amorphous materials including polymers were extensively studied for applications to photoreceptors for xerographic copiers, which are often called organic photoconductive coatings (OPCs). This is because the electrical properties of organic materials meet the requirements for the photoreceptors. In fact, the first organic photoreceptor for the copiers was developed with a mixture of polyvinylcarbazole (PVK) and trinitrofluorenone (TNF) in 1970. This application was extended to photoreceptors for laser printers in the following decades and has scored a great success today.

As described above, the electrical properties in the organic materials have been studied at two extremes, namely, single crystals and amorphous aggregates, and have been understood within the frameworks of the band conduction with a narrow bandwidth and the hopping conduction in the distributed localized states,

Self-Organized Organic Semiconductors: From Materials to Device Applications, First Edition.
Edited by Quan Li.

respectively. However, little attention has been paid to the mesophase materials whose molecular alignment is just in between these two extremes.

As for the first attention to the electrical properties of liquid crystals, there was a historical coincidence, and it determined the direction of research interests thereafter. Kusabayashi and Lebes were interested in the electrical properties of liquid crystals from their scientific view extended from those of organic crystals, and studied them at their phase transitions from phase to phase. They observed different current-voltage characteristics in various liquid crystals including smectic liquid crystals [4], but they did not come to any conclusive answer to their interests. Almost at the same time, Heilmeier at RCA proposed a liquid crystal display [5], which is quite different from the liquid crystal displays that we use in PC and TV monitors and was based on the light scattering caused by ion drift in the oriented nematic phase, that is, the so-called dynamic scattering mode, and studied the electrical properties of a nematic liquid crystal, namely, p-azoxyanisole, with time-of-flight experiments [6]. These studies inspired many researchers and provided a new interest in the electrical properties of liquid crystals that led to new applications in the 1970s [7]. All the experimental results at that time indicated that the electrical properties in liquid crystals were governed by ionic conduction. As a result, the general recognition that the conduction mechanism in the liquid crystal was ionic became accepted naturally before the electronic conduction in liquid crystals was discovered in the 1990s [8–10].

After the industrial success of organic photoreceptors for photocopiers and laser printers, organic amorphous semiconductors have found another industrial application, namely, organic light-emitting diodes (OLEDs) in the 1990s. Furthermore, the third potential application, organic field-effect transistors (OFETs) with polycrystalline thin films of organic materials having an extended π-conjugated aromatic system, has been proposed as described above, and the research and development for their industrial applications are now in progress. In the future, the organic solar cells will follow as the fourth application.

On the basis of present understanding about electrical properties of organic aggregates including liquid crystals, they are categorized in terms of mobility as illustrated in Figure 3-1.

The intrinsic conductivity of typical organic materials including OPC, OFET, and OLED materials is quite low, typically less than 10^{-12} Scm^{-1}, which is categorized as an insulator. However, this does not mean that carriers are immobile in these materials. The low conductivity is attributed to a small concentration of mobile carriers in the thermal equilibrium at the ambient atmosphere. For example, pentacene, which is the best-known material for OFETs and has high mobility over 1 cm^2/Vs in a polycrystalline thin film, exhibits low conductivity of less than 10^{-12} Scm^{-1} at room temperatures because of a very low concentration of free carriers less than 10^6 cm^{-3}. Therefore, mobility is a good index for electrical properties of organic materials. The mobility in organic aggregates depends on how molecules aggregate as illustrated in Figure 3-1, which is determined by electrical interaction of molecules in the aggregates. Because of weak van der Waals interaction of molecules in the organic aggregates, the bandwidth both in

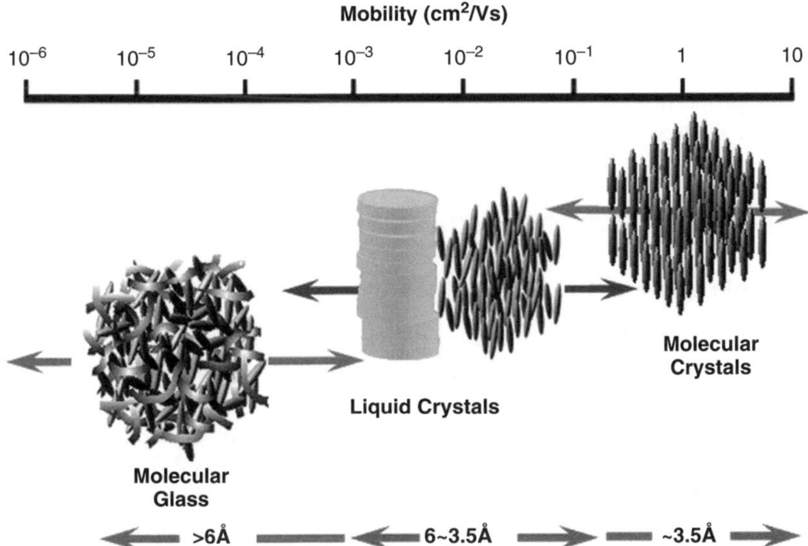

Figure 3-1 Typical mobility and intermolecular distance in various organic molecular aggregates. (A full color version of this figure appears in the color plate section.)

conduction and valence bands in a crystal becomes narrow and comparable to the thermal energy at room temperatures. Therefore, the conduction becomes localized rather than coherent even in single crystals. Thus, the mobility is roughly determined by a particular state of matter, as illustrated in Figure 3-1, and so by the average molecular distance in a particular aggregate.

Before the electronic conduction was established in liquid crystals in the 1990s, the conduction mechanism in the liquid crystals had been believed to be ionic for a long time. This is due to the historical studies on electrical properties of liquid crystals in the 1970s and later [11, 12], as described previously. In fact, the ionic conduction is often observed in liquid crystals, but it is not an intrinsic nature of liquid crystals but an impurity effect enhanced by a specific structure microphase separated in mesophases as described below.

The charge-carrier transport properties in liquid crystals are somewhat different from those of the totally disordered molecular systems of amorphous materials and from those of the totally ordered molecular systems of crystals because of disorder of molecular alignment in the ordered aggregates. Therefore, the liquid crystal is an excellent materials system to investigate how the molecular order contributes to the conduction mechanism and to have continuous understanding of conduction mechanism from amorphous materials to crystals in a framework of molecular order in organic aggregates. In addition, liquid crystalline materials exhibit very unique charge-carrier transport properties that can be realized in neither amorphous nor crystalline materials and have a high potential for device applications as a new class of organic semiconductors. Even though the mobility in liquid crystals cannot be superior to those of crystalline materials, as J. Simon forecast in his book, "Liquid crystalline molecular semiconductors can therefore be envisaged; the corresponding mobility will be, however, limited to be about 1 cm^2/Vs." In this context, this article describes charge-carrier transport properties of liquid crystals in terms of both experimental results and theoretical framework.

3.2. GENERAL FEATURES OF CARRIER TRANSPORT

3.2.1. Ionic and Electronic Conductions in Liquid Crystals

In the previous section, we mentioned that the conduction in mesophases had been thought to be ionic for a long time before the electronic conduction in both smectic and discotic liquid crystals was established in 1990s. The liquidlike nature of the mesophases and some experimental reports made it easy to accept this misleading idea. The ionic conduction in liquid crystals is an extrinsic effect caused by chemical impurities. Even in the most liquidlike nematic phase of small molecules, it is not always caused by dissociation of ionic impurities. The ionized neutral impurities formed by charge trapping or photoionization also cause the ionic conduction. As in the case of crystalline materials, the electronic conduction in mesophases is quite sensitive to chemical impurities, whose highest occupied molecular orbital (HOMO) and/or lowest unoccupied molecular orbital (LUMO) are located in the energy gap of those of the liquid crystalline molecule. According to impurity-doping experiments with highly purified liquid crystals, a trace amount of chemical impurity less than tens of parts per million degrades the electronic conduction into an ionic conduction totally as is shown in Figure 3-2 [13, 14]. In addition, there is another reason why the ionic conduction is so often observed in the liquid crystals. This is because there are easy conduction channels for ions in the mesophases due to the microphase separated structure of mesophases, where flexible hydrocarbon chains of liquid crystal molecules aggregate to form a liquidlike region, as illustrated in Figure 3-3. The impure molecules ionized by either charge trapping or photoionization diffuse into the liquidlike channel and start to drift as mobile ions in the mesophases. This is the reason why the ionic conduction is so easily observed in the liquid crystals and why the early reports misled us to the wrong conclusion that the ionic conduction is an intrinsic nature of liquid crystals.

The ion conduction never happens as easily to nonliquid crystalline materials because of the high viscosity in solids. In fact, the trap-controlled electronic conduction is often observed in highly ordered mesophases such as SmB and SmE phases when the liquid crystals are contaminated. The mobility in this case follows the Hoesterey–Letson formalism based on the multiple trapping model basically [15], as is reported in the single crystal of anthracene [16]. This could be the case for the highly ordered smectic phases such as SmF, SmG, and SmH.

As described above, the chemical impurities in a liquid crystal indeed affect the carrier transport properties seriously, but they give another advantage over the conventional organic semiconductors because the purity of the materials can

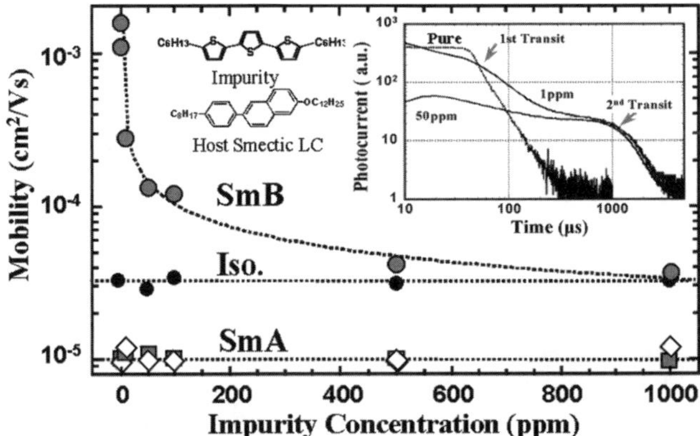

Figure 3-2   Mobility in various phases of a 2-phenylnapthalene derivative doped with chemical impurity of a terthiophene derivative as a function of its concentration. The inset shows transient photocurrents in nondoped and doped 2-phenylnaphthalene derivative with a terthiophene derivative. The first and the second transits are attributed to the electronic and ionic conduction, respectively.

be easily evaluated in liquid crystalline materials by means of detecting the ionic conduction quantitatively.

3.2.2. Anisotropic and Low-dimensional Conduction

Liquid crystalline molecules have an anisotropic molecular shape consisting of an extended π-conjugated moiety called a core and long hydrocarbon chains attached to it, namely, rodlike or disclike as shown in Figure 3-3. Because of this anisotropy, liquid crystalline molecules aggregate into molecular layers or molecular columns, which are called smectic and discotic columnar phases, respectively. The intermolecular distance in smectic phases is quite different between the inter- and intralayers because of a long molecular length, and so does the intermolecular distance differ in discotic columnar phases between inter- and intracolumns. In fact, the intermolecular distance ranges from 4 to 6 Å within a smectic layer and at around 3.5 Å within a discotic column, while the inter-molecular distance in the intersmectic layers and columns is about the molecular length basically. Therefore, the mobility is negligibly small in the intersmectic layers and discotic columns, and the conduction in smectic and columnar phases becomes two- and one-dimensional, respectively.

3.2.3. Ambipolar Carrier Transport

In amorphous aggregates, ambipolar transport, where both electron and hole contribute to the conduction, is seldom observed. Hole conduction is easy to find, but electron transport is very much limited in the materials having electron affinity higher than oxygen. This is due to deep trap states for electrons caused

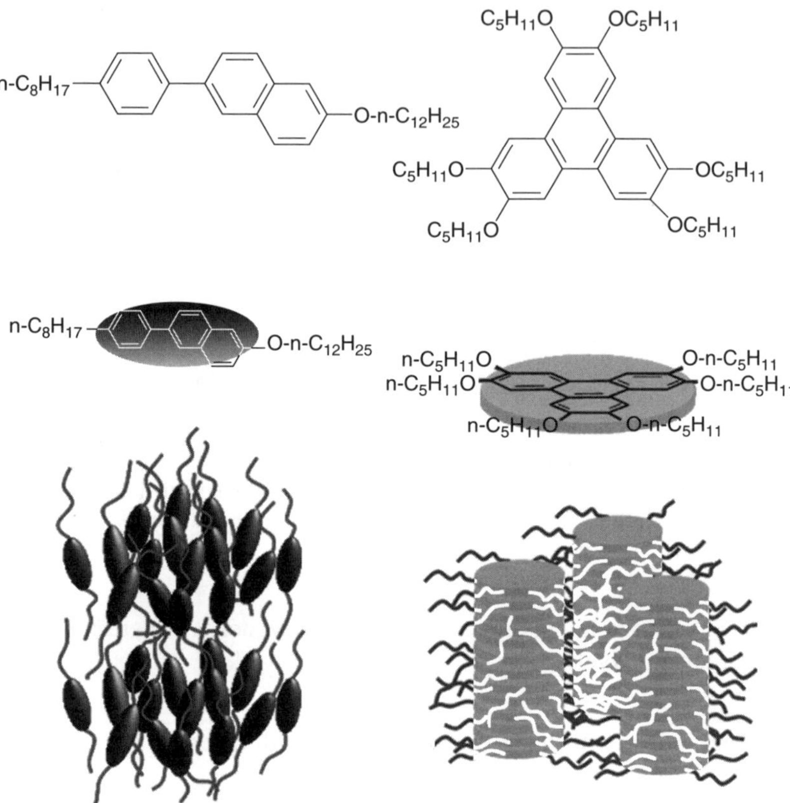

Figure 3-3   Typical liquid crystalline molecules and their self-organized aggregates. Rod-like molecules self-organize in layers in the smectic phases (left), while disc-like molecules do in columns (right), as illustrated at a certain temperature range. (A full color version of this figure appears in the color plate section.)

by oxygen contamination. The carrier transport in liquid crystals, however, is ambipolar irrespective of smectic or discotic mesophase [17–21], as in the case of the charge-carrier transport in single crystals of organic materials [22]. Judging from the fact that electron transport is observed even in the single crystal of phthalocyanine, which is a typical *p*-type material, the ambipolar carrier transport often observed in single crystals is probably due to both high purity of the material and closely packed molecules in a crystal that does not allow oxygen molecules to diffuse into the bulk. Therefore, it is very plausible that the ambipolar charge transport often observed in the mesophases of extensively purified discotic and smectic liquid crystals [23] is due to the same reason as in the single crystals. This is one of the unique features in the carrier transport in the mesophases. The electron and hole mobilities of various liquid crystals reported are summarized in Table 3-1 and Table 3-2 (a, b and c).

Table 3-1　Mobility in Various Smectic Mesophases Determined by Time-of-Flight (TOF) Measurements.

Materials	Hole Mobility (cm^2/Vs)	Electron Mobility (cm^2/Vs)
Biphenyls		
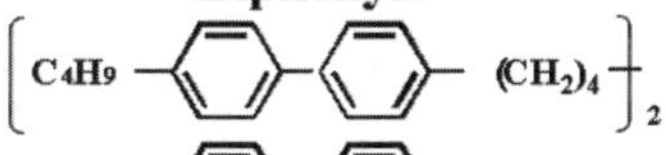 $\left[C_4H_9 - \bigcirc\!\!-\!\!\bigcirc - (CH_2)_4 \right]_2$	6×10^{-3} (SmB)	—
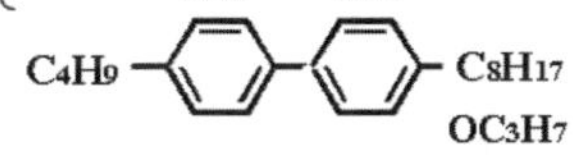 $C_4H_9 - \bigcirc\!\!-\!\!\bigcirc - C_8H_{17}$, OC_3H_7	3×10^{-3} (SmB) 1×10^{-3} (SmE)	—
2-Phenylbenzothiazoles		
$C_7H_{15}O - \bigcirc - \text{(benzothiazole)} - SC_{12}H_{25}$	1×10^{-4} (SmA)	1×10^{-4} (SmA)
2-Phenylnaphthalenes		
$C_8H_{17} - \bigcirc - \text{(naphthalene)} - OC_{12}H_{25}$	2.5×10^{-4} (SmA) 1.7×10^{-3} (SmB) 1.0×10^{-2} (SmE)	2.5×10^{-4} (SmA) 1.6×10^{-3} (SmB) 1.0×10^{-2} (SmE)
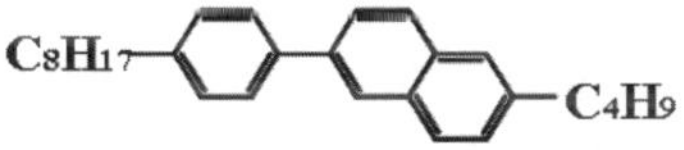 **-OC₄H₉** $C_8H_{17} - \bigcirc - \text{(naphthalene)} - C_4H_9$	8×10^{-4} (SmA) 8×10^{-3} (SmB) 4.7×10^{-2} (SmE)	—
Terphenyls 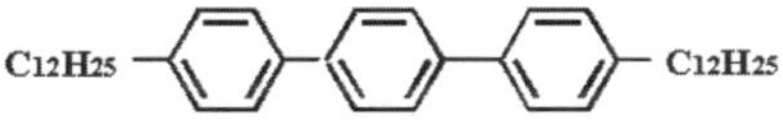 $C_{12}H_{25} - \bigcirc\!\!-\!\!\bigcirc\!\!-\!\!\bigcirc - C_{12}H_{25}$	0.01 (SmX$_1$) 0.04 (SmX$_2$) 0.13 (SmX$_3$)	—
Terthiophenes  $C_8H_{17} - \text{(terthiophene)} - C_8H_{17}$	8×10^{-4} (SmC) 3.1×10^{-3} (SmF) 2.4×10^{-2} (SmG)	5×10^{-4} (SmC) 2.5×10^{-3} (SmF) 2.4×10^{-2} (SmG)
Quaterthiophenes 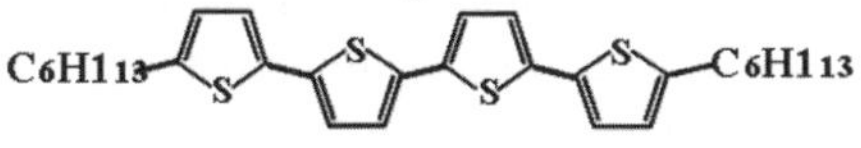 $C_6H_{13} - \text{(quaterthiophene)} - C_6H_{13}$ $-\!\!\equiv\!\!- C_4H_9$	6×10^{-2} (SmG) 0.1 (SmG)	— —
Benzothienobenzothiophenes 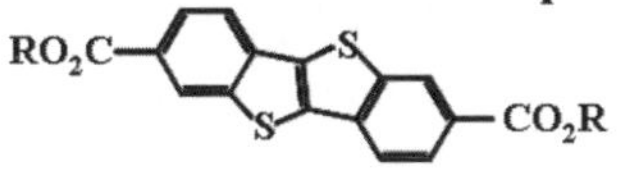 $RO_2C - \text{(BTBT)} - CO_2R$	2.2×10^{-3} (Lamello-columnar)	2.7×10^{-3} (Lamello-columnar)

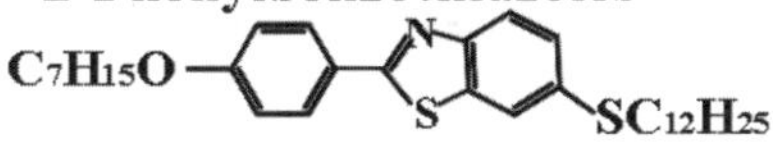

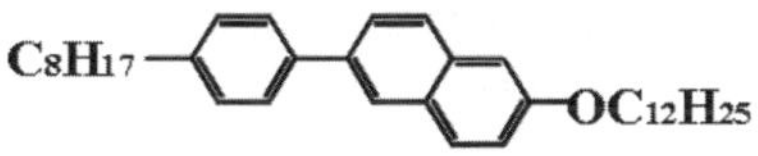

Table 3-1   (*Continued*)

Materials	Hole Mobility (cm^2/Vs)	Electron Mobility (cm^2/Vs)
Oligofluorenes	0.1 (Nematic Glass)	0.1 (Nematic Glass)
Polyacrylates	2×10^{-4} (SmA) 1×10^{-3} (Sm Glass)	—

3.2.4. Mesophase Structure

According to molecular alignments, that is, the orientational and positional orders of molecules, in molecular aggregates of liquid crystalline molecules there are a variety of mesophases in smectic and discotic liquid crystals. For discotic columnar mesophases, the geometrical order of columns gives a variation of the columnar phase in addition to the molecular order of intracolumn such as columnar tetragonal (Col_{tetr}) phase, columnar rectangular (Col_r) phase, and columnar hexagonal (Col_h) phase, in addition to the distinction of molecular order within the column as *ordered*, *disordered*, and *helical* shown in Figure 3-4.

For smectic mesophases, there are two series of mesophases in terms of molecular orientation. The first series of smectic mesophases are those in which the long axis of liquid crystal molecules is perpendicular to the smectic layer, while the second series of smectic mesophases are those in which the long axis is tilted to the molecular layer. For these two series, the geometrical order of molecules in the smectic layer gives an extended variation as indicated in Figure 3-5. The names of mesophases are given from A to K in the order of their historical discovery. For example, the smectic mesophases in which the liquid crystal molecules organize in a rectangular and hexagonal fashion are called smectic E (SmE) and smectic B (SmB), respectively, while the smectic layer with no positional order is called smectic A (SmA). In the tilted series, the mesophases having a rectangular and hexagonal arrangement are called smectic F (SmF) and smectic G (SmG), respectively, while that with no positional order in the smectic layer is called smectic C (SmC) as illustrated in Figure 3-5.

Experimental results accumulated in a series of liquid crystals having the same core moiety revealed the enhancement of mobility in a stepwise manner from phase to phase as the molecular order in mesophases is increased because of shortening of intermolecular distance. Therefore, mobility is enhanced according to the following order for smectic mesophases: SmA, SmC < SmB_{hex}, SmF < SmB_{cryst}, SmE, SmG. It is found that the mobility hardly depends on the orientation of molecular axis, that is, perpendicular or

Table 3-2a   Mobility in Various Discotic Columnar Phases Determined by Time-of-Flight (TOF) Measurements.

Materials	R	Hole Mobility (cm^2/ Vs)	Electron Mobility (cm^2/ Vs)
	C_4H_9	2.5×10^{-2} (Col_{hp})	2.3×10^{-2} (Col_{hp})
	C_5H_{11}	1.9×10^{-2} (Col_{ho})	1.7×10^{-2} (Col_{ho})
	C_6H_{13}	4×10^{-4} (Col_{ho})	4×10^{-4} (Col_{ho})
	SC_6H_{13}	0.08 (Helical)	0.08 (Helical)
	C_8H_{17}	0.1 (Col_h) 0.2 (Col_r)	0.2 (Col_h) 0.3 (Col_r)
	OC_9H_{19} $M = Zn$	1×10^{-2} (Col)	8×10^{-3} (Col)

tilted to the molecular layer. Figure 3-6 shows mobility in various mesophases of 2-phenylnaphthalene and terthiophene derivatives as a function of temperature. Judging from these results, it seems that the mobility is primarily determined by mesophase structure and the structural variation in liquid crystals gives minor contribution to the mobility. In fact, the mobility for SmA and SmC phases is on the order of 10^{-4} cm^2/Vs, those for SmB_{hex} and SmF phases on the order of 10^{-3} cm^2/Vs, and those for $SmB_{cryst,}$ SmE, and SmG phases on the order of 10^{-2} cm^2/Vs. This indicates that the intermolecular distance in the smectic layer determines the mobility in each phase primarily.

Table 3-2b   Mobility in Discotic Columnar Phases of Triphenylene Derivatives Determined by Pulse-Radiolysis Time-Resolved Microwave Conductivity (PR-TRMC) Measurements.

Materials	R		Phase	Mobility $(m^2/\,Vs)$
Triphenylenes	OC_4H_9		Col_{hp}	2.5×10^{-2}
			Cryst	6.9×10^{-3}
	OC_5H_{11}		Col_h	1.0×10^{-2}
			Cryst	3.3×10^{-1}
	OC_6H_{13}		Col_{hp}	2.0×10^{-2}
			Cryst	1.2×10^{-2}
	SC_6H_{13}		Col_h	1.0×10^{-2}
			Helical	8.7×10^{-2}
			Cryst	3.3×10^{-1}
	SC_8H_{17}		Col_h	1.5×10^{-2}
			Cryst	2.0×10^{-1}
Triphenylene dimers	OC_4H_9	$n = 10$	Col_p	9.5×10^{-3}
			Cryst	5.5×10^{-3}
	OC_4H_9	$n = 12$	Col_p	1.4×10^{-2}
			Cryst	5.8×10^{-3}
Triphenylene trimers	OC_4H_9	$n = 4$	Col_{hp}	6.0×10^{-2}
	OC_6H_{13}	$n = 3$	Col_h	1.7×10^{-2}
		$n = 6$	Col_{obl}	1.8×10^{-1}

For columnar mesophaeses, the mobility is increased as the molecular order is sophisticated in inter- and intracolumns as in the case of smectic mesophases. For example, the mobility is on the order of 10^{-3} cm^2/Vs in a columnar ordered phase, on the order of 10^{-2} cm^2/Vs in a columnar plastic phase, and on the order of 10^{-2} cm^2/Vs or higher in a columnar helical phase. The mobility, however, is not decided by intermolecular distance in a column because the intermolecular distance is around 3.5 Å irrespective of columnar mesophases. Taking account of the one-dimensional transport in columnar mesophases, where a carrier has to transport in a column without a detour along a given electric field, it is very likely that the disorder of molecular alignment in a column resulting from dynamic translational motion of the liquid crystalline molecules affects the charge-carrier transport properties in columnar phases [24]. From this point of view, there are several experimental facts to support this idea [25–27], but we have to wait to

Table 3-2c Mobility in Discotic Columnar Phases Determined by Pulse-Radiolysis Time-Resolved Microwave Conductivity (PR-TRMC) Measurements.

Materials	R		Phase	Mobility (m^2/Vs)
Perylene diazadiimides	H $C_{18}H_{37}$, H, $C_{18}H_{37}$ C_6H_{13}, H, C_6H_{13} C_6H_{13}, C_6H_{13}, H		Col Cryst.	$0.1-1.0 \times 10^{-1}$
Perylene diimides	$C_{18}H_{37}$		SmX Cryst.	1.1×10^{-1} $1.1-2.1 \times 10^{-1}$
Hexabenzocolonenes	$OC_{12}H_{25}$		Col_h Cryst.	3.8×10^{-2} 8.0×10^{-1}
	$PhOC_{12}H_{13}$		Col_h	3.5×10^{-1}
Phthalocyanines	$OC_{10}H_{21}$		Col_h Cryst.	3.8×10^{-1} 1.3×10^{-1}
	$OC_{12}H_{25}$		Col_h Cryst.	3.5×10^{-1} 2.1×10^{-1}
	$SC_{12}H_{25}$		Col_h Cryst.	2.0×10^{-1} 2.5×10^{-1}
	$PhOC_{12}H_{13}$		Col_h Cryst.	1.8×10^{-1} 3.4×10^{-1}
Porphyrins	OC_9H_{19}	$M = Zn$	Col Cryst.	6.0×10^{-3} 2.6×10^{-2}
	$OC_{10}H_{21}$	$M = Pd$	Col Cryst.	5.60×10^{-3} 3.9×10^{-1}
		$M = Cu$	Col_h Cryst.	6.5×10^{-2} 3.6×10^{-1}
		$M = Co$	Col_h Cryst.	5.3×10^{-2} 2.7×10^{-1}
		$M = Zn$	Col_h Cryst.	6.1×10^{-2} 3.0×10^{-1}

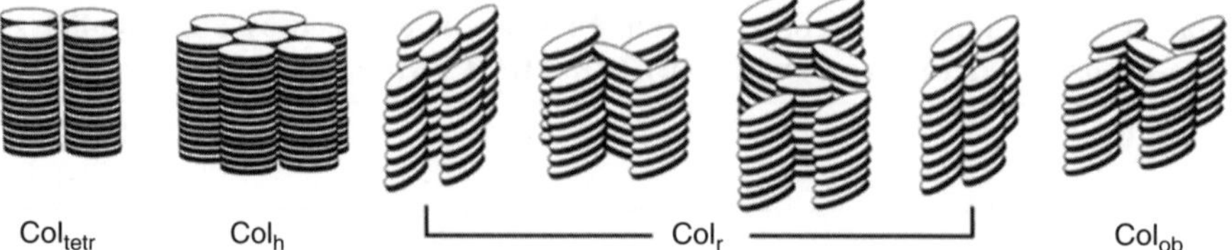

Figure 3-4   Aggregate structures in various discotic columnar phases.

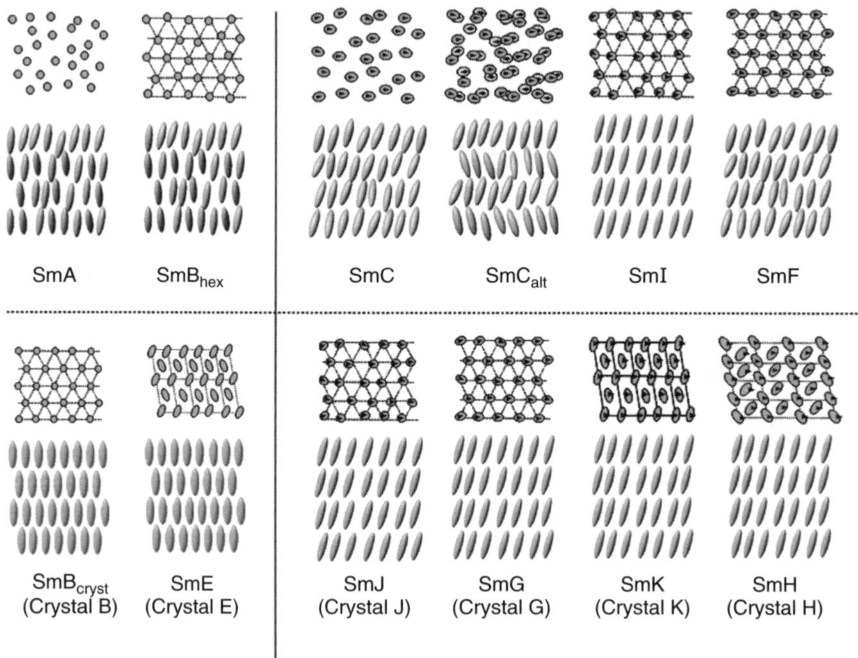

Figure 3-5   Aggregate structures illustrated with side and top views in various smectic phases. The molecules sit perpendicular to the smectic layers in left-hand side groups, while they sit tilted toward the smectic layers in the right-hand side groups. There is no correlation of the molecules among the layers in the upper groups, while there is positional correlation of molecules among the layers in the bottom groups.

draw a conclusion about what dominates the carrier mobility in the columnar mesophases because of a lack of accumulated data on the carrier mobility in various discotic liquid crystals.

3.2.5. Temperature and Electric Field Dependence

The charge-carrier transport properties in mesophases are quite unique in contrast to those of amorphous and crystalline materials. Mobility hardly depends on temperature at room temperatures or higher, as shown in Figure 3-7; furthermore, it

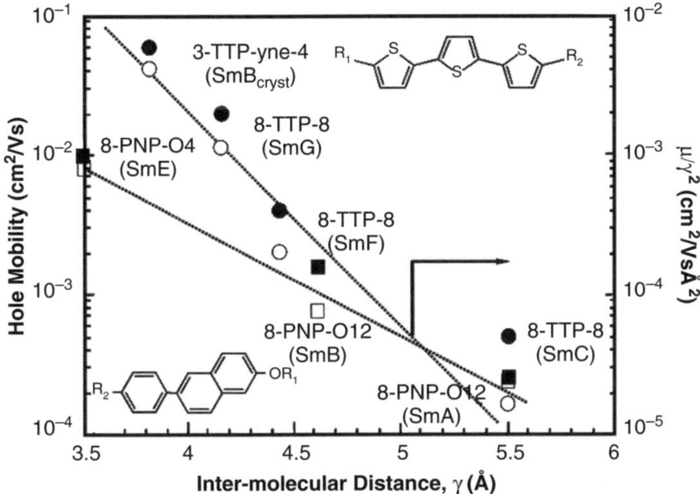

Figure 3-6   charge-carrier mobility μ as a function of inter-molecular distance, γ in 2-phenylnaphthalene (PNP) and terthiophene (TTP) derivatives. The numbers indicate the number of carbons in a side chain.

does not depend on electric field either, as shown in Figure 3-8. These behaviors are common in various mesophases including discotic columnar phases at the temperature range above room temperatures [8, 25–27]. On the other hand, it is found that the mobility in a smectic quaterthiophene derivative of 5-hexyl-5′-hexenyl-2.2′:5′2″-terthiophene having a wide temperature range below room temperatures for a smectic mesophase does depend on both temperature and electric field [28, 29]. These unique carrier transport properties are not only beneficial for device application of mesophase materials but also useful for understanding the conduction mechanism in the mesophase materials. This is discussed in detail below.

As for the charge-carrier transport properties in the nematic phase, there are few reports available because of the very recent discovery of electronic conduction in it [30, 31]. Considering the molecular alignment of the nematic phase that has no positional order as in the amorphous aggregates, it is very plausible that the conduction takes place in the distributed localized states similar to those for amorphous aggregates, and that the conduction can be three-dimensional. Therefore, it is very likely that the intrinsic carrier transport properties would be determined by carrier-dipole interaction as in the case of amorphous materials, namely, the Pool–Frenckel type of carrier transport as reported in the chiral nematic phase of a phenylquaterthiophene derivative [32].

3.2.6. Structural Defects

Unlike isotropic materials such as amorphous aggregates, ordered materials cannot be free from structural defects originating from misorientation of molecules

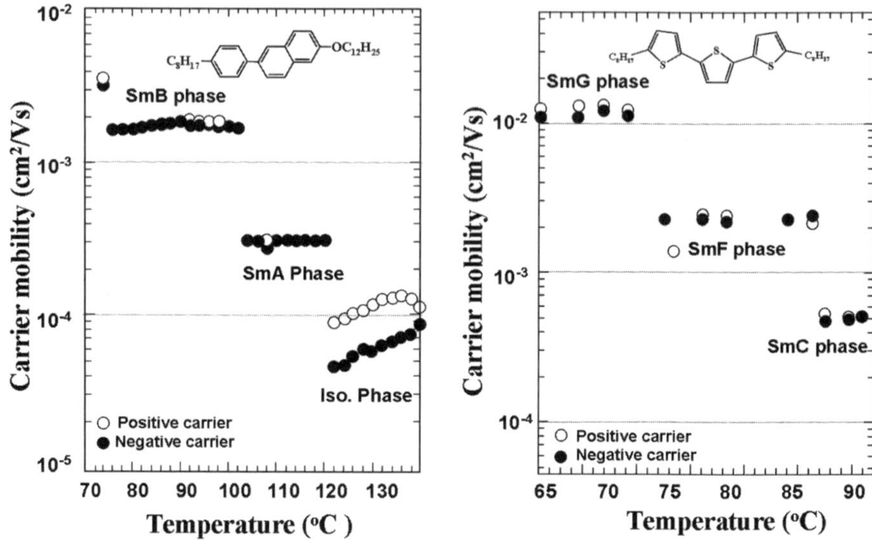

Figure 3-7 charge-carrier mobility as a function of temperature in various smectic mesophases of typical smectic liquid crystals, 2-phenylnaphthalene and terthiophene derivatives.

such as dislocations and inhomogeneous boundaries except for single crystals. These defects often accumulate chemical impurities taken in during synthesis and adsorb oxygen and water, causing shallow or deep trap states for carriers and resulting in deteriorating carrier transport properties. Indeed, polycrystalline thin films, for example, are so defective that the transient photocurrent in time-of-flight experiments becomes too dispersive to determine mobility, or often exhibits an exponential decay.

In liquid crystalline mesophases, there indeed exist domain boundaries in polydomain textures, in addition to disclinations in a domain. However, judging from the facts that neither mobility nor $\mu\tau$-product, which is the key parameter to determine the photoconductive properties in organic semiconductors, depends on a size of domains in a polydomain texture: these structural defects hardly affect the carrier transport properties in smectic mesophases [33–35]. It is plausible that the flexible nature of mesophase aggregates, or a soft structure of mesophases, makes the carrier transport at defective sites possible. As yet, the exact reason why the structural defects in smectic mesophases are less harmful to the carrier transport has not been explained. This is another outstanding feature of carrier transport in the mesophase materials, which distinguishes the mesophases from the crystalline materials as described above.

For the discotic columnar phases, there are few reports on the effect of structural defects on the charge-carrier transport so far. Unlike the smectic mesophases,

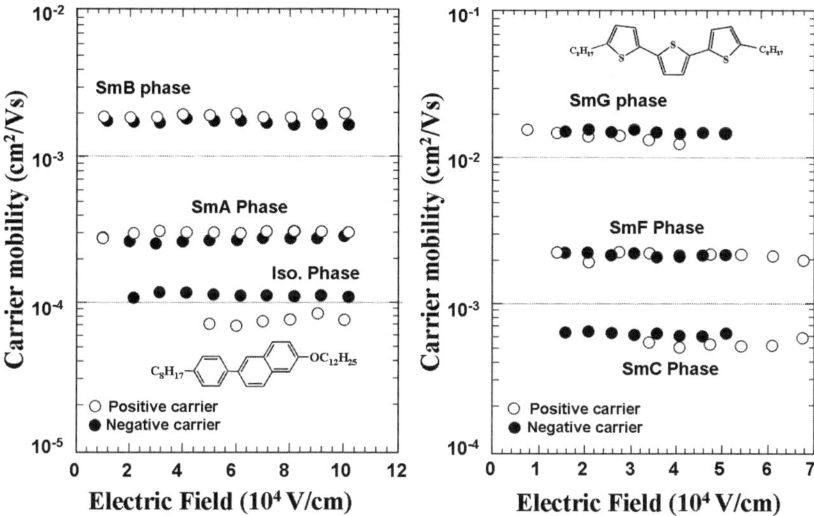

Figure 3-8 Charge-carrier mobility as a function of electric field in various smectic mesophases of typical smectic liquid crystals, 2-phenylnaphthalene and terthiophene derivatives.

where two-dimensional transport takes place, it is likely that the structural defects in a column may affect the carrier transport properties seriously, because a carrier has to pass throughout a column without detouring to adjacent columns: From this point of view, relatively low mobility in the columnar ordered phase of triphenylene derivatives, where the intermolecular distance is as small as 3.5 Å, might be explained by the structural defects or disorder of molecular alignment in a column as described above.

3.3. CHARGE TRANSPORT MODEL FOR LIQUID CRYSTALS

3.3.1. Introduction

Charge-carrier transport in liquid crystals is one of the current topics of theoretical study on organic semiconductors. It is different from the charge transport in amorphous solids or crystals. This is because the liquid crystalline phase is the mesophase between the liquid phase and the crystal phase, which is an ordered molecular aggregate lacking in long-range order. We often apply existing models developed to describe the charge-carrier transport properties of amorphous solids or crystals when analyzing the carrier transport properties in liquid crystals [36, 37]. Sometimes these are successful to explain some experimental results partly. In fact, in liquid crystals, we often observe temperature- and field-independent carrier transport at a temperature range above room temperatures, as is observed in the crystalline materials [38], while we observe dispersive carrier transport below room temperatures [39], as is often observed in amorphous solids. These behaviors indicate that the carrier transport in liquid crystals comprises those of the amorphous and crystalline solids, just as in the case of the state of matter. All these behaviors are attributed to the features of mesophase as described above.

Before considering a charge transport model for liquid crystals, let us look over the charge transport process in two different scales: One is the electron transfer (ET) process from molecule to molecule, namely, the Marcus model, which describes *the microscopic view* of the charge transport in molecular scale; the other is the charge transport process in a whole molecular system of liquid crystal, that is, *the macroscopic view* of the charge transport in molecular aggregate. In order to understand the charge-carrier transport properties in liquid crystals we need both the micro- and macroscopic views, and any model combines these views.

3.3.2. Microscopic View

It is very important to know which electronic state is dominant, *delocalized (bandlike)* or *localized (hopping) states*, when we try to understand the conduction mechanism in a particular material system. Judging from the charge transport behaviors as described above and the reported mobility less than 1 cm^2/Vs in the mesophase, it is very likely that the conduction regime in the liquid crystals is dominated by localized states rather than by bandlike states. Therefore, the conduction can be described not by the band regime but by the hopping regime.

In this section, we start from the brief introduction of the small polaronic transport using the Marcus-type scheme for the electron transfer (ET) process, which is well established in chemical and biological ET processes. This eventually provides almost same model as the Holstein-type explanation of the small polaron regime [40].

3.3.2.1. Fermi's Golden Rule.

In the framework of Fermi's Golden Rule, the transition probability per unit time in the long time limit ($t \gg 1/\omega_{mi,nf}$) from a vibronic level m of the electronic state i to a vibronic level n of the electronic state f can be denoted by first-order perturbation theory [41–43]

$$W_{fi} = \frac{2\pi}{\hbar^2}\left|V_{mi,nf}\right|^2 \delta(\omega_{mi,nf}) = \frac{2\pi}{\hbar}\left|V_{mi,nf}\right|^2 \delta(E_{nf} - E_{mi}) \qquad (3.1)$$

where $2\pi\hbar$ is the Planck constant and $V_{mi,nf} = \langle mi|V|nf \rangle$ is a matrix element of the electronic coupling, or a transfer integral from a vibronic state $|mi\rangle$ to a vibronic state $|nf\rangle$. $\hbar\omega_{mi,nf} = E_{nf} - E_{mi}$ is an energy difference between $|mi\rangle$ and $|nf\rangle$. E_{mi} and E_{nf} are the (zero order) energy states, respectively. The electronic transfer (ET) rate k_{mi} from an initial vibronic state $|mi\rangle$ of reactants to all of the finally vibronic manifold $\{|nf\rangle\}$ is derived by the summation of all of n states.

$$k_{mi} = \frac{2\pi}{\hbar}\sum_{n}\left|V_{mi,nf}\right|^2 \delta(E_{nf} - E_{mi}) \qquad (3.2)$$

This ET process is nonadiabatic, and the microscopic ET rates are thermally averaged. The ET rate k_{ET} from the initial manifold to the final manifold is then obtained by summing k_{mi} over all the initial states $|mi\rangle$ where each is weighted by the Boltzmann probability $P(E_{mi})$

$$k_{\text{ET}} = \sum_m k_{mi} P(E_{mi}) \tag{3.3}$$

where $P(E_{mi})$ is

$$P(E_{mi}) = \frac{1}{Z} e^{-E_{mi}/k_B T} \text{ and } Z = \sum_m e^{-E_{mi}/k_B T} \tag{3.4}$$

3.3.2.2. Franck–Condon Factor.

Applying the Franck–Condon approximation $V_{mi,nf} \cong V_{if}\langle mi|nf\rangle$, which means that V is independent of nuclear coordinate, the nonadiabatic ET rate is obtained by substituting Eq. (3.2) into Eq. (3.3).

$$k_{\text{ET}} = \frac{2\pi}{\hbar} \sum_m \sum_n |V_{mi,nf}|^2 P(E_{mi})\delta(E_{nf} - E_{mi}) \tag{3.5}$$

$$\cong \frac{2\pi}{\hbar} |V_{if}|^2 \sum_m \sum_n F_{mn} P(E_{mi})\delta(E_{nf} - E_{mi}) \tag{3.6}$$

$$= \frac{2\pi}{\hbar} |V_{if}|^2 (\text{FCWD}) \tag{3.7}$$

where $F_{mn} = |\langle mi|nf\rangle|^2$ is the Franck–Condon overlap and FCWD is the Franck–Condon weighted density-of-states. As E_{mi} increases, F_{mi} increases while $P(E_{mi})$ decreases. The summation of Eq. (3.6) has included all intramolecular vibrations and surrounding vibrations such as solvent vibronic modes. However, the frequency of solvent vibrations is relatively low, and so classical treatment can be applied. On the other hand, the intramolecular vibrations are relatively high, and then the quantum treatment is applied. Consequently the ET rate of Eq. (3.7) is reproduced as the semiclassical Marcus equations [44–46] (Fig. 3-9)

$$k_{\text{ET}} = \frac{2\pi}{\hbar} \frac{|V_{if}|^2}{\sqrt{4\pi\lambda_{\text{out}}k_B T}} \sum_m \sum_n F_{mn} P(E_{mi})$$

$$\exp\left[-\frac{(\Delta G_0 + E_{nf} - E_{mi} + \lambda_{\text{out}})^2}{4\lambda_{\text{out}}k_B T} \right] \tag{3.8}$$

where λ_{out} is a solvent reorganization energy.

In some models [40, 47, 48], it is assumed that the contribution of the relevant high-frequency vibrations of states $|mi\rangle$ can be reduced to a single high vibration with averaged frequency $\langle\omega_c\rangle$. Thus, Eq. (3.8) is simplified to be

$$k_{\text{ET}} = \frac{2\pi}{\hbar}\frac{|V_{if}|^2}{\sqrt{4\pi\lambda_{\text{out}}k_B T}}\sum_{v'}\frac{S_c^{v'}e^{-S_c}}{v'!}\exp\left[-\frac{(\lambda_{\text{out}}+\Delta G_0+v'\hbar\langle\omega_c\rangle)^2}{4\lambda_{\text{out}}k_B T}\right] \quad (3.9)$$

where v' is an integer and S_c is vibrational coupling constant or Huang–Rhys factor. In the regime of ET rate denoted by Eq. (3.9), intramolecular reorganization energy λ_{in} is treated as the summation of harmonic oscillations. The coupled vibrations characterized by frequency ω arise from structural differences between the equilibrated configurations of the initial state i and final state f.

$$\lambda_{\text{in}} = \frac{1}{2}\sum_j \lambda_j = \frac{1}{2}\sum_j \overline{f}_j \Delta q_{\text{eq}}^2 \quad (3.10)$$

where Δq_{eq} is the equilibrium displacement along the normal mode (NM) j between the equilibrated positions of the two states. This can be related to S_j as defined in Eq. (3.11)

$$S_j = \frac{\lambda_j}{\hbar\omega_j} = \frac{\overline{f}_j \Delta q_{\text{eq}}^2}{2\hbar\omega_j} \quad (3.11)$$

The effects of quantum behavior can be subsumed by treating only a few modes with the mode averaged. Those few modes can be represented as averages of

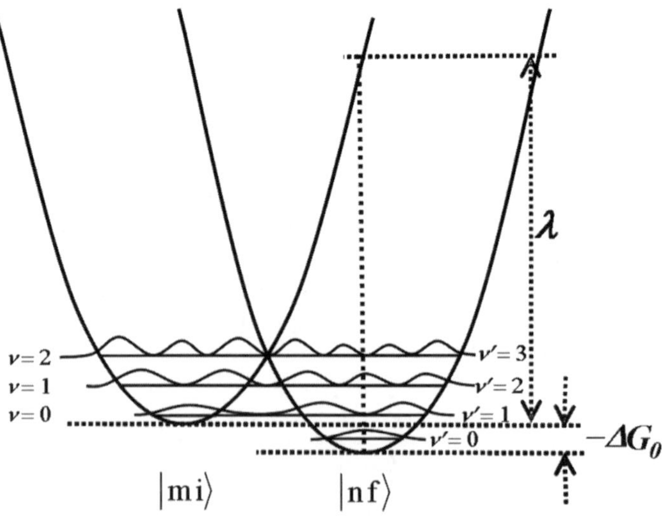

Figure 3-9 Schematic representation of potential- energy diagram for the quantum mechanical view of electron transfer. The vibrational wave functions are shown symbolically for importance of vibrational overlap.

many manifolds. Thus the S_c is defined as:

$$S_c = \sum_j S_j = \sum_j \frac{\lambda_j}{\hbar \omega_j} \tag{3.12}$$

The averaged frequency is derived as

$$\hbar \langle \omega_c \rangle = \frac{\sum_j \hbar \omega_j S_j}{\sum_j S_j} = \frac{\sum_j \lambda_j}{\sum_j S_j} = \frac{\lambda_{\text{in}}}{S_c} \tag{3.13}$$

Equation (3.8) maximizes at $-\Delta G_0 \sim \lambda_{\text{in}} + \lambda_{\text{out}} \equiv \lambda$ in the normal region, where $-\Delta G_0 < \lambda$, Eq. (3.8) is represented very well by the simplified expression

$$k_{\text{ET}} = \frac{2\pi}{\hbar} \frac{|V_{\text{if}}|^2}{\sqrt{4\pi \lambda k_B T}} \exp\left[-\frac{(\lambda + \Delta G_0)^2}{4\lambda k_B T} \right] \tag{3.14}$$

This constitutes the *high temperature limit* of semiclassical Marcus–Hush theory [49]. All vibrational modes can be treated as classical, which means that the spacing between the vibrational states is small compared with thermal energies, $k_B T$. On the other hand, Eq. (3.9) is treated as quantum mechanics where the spacing between the vibrational states is large compared with thermal energies. Thus Eqs. (3.15-a) and (3.15-b) provide a guide to select the hopping regime as either *quantum* or *classical*

$$low\ temperature\ limit: \hbar \omega_j \gg k_B T \quad \text{quantum} \tag{3.15-a}$$

$$high\ temperature\ limit: \hbar \omega_j \ll k_B T \quad \text{classical} \tag{3.15-b}$$

Rate equation (3.9) under the low temperature limit consists of the product of two parts, namely, the Franck–Condon (FC) term and another term.

$$k_{\text{ET}} = \frac{2\pi}{\hbar} |V_{if}|^2 \frac{1}{\sqrt{4\pi \lambda_{\text{out}} k_B T}} (\text{FC}) \tag{3.16}$$

$$(\text{FC}) = \sum_{v'} \frac{e^{-S} S^{v'}}{v'!} \exp\left[-\frac{(\lambda_{\text{out}} + v' \hbar \omega + \Delta G_0)^2}{4\lambda_{\text{out}} k_B T} \right] \tag{3.17}$$

The front term of the FC provides the frequency of electron transfer for nonexistence of thermal activation factor. The Franck–Condon factor is composed of all available overlap integral between the initial vibrational state v and the final state v'. Thus each v' indicates a different channel of ET from $v = 0$ to $v = v'$. Each different exponential term of the FC in the sum is the distribution of molecules having the energy for ET through the $v = 0 \rightarrow v'$ channel.

Two ET rates, classical Eq. (3.14) and quantum Eq. (3.9), afford the same behavior very well in the normal region where $\lambda > -\Delta G_0$, and thus Eq. (3.14) is often used even when the condition is not the high temperature limit.

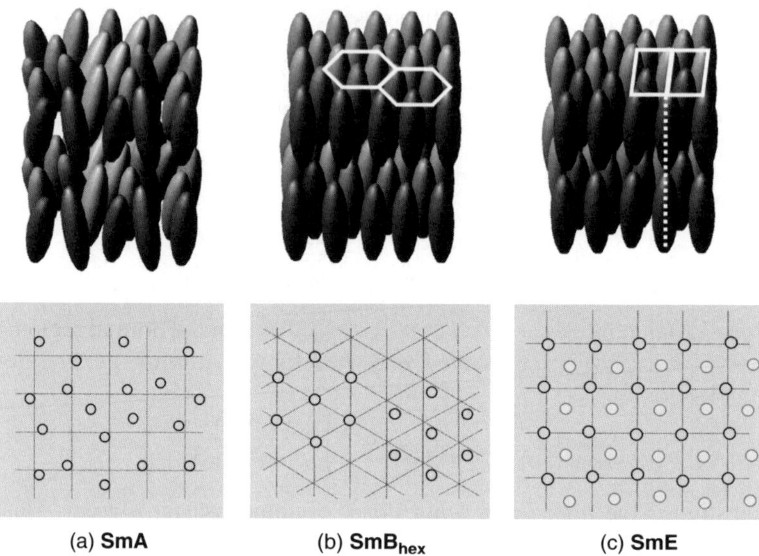

Figure 3-10   Smectic mesophases and their molecular alignments. (A full color version of this figure appears in the color plate section.)

3.3.2.3. Reorganization Energy.

The internal reorganization energy λ_{in} is defined in Eq. (3.10). In the ET process between adjacent localized states of the same molecules, the equilibrium of displacement Δq_{eq} arises due to the difference of an equilibrated geometry between the neutral and the charged molecule. Thus the molecule forced to a large structural rearrangement by an induced charge has a large amount of the reorganization energy or the polaron binding energy. The total internal reorganization energy λ_{in} can be decided only by the total-energy difference between the energies of the system before and after ET, with molecular geometry at equilibrium for final state [50] (Fig. 3-9).

$$\lambda_{in}^{\pm} = \left[E^{\pm}(g^0) - E^{\pm}(g^{\pm}) \right] + \left[E^0(g^{\pm}) - E^0(g^0) \right] \tag{3.18}$$

where $\lambda_{in}^{+}(\lambda_{in}^{-})$ is the internal reorganization energy of a hole (electron). $E^{+}(g^0)$ is the energy of a single molecule with an excess hole forming nuclear geometry corresponding to that for the neutral molecule at equilibrium, etc. Based on Eq (3.18), the internal reorganization energy can be numerically calculated using molecular orbital calculations, such as the Hartree–Fock or DFT methods [51–55].

3.3.2.4. Transfer Integral.

The electronic coupling or transfer integral from the initial state $|i\rangle$ to the final state $|f\rangle$ can be described as the matrix element of V; then this can be calculated as the matrix element $\langle i|h|f \rangle$, where h denotes the Hartree–Fock Hamiltonian or the Kohn–Sham Hamiltonian. The effective charge transfer integrals V_{eff} are defined as [56]:

$$V_{\text{eff}} = \langle i|h|f \rangle - \frac{E_i + E_f}{2} \langle i|f \rangle \qquad (3.19)$$

This recasting is due to removal of the spatial overlap of the neighboring-site orbitals from overestimation of the transfer integral in Hamiltonian **h**. Matrix elements of the transfer integral $V_{if} = \langle i|h|f \rangle$, the overlap integral $S_{if} = \langle i|f \rangle$, and site energies $E_i = \langle i|h|i \rangle$, $E_f = \langle f|h|f \rangle$ can be directly calculated as matrices of dimer in basis set of the molecular orbitals of monomers (fragment orbitals). The Amsterdam Density Functional theory program [57] can numerically calculate **S**, **E**, and **C**, and the Gaussian program [58] can derive these transformed matrices from the atomic orbital basis set to the fragment orbital set [59–61], where **S**, **E**, and **C** are the overlap matrix, the diagonal-eigenvalue matrix, and the eigenvector matrix of the dimer. Thus the charge transfer integral can be obtained as the matrix elements of the Kohn–Sham Hamiltonian $\langle i|h_{\text{KS}}|f \rangle$ in the basis of fragment orbital [62–64]:

$$\mathbf{h}_{\text{KS}} = \mathbf{SCEC}^{-1} \qquad (3.20)$$

Using the derived matrix elements Eq. (3.20), effective transfer integrals are obtained from the calculation of Eq. (3.19). This calculation shows that the transfer integrals basically exhibit exponential decay along the radius direction depending on the molecular distance [50, 59]. However, the values of the transfer integral have variation depending on the dihedral angle between two molecules. Thus the precise face-to-face position of molecules is very important in closely packed material such as crystals, where it dominates the potentiality of the charge transport in crystals. In liquid crystal or other meso-ordered materials this variation provides variation of the hopping rate from molecule to molecule with a certain deviation of a transfer-rate distribution limited by molecular structure.

3.3.3. Macroscopic View of Charge Transport

In the hopping regime, if the hopping energy and transfer integral at each site are the same in the ordered material system, the ET rate is constant. In this case, the Einstein relation is held between the mobility μ and diffusivity $D = a^2 \nu$, where the ν is simply a constant of the ET rate. In most self-organized molecular systems including liquid crystalline phases, however, they have some static structural disorder, and so the hopping rate varies from site to site because of variation in the hopping energy and the transfer integral. As for the molecular aggregates of liquid crystals, each molecule has several kinds of dynamic molecular motion such as translational and rotational motions, but these motions are slow compared with the hopping rate. Therefore, it is quite likely that charge electronic states are localized to the scale of the molecule and the thermal activation process of carriers dominates the hopping rate.

Consequently, the hopping transport in these systems does not provide a simple Einstein relation with diffusivity $D = a^2 v$. For this case, we must take into account the structure of molecular aggregates, that is, degrees of molecular alignment and molecular orientation as a molecular system. In addition, we must pay attention to the anisotropic conduction in the liquid crystalline mesophase because of the anisotropic structure of the liquid crystalline molecule.

In this section, we describe the molecular order and orientation in typical liquid crystalline mesophases and how they are involved in the macroscopic model for the charge-carrier transport in the liquid crystals.

3.3.3.1. *Liquid Crystalline Phase and Order Parameters.*   The liquid crystalline mesophases are characterized as ordered molecular aggregates without long-range orders, which are the intermediate matter of state between liquids and crystals. We emphasize that the liquid crystalline phase is a unique phase in terms of a state of matter but not so different from both the crystalline and the liquid phases; thus its feature as a molecular aggregate can be described by the order parameters for orientational and/or translational orders, as characterized by a complete set of orders for the crystal states and by none of the order parameters for the liquid or amorphous state: The variation of liquid crystalline mesophases is described by different sets of order parameters. Detailed descriptions of each phase and related order parameters are provided in many texts [64–67]. Here we focus on typical liquid crystalline phases of which we are interested in the charge conduction, such as series of smectic mesophases of smectic A (SmA), smectic B (SmB), and smectic E (SmE) and discotic columnar phases. First, we describe the relation between the ordered molecular alignment and the structure of each phase, and then we describe the macroscopic models to understand the charge-carrier transport properties in the liquid crystals.

3.3.3.2. *Orientational Order and Nematic Phase.*   Order parameters are categorized into two kinds: one is the orientational order parameter, and the other is the translational order parameter. One of the common features in liquid crystalline molecules is that their molecular long axis and short axis are well defined. Provided that the primary axis for the molecular long axis is well defined, the scalar order parameter is given by the average

$$S = \langle P_2(\cos\theta)\rangle = \frac{1}{2}\langle 3\cos\theta - 1\rangle \qquad (3.21)$$

where $P_2(x)$ is the second Legendre polynomial and the brackets denote statistical averaging. θ is the angle between the long axis and the director. Thus the order parameter defines dispersion of molecular orientation: $S = 1$ means that the molecular long axis of each molecule in an aggregate is completely oriented to the director. In contrast, $S = 0$ means the orientation of each molecule is random.

In common, the direction of the π-conjugated wave function is perpendicular to the molecular long axis. Hence, the direction of the wave function must be characterized by order parameter S. If S is close to 1, π-wave functions are strongly oriented in the direction perpendicular to the director. By contrast, the wave functions are weakly limited when S is closed to 0.

As described in the historical studies on electrical properties in liquid crystals, the conduction in the nematic phase of small molecules had been considered to be ionic before it was established in highly purified phenylbenzothiazole derivatives reported in 2009 [68]. The electronic conduction has been reported in a few examples, and the intrinsic charge transport properties in the nematic phase have not been clarified yet.

Because of the lack of translational or positional order in the nematic phase, its molecular alignment is little different from that of the isotropic (liquid) phase except for molecular orientation. Therefore, it is expected that the distribution of electronic coupling between two sites responsible for charge transfer is anisotropic but that their values are not different from those in the isotropic phase. We see that a different behavior of charge conduction between nematic and isotropic phases originates from the constitution of anisotropy in nematic phase: When the translational order parameter S is close to 1, the coupling is very weak along the direction of the nematic director; in contrast, the coupling is strong perpendicular to the director; consequently, anisotropy in conduction arises.

3.3.3.3. *Translational Order and Smectic Phase.*   In smectic phases, rod-like molecules having π-conjugated core moieties and long hydrocarbon chains often aggregate in layers in a self-organizing manner. Periodic modulation in electron density along the layer normal is applied for the expression of layered structure in smectic liquid crystal. The distribution function is defined as

$$f(z) = \frac{1}{d} + \sum_{n=0}^{\infty} \tau_n \cos\left(\frac{2\pi n z}{d}\right) \tag{3.22}$$

along the z-axis, where d is the nearest interlayer distance. The first harmonic of the Fourier coefficient in Eq. (3.22)

$$\tau_1 = \left\langle \cos\left(\frac{2\pi z}{d}\right)\right\rangle = \int_{-d/2}^{d/2} dz \cos\left(\frac{2\pi z}{d}\right) f(z) \tag{3.23}$$

is often referred to as "the smectic order parameter," which gives the strength of sinusoidal order of the layer. Locations of molecules are highly periodic and aligned on the layer if τ_1 is large. On the other hand, centroid alignment of molecules is nearly random if τ_1 is close to 0. Molecules aligned in the layer promise quasi-two-dimensional conduction in smectic liquid crystals because of large electronic coupling along the direction of the layer.

In the smectic phase, there exist subcategorized phases according to the molecular alignment of the molecules in a smectic layer. Here we pick up SmA, SmB_{hex}, SmB_{cryst}, and SmE phases. In the SmA phase, molecules in the layers align randomly, so that it can be a two-dimensional liquid. On the other hand, they behave elastically for the other direction. In the SmB_{hex} phase, the molecules in a layer have a long-range order of bonding orientation, which means that the direction of the bond between molecules has long-range order although they have a short-range order of translation in the layer. The translational order of molecules is broken a few molecules away. On the other hand, in the SmB_{cryst} and SmE phases, they have long-range order with translational order, in addition to hexatic and rectangular order in the layer, respectively.

SmB_{cryst} and SmE phases have long-range order in periodical molecular alignment. This means the disappearance of *Landau-Peierls instability* of SmA phase, where the thermal fluctuation breaks the long range order along z axis. Thus the radial distribution function along the layer normal decays $\sim r^{-\eta}$, where r is the distance between molecules in a layer and η is the exponent depending on temperature. This instability disappears in SmB_{cryst} and SmE phase. However, the thermal fluctuation occurs in SmB_{cryst} and SmE due to the weak elastic constant. Therefore, their periodic alignment is obscure in the layers (see Fig. 3-10 and Fig. 3-5).

3.3.3.4. Discotic Liquid Crystal.

Disclike molecules consisting of π-conjugated aromatic moieties can easily form a planar conformation because the orbital overlap between adjacent molecules becomes optimal in a planar conformation. Peripheral substitution of the disclike aromatic molecule with long hydrocarbon chains is helpful to cause mesomorphic properties when the molecules aggregate. The most popular mesophases are discotic columnar phases, in which the disclike molecules aggregate in molecular columns to form one-dimensional conducting pathways along the axis of the column. On the other hand, the columns are packed in two dimensions and give various discotic columnar phases such as hexagonal columnar phase and rectangular columnar phases according to column alignment. The charge-carrier conduction among the columns, however, is very much limited because of the long hydrocarbon chains attached to the disclike core moiety, resulting in the anisotropic conduction in columnar phases, that is, one-dimensional conduction along with the column axis.

3.3.3.5. Dimensions and Charge Transport in Liquid Crystal.

As we see in the next section, charge transport can be described by introducing the concept of the *Gaussian disorder model* (GDM), [69] which has two kinds of static structural disorder. One is related to the variation of transfer integral depending on face-to-face condition randomness between neighbor molecules. The other is the energetic variation of the charge hopping states originated by spatial randomness of coulomb potential by randomly oriented and distributed charge distributions such as the dipole moment that exists in surrounding molecules. Focusing on the charge transport in the smectic phase, typical order parameter are $S_2 \sim 0.6-0.8$ in nematic and $\tau_1 \sim 0.7-0.8$ in smectic [65, 67]. But no long-range order can be seen along the smectic layer direction in SmA and SmB_{hex} phases. Thus in these phases we can apply the two-dimensional GDM, which is almost the same as three-dimensional GDM except for dimension. We consider the model for Monte Carlo simulation in the next section, while we can conclude that the hopping model can be applicable in such phases. SmA and SmB_{hex} phases have no long-range order along the smectic layer where the charge hopping occurs. Thus each hop occurs based on different transfer integrals and different energetic hopping states. Thus it is difficult to delocalize, and hopping conduction is dominant.

3.3.4. Modeling of Charge Transport

As we saw in the previous section, the presence of static structural disorder such as spatially varying transfer integrals stimulates the localization of hopping sites. At the same time the static electronic distribution, for example, random alignment and orientation of dipole moments, provides spatially varying electronic states of hopping sites.

In amorphous organic solids, the Gaussian disorder model (GDM) proposed by Bässler [69] is well-accepted and conveniently used for the analysis of the charge transport behavior in those materials. The GDM of photoinjected charge transport is a model in which charges hop among molecules with different energies coming from a Gaussian distribution. From the viewpoint that the dominant regime of charge migration in our self-organized molecular system is hopping between localized states of adjacent molecules, the Bässler model should be an acceptable model also in the case of smectic liquid crystals or a self-organized molecular system. The static structural disorder as in amorphous materials exists in liquid crystals, but they are controlled by an order parameter as shown in the previous section. In liquid crystalline materials, then, this kind of disorder is smaller than those of amorphous materials. In this section we introduce the GDM into Monte Carlo simulation to describe the electronic transport in smectic liquid crystals and thin films of self-organized semiconductors forming a two-dimensional lattice structure.

3.3.4.1. Gaussian Disorder Model.

The charge transport in amorphous materials often shows mobility with strong field and strong temperature dependence. Off-diagonal disorder, due to Gaussian-type deviation with respect to position and orientation from the simulated three-dimensional periodic lattice points, comes from the disorder of the position and orientation of each molecule in amorphous organic material [70]. This provides the random variation of transfer integral between hopping sites. Diagonal disorder, due to Gaussian-type deviation with respect to energetic level, comes from energetic disorder of the HOMO (for holes) or LUMO (for electrons) levels. Unfortunately, however, there is no direct and easy way to find the origin of the diagonal disorder because we cannot measure the inter- and intramolecular contributions to the transfer activation energy. Direct probing by absorption spectroscopy can hardly give information about the density of states (DOS) function for charge transport state. The origin of the disorder may be ascribed to the superposition of ubiquitous contributions of electrostatic potential caused by randomly distributed permanent dipoles [71–73]. We assume that the energy-localized state at ε_i is characterized by a Gaussian-distributed DOS

$$g(\varepsilon_i) = \frac{1}{\sqrt{2\pi}\sigma} \exp\left[-\frac{(\varepsilon_i - \varepsilon_0)^2}{2\sigma} \right] \qquad (3.24)$$

as is used in the case of disordered organic solids, where σ is the energetic width characterizing the fluctuation of Gaussian DOS in the polarization energy and charge-dipole interactions. The long-range nature of the carrier-dipole interaction has been shown to play a crucial role, for it creates a spatial correlation in the DOS, so that the potential change in neighboring sites is smooth. In liquid crystals, however, the random variation of dipole moment is controlled by an order parameter, and the energetic disorder caused by the carrier-dipole interactions is small because the local Coulomb potential variation originated from dipole moments is controlled by order parameters.

We consider a system where the carriers hop among the molecules in the smectic mesophase. All the molecules sit in layers with thermal fluctuation, and the molecular alignment in the layers determines each mesophase. The interlayer molecular distance of about $\sim$40 Å is much greater than the intralayer distance of $\sim$5 Å in smectic B or E phases of 8-PNP-O12 and 8-PNP-O4 [74].

In amorphous semiconductors, a reasonable transition rate for hopping processes was suggested by Miller and Abrahams [75]. It is the product of prefactor v_0, a carrier wave-function overlap factor, and a Boltzmann factor for jumps upward in energy, that is, the jump rate between sites i and j, with the distance between the sites being r_{ij}, is

$$v_{ij} = v_0 \exp\left(-\frac{2r_{ij}}{\alpha}\right) \begin{cases} \exp\left[-\dfrac{\varepsilon_j - \varepsilon_i - e\mathbf{E} \cdot \mathbf{r}_{ij}}{kT}\right] & (\text{for } \varepsilon_j - \varepsilon_i - e\mathbf{E} \cdot \mathbf{r}_{ij} > 0) \\ 1 & (\text{for } \varepsilon_j - \varepsilon_i - e\mathbf{E} \cdot \mathbf{r}_{ij} < 0) \end{cases}$$

$$(3.25)$$

Here, α is a specific decay length of the wave function in the localized states, $\mathbf{r}_{ij}$ is the vector from site i to site j, and E is the applied electric field. The quantity $e\mathbf{E} \cdot \mathbf{r}_{ij}$ is the difference of electrostatic potential between the ith and jth hopping sites. The only source of activation energy comes from the difference in site energies, and the rate is only thermally activated for an "uphill" hop (Fig. 3-11).

The probability P_{ij} for jumping from a site i to a site j within a two-dimensional lattice is

$$P_{ij} = \frac{v_{ij}}{\sum_k v_{ik}}$$

$$(3.26)$$

and the dwell time of a carrier at a site i is described by the following equation:

$$t_{ij} = -\frac{\ln(x)}{\sum_k v_{ik}}$$

$$(3.27)$$

Here, x is a random number taken from a uniform distribution between 0 and 1. The sum is taken over all connected sites k except for the site i from which the hop commences.

3.3.4.2. *Scheme of Monte Carlo Simulation.*

In the case of the charge conduction model in smectic phases, two-dimensional simulations on a model layer were considered. The model consists of 100×5000 sites aligned as a

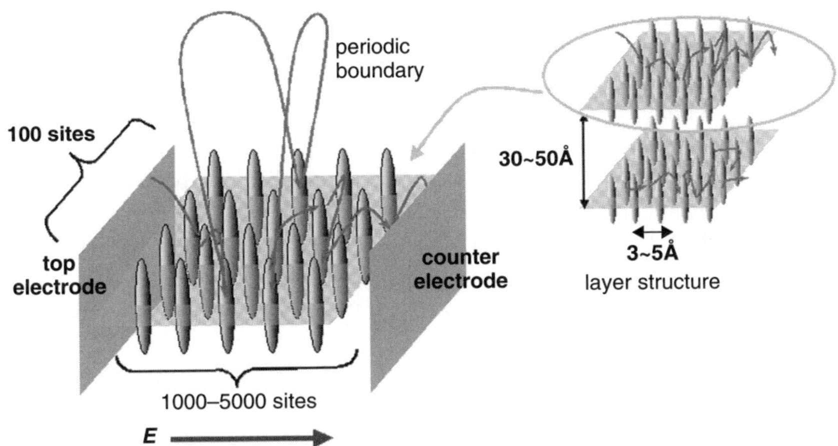

Figure 3-11 Scheme of GDM simulation. The simulation measures the transit time from the top electrode to the counter electrode on the two-dimensional layer. (A full color version of this figure appears in the color plate section.)

hexagonal lattice and a rectangular lattice, to compare with the experimental results in SmB and SmE phases, respectively [76, 77]. To extend the lattice size in the direction perpendicular to the applied field we applied a periodic boundary condition. For the site distance $a = \sim 3{-}5$ Å, 5000 sites corresponds to a sample length of around 2 μm, which is typical of the thickness of the actual samples examined in time-of-flight experiments (Fig. 3-11). The parameters were chosen as $\nu_0 = 6 \times 10^{12}$ Hz and $\alpha = 2.3$ Å. The distance between neighboring sites, a, is to be 4.6 Å in SmB and 3.5 Å in SmE phases.

Before the simulation, each of the hopping sites is assigned an energetic shift ε_i and a positional parameter Γ_i. In the simulation we allow a charge to start at arbitrary selected sites located near the top electrode. The charge then hops randomly around the sites according to the probability P_{ij} of transfer from i to j (Eq. 3.26) during the time t_{ij} (Eq. 3.27) and disappears on arriving at one of the sites adjacent to the counter electrode. This process is repeated $N = 10{,}000$ times. For each random walk a transit time t_n is calculated. Average velocity $< v >$ is thus derived as

$$< v > = \frac{1}{N} \sum_{n=1}^{N} \frac{L}{t_n} \tag{3.28}$$

where L is the sample length. The mobility μ_a is calculated as

$$\mu_a = \frac{< v >}{E} = \frac{1}{NE} \sum_{n=1}^{N} \frac{L}{t_n} \tag{3.29}$$

3.3.4.3. *Simulation of Field Dependence.*   The computational simulations were done in SmB and SmE lattice structures. First, we examined the simulation for the SmB phase and compared this with the measurements on 8-PNP-O-12 [76]. The behavior of the calculated charge-carrier mobility as a function of the electric field is shown in Figure 3-12, for varying values of σ. We can recognize the existence of a region having a Poole–Frenkel behavior as well as a region of field-independent mobility. In fact, there are three separate regions: (i) a high-field region $\sigma \ll eEa$, where the mobility decreases with increasing electric field; (ii) a region with a field-dependent mobility (the Poole–Frenkel region) in the range $\sigma \sim eEa$; and (iii) a region with a field-independent mobility ($\sigma \gg eEa$).

In region (ii) where the electric field strength is in the range of $\sim 10^5 - 10^6$ V/cm, the drift and diffusion of the carriers depends on E. To parameterize the Poole–Frenkel behavior, we apply a trial function of the following form:

$$\mu = \mu_0 \exp\lfloor (c\hat{\sigma})^n + \gamma' \sqrt{eaE/\sigma} \rfloor \tag{3.30}$$

$$\gamma' = C'(\hat{\sigma}^m - \Gamma) \tag{3.31}$$

where $\hat{\sigma} = \sigma/kT$. This form was first introduced by Novikov et al. [78] to describe the results in the case of correlated disorder. Figure 3-13 shows the mobility vs. $\sqrt{eaE/\sigma}$ for calculations pertaining to the SmB phase. This behavior is also typical for three-dimensional (Bässler et al. [69]) and one-dimensional (Bleyl et al. [79], Novikov et al. [78], and Kohary et al. [80]) systems. We have

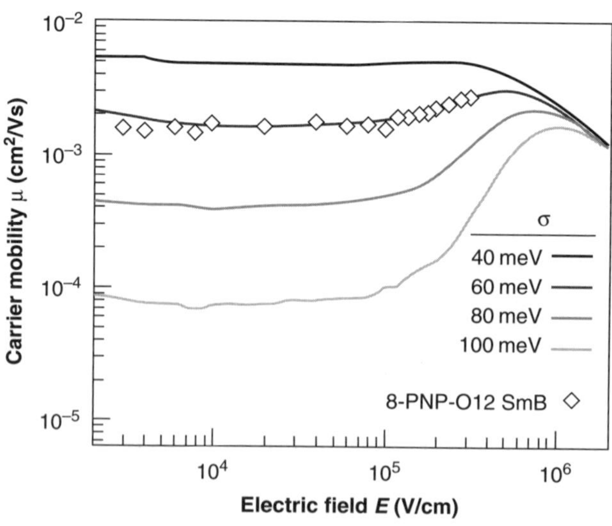

Figure 3-12   Field dependence of carrier mobility in SmB. The lines show the simulation results. Open diamonds are experimental data. (A full color version of this figure appears in the color plate section.)

Table 3-3   Coefficients c in Eq. (3.30) versus Decay Length α.

α(Å)	1.6	1.8	2.0	2.2	2.3	2.6	2.8
c	0.83	0.84	0.82	0.80	0.80	0.80	0.80

obtained $\mu_{zf} = \mu_0 \exp[-(c\hat{\sigma})^2]$ via extrapolation to zero field. We plot the log μ_{zf} vs. $\hat{\sigma}^2$ to obtain $\mu_0 = 1.44 \times 10^{-2}$ cm^2/Vs, $c = 0.8$, and $n = 2$. However, the value of c varies only slightly with the decay length α or the distance between neighboring sites r, as indicated in Table 3-3. To get C', μ, and Γ, we calculated γ', the slope of $\log(\mu)$ vs. $(eaE/\sigma)^{1/2}$ and plotted γ' against $\hat{\sigma}^{2.5}$ as shown in the inset of Figure 3-13. We obtained: $C' = 0.54$, m = 2.5 and = 0.87. We also performed simulations for a two-dimensional hexagonal lattice applied to the SmB phase, and fit the mobility to the expression

$$\mu_{\text{SmB}} = \mu_0 \exp\left[-\left(0.80\frac{\sigma}{kT}\right)^2\right] \exp\left\{C'\left[\left(\frac{\sigma}{kT}\right)^{2.5} - \Gamma\right]\sqrt{\frac{eaE}{\sigma}}\right\} \quad (3.32)$$

This equation describes asymptotes to the simulated data in Figure 3-13. The range in which Eqs. (3.30) and (3.31) hold is very narrow. On the other hand,

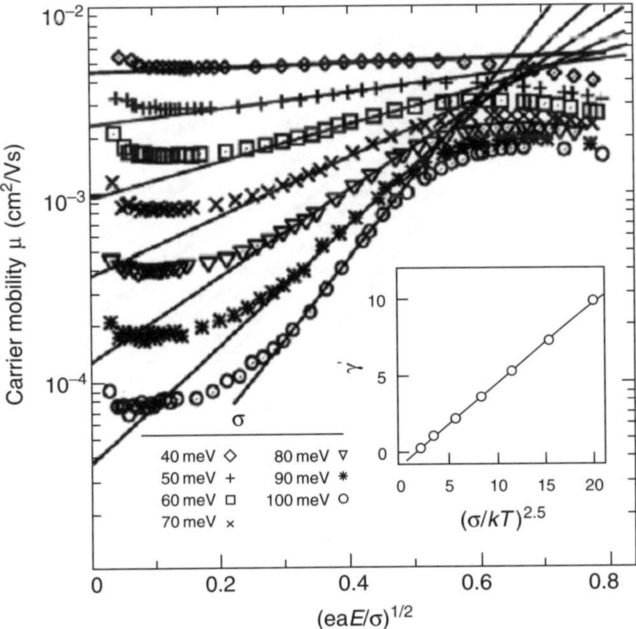

Figure 3-13   Simulation results showing field dependence of the carrier mobility in the SmB phase. The inset shows the slope of $\log(g)$ vs. s2.5.

we think that Eq. (3.32) will be accepted in a wide-field range as pointed out in Ref. 20, if we adopt the correlation disorder model (CDM), which means that if the spatial correlation of Coulomb potential exists on a close hopping site to the charge, then the hopping sites is smoothly change. However, CDM is well accepted in the amorphous semiconductor, where the energetic disorder originated by the charge-dipole interactions is very dominant [81]. We cannot find this kind of strong disorder in our liquid crystal, so we did not think of adopting CDM instead of GDM in this chapter. Thus, in GDM, we can say that region (ii) is the region in which field-dependent mobility is shown and Eqs. (3.30) and (3.31) are shown around the inflection point of this region.

Figure 3-12 shows that region (ii) with field-dependent mobility becomes pronounced with increasing σ. These results suggest that for smectic liquid crystal, where only a small value of σ is expected because of orientation effect, the Poole–Frenkel-like behavior will disappear in the field range of $\sim 10^4 - 10^5$ V/cm often employed in TOF experiments.

In region (iii) where the electric field is in the range $<10^5$ V/cm and the mobility is independent of electric field, we found that the model with $\sigma = 60$ meV showed good agreement with experiment [82], as shown with diamonds in Figure 3-12. The width of the disorder $\sigma = 60$ meV is considerably smaller

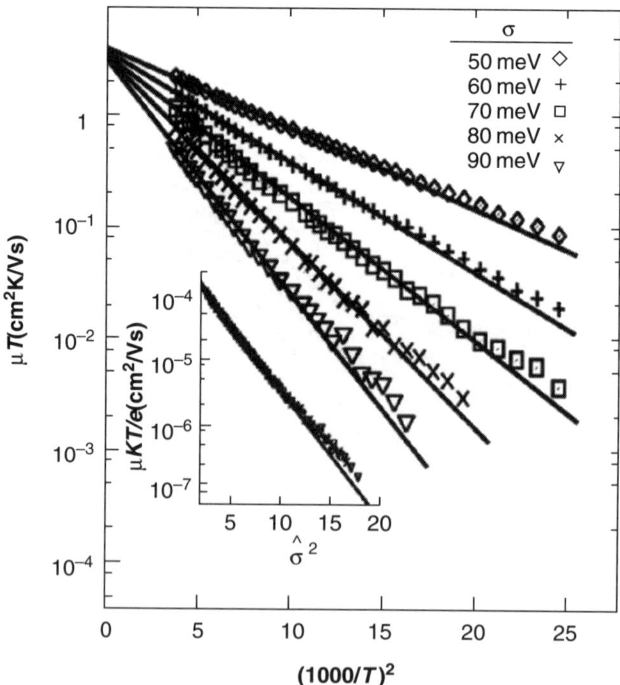

Figure 3-14 Temperature dependence of the carrier mobility μT vs. $(1000/T)^2$. The inset shows $\log(\mu k_B T/e)$.

than a typical value $\sigma = \sim 100\text{--}120$ meV in the organic disordered solids [71]. As shown in Figure 3-12, different behaviors of the mobility appear in regions (ii) and (iii), where the diffusion process plays a significant role in the hopping process, depending on σ and E. These results are shown in Figure 3-14, where $\log(\mu T)$ is plotted vs. $(1000/T)^2$ at constant field $E = 8 \times 10^4$ V/cm belonging to region (iii). We find that the simulated results are consistent with the relation:

$$\mu = \frac{e}{kT} A \exp[-(B\hat{\sigma})^2] \tag{3.33}$$

This implies that the diffusion coefficient depends on $\hat{\sigma}^2$, as $D = A \exp[-(B\hat{\sigma})^2]$, assuming that the Einstein relation holds. This is confirmed from the inset of Figure 3-14. Here, all the points were calculated by using Eq. (3.33) with the same parameters: $A = 2.9 \times 10^{-4}$ cm^2/s and $B = 0.67$. Equation (3.33) shows that the diffusion process does not depend on field E in region (iii).

Our simulated results are consistent with the experimental results on 8-PNP-O12 [82]. The simulation, however, predicts the appearance of field-dependent mobility at the high fields E for 8-PNP-O12. To test this prediction, we have attempted a mobility measurement using 8-PNP-O12 over an extremely wide range of electric field including strengths over 10^5 V/cm. The temperature of the sample was maintained at $80°$C so as to maintain the SmB phase. The diamonds in Figure 3-13 show the experimentally measured mobility as a function of electric field in SmB. The transition from a field-independent mobility to a field-dependent mobility in the experiment is consistent with the simulated predictions using $\sigma = 60$ meV. The separation of region (iii) from region (ii) is dependent on σ, as is clearly seen in Figure 3-13. Region (ii) will extend with an increase in the σ value. The conventional disordered organic solids with a small σ value over 100 meV have a narrow region (ii) at lower electric fields, for example, 10^5 V/cm. This is the reason why the field-independent mobility has been overlooked thus for in the amorphous organic solids.

In summary, we have applied disorder formalism for charge transport in SmB and SmE phases, obtaining field dependencies which are as follows:

$$\mu_{\text{SmB}} = \mu_0 \exp\left[-\left(0.80\frac{\sigma}{kT}\right)^2\right] \exp\left\{C\left[\left(\frac{\sigma}{kT}\right)^2 - \Sigma_0\right]\sqrt{E}\right\} \tag{3.34}$$

The parameters of one-dimensional [79] and three-dimensional [69] disorder formalism are shown in Table 3-4. The parameters seem to have a dependence on the number of nearest-neighbor sites. With higher dimensions, a lower value of the parameter c is obtained. The reason for this behavior is that, when more nearest-neighbor sites are available, carriers can find a site to hop more easily, and they are able to avoid the higher energetic barriers. This lowers the activation energy. On the other hand, μ in Eq. (3.10) increases with increasing the number of nearest-neighbor sites, μ being concerned to the number of paths from the carrier-dwelling site to the nearest-neighbor sites.

3.3.5. Two-Dimensional Disorder Model Using Marcus Equation

Focusing on the temperature dependence of the mobility in the Miller–Abraham model in Eq. (3.25) is not accurate because the reorganization energy λ as shown in Marcus equation (3.14) is ignored. However, this is a good approximation in the case of the amorphous organic semiconductor because the energetic disorder σ (>100 meV) is large enough to be considered as the only dominant factor in the activation process. This is not the case in liquid crystals. For the precise treatment of hopping rate, the Marcus–Hush formalism of Eq. (3.14) should be applied for the rate of ET.

In the case of a single crystal of organic semiconductors, each site is quite uniform, so that ΔG_0, whose distribution can be considered to form a Gaussian DOS, can be ignored at each charge transfer event, that is, $\lambda \gg \Delta G_0$, resulting in the Arrhenius type of temperature dependence. Suppose that the difference of the potential between nearest-neighbor sites under the applied field is smaller than $k_B T$ or λ, that is, $eEa/k_B T$ or $eEa/\lambda \ll 1$; the mobility can be simply denoted by the following equation:

$$\mu = \frac{ea^2}{k_B T} \frac{J_{\text{eff}}^2}{\hbar} \sqrt{\frac{\pi}{\lambda k_B T}} \exp\left(-\frac{\lambda}{4 k_B T}\right) \tag{3.35}$$

where a is the distance between nearest neighbor sites, E is the applied external field, and e is the elementary electric charge. This means that the mobility exhibits an Arrhenius type of temperature dependence whose activation energy is a quarter of λ.

Alternatively, in the case where each site has a large electrostatic variation and ΔG_0 is dominant compared to the reorganization energy ($\Delta G_0 > \lambda$), the charge transfer rate is governed by ΔG_0. This is the case of amorphous organic semiconductors, which are characterized by energetic disorder at hopping sites. ΔG_0 is determined by the energetic disorder σ, which plays a dominant role for charge-carrier transport properties. Here, the energetic disorder σ is a standard deviation of the Gaussian density of states, which is usually assumed in the disorder model. Bässler's version of the disorder model can successfully explain the charge-carrier transport in amorphous organic semiconductors using the Miller–Abraham hopping rate. Thus the model ignores the reorganization energy.

For the materials characterized by molecular alignment with small disorder, however, none of the two factors, reorganization energy λ and energetic disorder σ, can be neglected when the charge transfer rate at each site is decided with the energetic disorder far smaller than that of disordered materials, for example, ~ 50 meV.

Usually, the typical value of the reorganization energy for such small molecules, as in organic semiconductors, is 0.2–0.4 eV. Thus the activation energy originating from the reorganization energy is 50–100 meV. As pointed out in references [83, 84] the average value of energy for charge distribution is located below the center of the Gaussian DOS by $\sigma^2/k_B T$ in the low field range

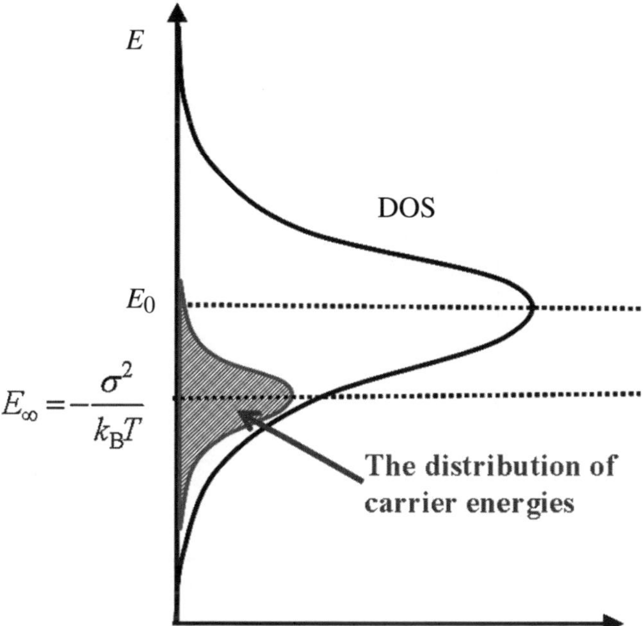

Figure 3-15 The distribution of carrier energies in a Gaussian DOS with standard deviation σ. The mean energy of the carrier distribution keeps the location below the center of the Gaussian DOS by $\sigma^2/k_\mathrm{B}T$ in the long time limit.

(see Fig. 3-15). Thus the apparent activation energy caused by σ is related to $\sigma^2/k_\mathrm{B}T$. The apparent activation energy increases with decrease of temperature. This causes a steep decrease of mobility with decrease of temperature. Here, we show these kinds of phenomena by using a Monte Carlo simulation and derive the disorder formalism in a low field domain.

We assume a Gaussian type of energetic distribution for the density of states (DOS) as we discussed previously. Furthermore, taking care of highly anisotropic conduction in liquid crystals or in a channel area of an organic field-effect transistor (OFET), we execute the Monte Carlo simulation in two-dimensional systems consisting of a two-dimensional lattice structure composed of 500 (taking periodical boundary condition at edge) $\times$ 5000 (taking direction from the top to the counterelectrode) sites whose lattice constant is 3.5 Å. This is comparable to 1.75-μm thickness of a cell. We examined 10,000 charges. Then we calculated the mobility: $\mu = L^2 <1/t_\mathrm{tr}>/V$. We set the effective transfer integral $J_\mathrm{eff} = 5$ meV, and reorganization energy λ is 200 meV, satisfying the nonadiabatic limit: $J_\mathrm{eff} \ll \lambda$. The applied field varies from 5×10^4 to 5×10^6 V/cm. Temperature ranges from 250 to 350 K, and energetic disorder changes from 10 to 50 meV.

Figure 3-16 shows the field dependence of the mobility parameter with temperature for $\sigma = 40$ meV by our simulation. The field domain in which the mobility does not depend on the field ranges up to $\sim 3 \times 10^5$ V/cm. In higher

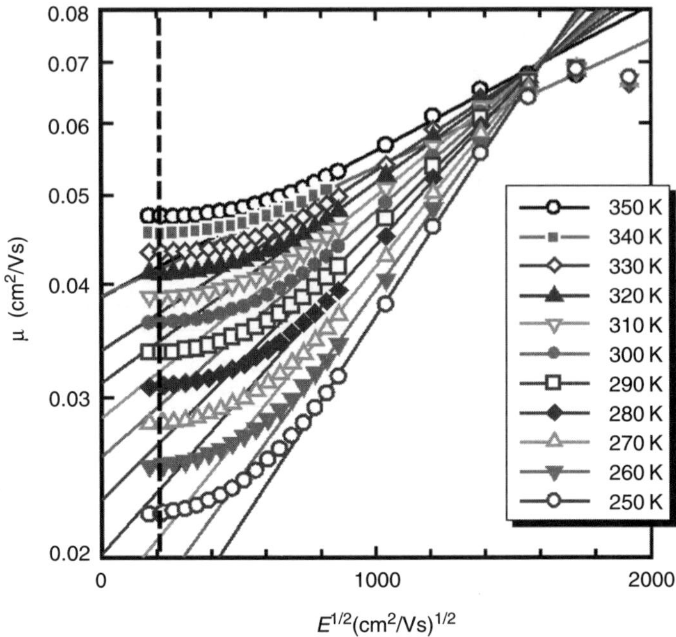

Figure 3-16 Simulation results for the field dependency of the mobility parameter with temperature. (A full color version of this figure appears in the color plate section.)

field domains, mobility seems to depend on the field around $\sim 10^6$ V/cm. We can apply the disorder model even to a system that has no field dependence on the mobility. Figure 3-16 shows the plots of $\log(\mu)$ vs. $E^{1/2}$, taking the intercepts at $E = 4.7 \times 10^4$ V/cm (dashed line in Fig. 3-16), which is in the field-independent domain of mobility μ_{lowE}. The temperature dependence of the mobility μ_{lowE} is shown in Figure 3-17(a). On the other hand, the resultant plots in the high-field domain from 3×10^5 to 3×10^6 V/cm show the Poole–Frenkel type of field dependences. Extrapolation to zero field derives "zero-field mobility" $\mu_{E \to 0}$ shown in Figure 3-17(b).

One important feature is that the plots are not dominated by a simple Arrhenius type of mobility. Even if we assume a reorganization energy that is comparable with the energetic disorder, behaviors of mobility are still in the framework of the disorder model. Fitting parameters accord well with the initial set of parameters, leading to the relation:

$$\mu = \frac{W_0}{T^{3/2}} \exp\left[-\frac{\Delta}{k_{\mathrm{B}}T} - \left(\frac{c\sigma}{k_{\mathrm{B}}T} \right)^2 \right] \tag{3.36}$$

The fittings are done at once for all σ s. The relation (3.36) is applied to both μ_{lowE} and $\mu_{E \to 0}$. The fitting parameters thus determined are shown in Table 3-4. For a temperature range higher than room temperature, the mobility exhibits

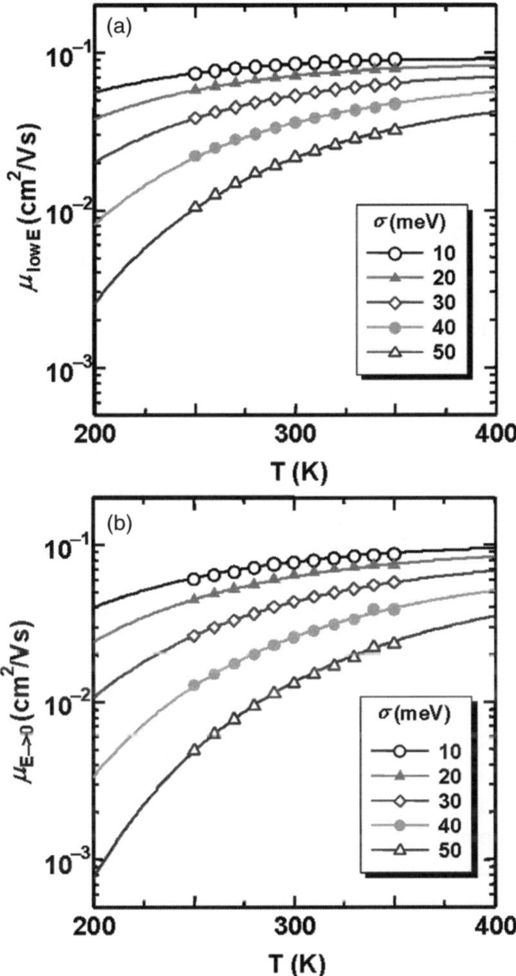

Figure 3-17 Temperature dependency of mobility parameter in energetic disorder σ. (a) Mobilities in the low field range ($E = 4.7 \times 10^4$ V/cm). (b) Zero-field mobilities extrapolated to the field from the plots in the high field range. Solid lines are fitting lines using our proposed disorder formalism denoted by Eq. (3.36).

Table 3-4   Disorder Parameters Concerning Dimension Number.

	1D[21]	SmE	SmA	SmB	3D[1]
C	0.9	0.74	0.75	0.80	0.67
M	1	1.5	1.5	2	2

very weak temperature dependence. For a temperature range lower than room temperature, the mobility has temperature dependence, exhibiting rapid decrease with decreasing temperature. These behaviors are caused by the energetic disorder in the second term in Eq. (3.36).

The model is applicable to recent high-ordered materials exhibiting high mobility in the hopping regime because the energetic disorder is expected to be smaller than 100 meV. We applied our model to the experimental data of mobility measured by TOF for one of the organic semiconductors, a terthiophene derivative, 2-alkyl-5″-hexynyl-3:5′-2′:2″-terthiophene (6-TTP-yne-4) that exhibits a liquid crystalline phase. We used the data in Ref. [85]. Field dependence of the mobility is shown in Figure 3-18(a). The temperature dependence of the extrapolated mobility to zero field mobility is shown in Figure 3-18(b).

Two formalisms are well fitted to the experimental data in each temperature range. The dashed curve shows the fitting curve of $\mu_{E\to 0}$, and the dot-dashed curve shows the fitting curve of μ_{lowE}. Fitting parameters W_0, Δ, and σ are shown in Table 3-5. We can see that fitting parameters from different temperature ranges almost corresponds to different fittings from each different temperature range. This means that our choice of fitting area is relevant for both fittings. A steep decrease of mobility with decrease of temperature can be explained by the disorder formalism of $\mu_{E\to 0}$. The derived disorder formalism can be well fit to the experimental data as shown in Figure 3-18(b). Derived fitting parameters are shown in Table 3-6.

In this simulation and analysis, we found that the temperature dependence of the mobility is dominated by a combination of reorganization energy and energetic disorder, and cannot be negligible for one or both parameters for such a material. The small order $\sim$50 meV of both parameters causes maximum value and weak temperature dependence of the mobility around room temperature. Thus the temperature dependence of mobility is not a simple Arrenius type, even if the energetic disorder is small.

3.4. CONCLUSION

Liquid crystals are very unique in terms of not only as a state of matter but also in terms of electrical properties as we described in this chapter.

In the framework of the hopping transport, the charge transport process in amorphous material is dominated by the energetic fluctuation of localized states, while that in crystalline material is dominated by the reorganization energy or polaron binding energy of molecules. On the other hand, both of these activated processes cannot be neglected in liquid crystals, which is just the same as the fact that the nature of molecular alignment in liquid crystalline material is in between the ordered nature in crystal and the disordered nature in amorphous materials. Better understanding of the carrier transport mechanism in liquid crystals would lead to the continuous understanding of the carrier transport mechanism from amorphous to crystalline materials.

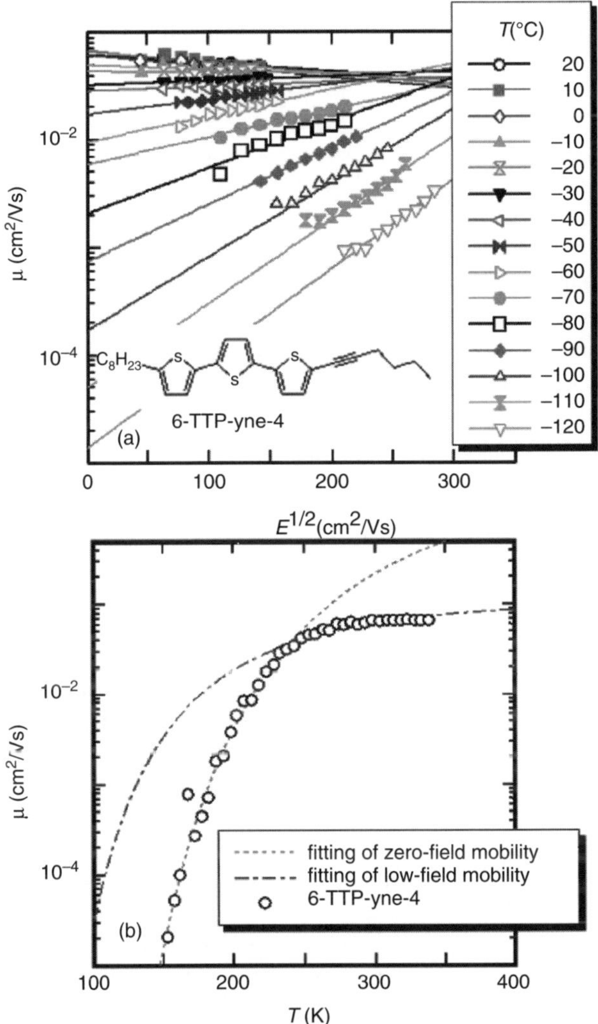

Figure 3-18   (a) Field dependency of mobility parameter in temperature for experimental results of 6-TTP-yne-4 (referenced in [85]) whose chemical structure is shown in the inset. (b) Temperature dependence of the zero-field mobility derived by extrapolation from (a). Solid lines are fittings using our disorder formalism explained in the text. (A full color version of this figure appears in the color plate section.)

Table 3-5   Fitting Parameters of Disorder Formalism (Eq. 3.36) Applied to μ_{lowE} and $\mu_{E\to0}$ (see Fig. 3-17).

	W_0 (cm²K$^{3/2}$ V^{-1}s^{-1})	Δ (meV)	C
μ_{lowE}	3.2×10^3	49	0.62
$\mu_{E->0}$	1.4×10^5	75	0.70

Table 3-6   Fitting Parameters of Disorder Formalism of Both μ_{lowE} and $\mu_{E\rightarrow0}$ to Experimental Data [85] (see Fig. 3-18(b)).

	W_0	Δ (meV)	σ (meV)
μ_{lowE}	1.7×10^3	8	51
$\mu_{E->0}$	1.7×10^6	82	53

From a materials point of view, liquid crystalline materials are very promising as a new class of organic semiconductors for large-area opto-electronic devices thanks to their high mobility comparable to crystalline materials, less defective domain boundaries, and the flexible nature of soft materials, as are the amorphous materials thanks to the isotropic nature of materials. Despite their high potential, the research is still in a primitive stage because of the relatively recent recognition of the organic semiconductor.

REFERENCES

1. H. Akamatu and H. Inokuchi. On the electrical conductivity of violanthrone. *J. Chem. Phys*. **1950**, *18*, 810.

2. N. Karl. *Conjugated Polymers and Low Molecular Weight Organic Solids Series: Springer Series* in Materials Science Vol. 41 Chapter 8, R. Farchioni and G. Grosso, Eds., **2001**.

3. S. F. Nelsona, Y.-Y. Lin, D. J. Gundlach, and T. N. Jackson. Temperature-independent transport in high-mobility pentacene transistors. *Appl. Phys. Lett*. **1998**, *72*, 1854–1856.

4. S. Kusabayashi and M. M. Labes. Conductivity in liquid crystals. *Mol. Cryst. Liq. Cryst*. **1969**, *7*, 395–405.

5. G. H. Heilmeier, L. A. Zanoni, and L. A. Burton. Dynamic scattering: A new electrooptic effect in certain classes of nematic liquid crystals. *Proc. IEEE* **1968**, *56*, 1162–1171.

6. G. H. Heilmeier and P. M. Heyman. Note on transient current measurement in liquid crystals and related systems. *Phys. Rev. Lett*. **1967**, *18*, 583–585.

7. G. Drefel and A. Lipnski. Charge carrier mobility measurements in nematic liquid crystal. *Mol. Cryst. Liq. Cryst*. **1979**, *55*, 89–100.

8. D. Adam, F. Closs, T. Frey, D. Funhoff, D. Haarer, H. Ringsdorf, P. Schuhmacher, and K. Siemensmeyer. Transient photoconductivity in a discotic liquid crystal. *Phys. Rev. Lett*. **1993**, *70*, 457.

9. M. Funahashi and J. Hanna. Fast hole transport in a new calamitic liquid crystal of 2-(4′-hepthyloxyphenyl)-6-dodecylthiobenzothiazole. *Phys. Rev. Lett*. **1997**, *78*, 2184–2187.

10. K. Tokunaga, H. Iino, and J. Hanna. Reinvestigation of carrier transport properties in liquid crystalline 2-phenylbenzothiazole derivatives. *J. Phys. Chem. B* **2007**, *111*, 12041–12044.

11. K. Okamoto, S. Nakajima, M. Ueda, A. Itaya, and S. Kusabayashi. Electrical dark- and photo-conductivities of 2-(p-decyloxybenzilideneamino)-9-fluorenone in nematic state. *Bull. Chem. Soc. Jpn.* **1983**, *56*, 3830.

12. S. Murakami, H. Naito, M. Okuda, and A. Sugimura. Transient photocurrent in amorphous selenium and nematic liquid crystal double layers. *J. Appl. Phys.* **1995**, *78*, 4533.

13. M. Funahashi and J. Hanna. Impurity effect on charge carrier transport in smectic liquid crystals. *Chem. Phys. Lett.* **2004**, *397*, 319–323.

14. H. Ahn, A. Ohno, and J. Hanna. Detection of trace amount of impurity in smectic liquid crystals. *Jpn. J. Appl. Phys.* **2005**, *44*, 3764–3768.

15. H. Ahn, A. Ohno, and J. Hanna. Impurity effects on charge carrier transport in smectic liquid crystal: The effect of conducting path consisting of different mesophases. *J. Appl. Phys.* **2007**, *102*, 093718.

16. G. E. Hoestery and G. M. Letson. The trapping of photocarriers in antracene by anthraquinone, anthrone and naphthacene. *J. Chem. Phys. Solids* **1963**, *24*, 1609–1615.

17. M. Funahashi and J. Hanna. Fast ambipolar carrier transport in smectic phases of phenylnaphthalene liquid crystal. *Appl. Phys. Lett.* **1997**, *71*, 602.

18. M. Funahashi and J. Hanna. High ambipolar carrier mobility in self-organizing terthiophene derivative. *Appl. Phys. Lett.* **2000**, *76*, 2574.

19. S. Mery, D. Haristoy, J. F. Nicoud, D. Guillon, S. Diele, H. Monobe, and Y. Shimizu. Bipolar carrier transport in a lamello-columnar mesophase of a sanidic liquid crystal. *J. Mater. Chem.* **2002**, *12*, 37–41.

20. H. Iino, J. Hanna, D. Haarer, and R. Bushby. Fast electron transport in discotic columnar phases of triphenylene derivatives. *Jpn. J. Appl. Phys.* **2006**, *45*, 430–433.

21. H. Iino, Y. Takayashiki, J. Hanna, and R. Bushby. Fast ambipolar carrier transport and easy homeotropic alignment in a metal-free phthalocyanine derivative. *Jpn. J. Appl. Phys.* **2005**, *44*, L1310–L1312.

22. L. B. Shein. Temperature independent drift mobility along the molecular direction of As_2S_3. *Phys. Rev. B* **1977**, *15*, 1024.

23. H. Iino and J. Hanna. Ambipolar charge carrier transport in liquid crystals. *Optoelectronics Rev.* **2005**, *14*, 295–302.

24. E. O. Arikainen, N. Boden, R. Bushby, J. Clements, B. Movaghar, and A. Wood. Effect of side-chain length on charge transport properties of discotic liquid crystal and their implication for the transport mechanism. *J. Mater. Chem.* **1995**, *5*, 2161–2165.

25. N. Boden, R. J. Bushby, O. R. Lozman, Z. Lu, A. McNeill, B. Movaghar, K. Donovan, and T. Kreouzis. Enhanced conductivity in the discotic mesophases. *Mol. Cryst. Liq. Cryst.* **2004**, *410*, 13–21.

26. H. Monobe, Y. Shimizu, S. Okamoto, and H. Enomoto. Ambipolar charge carrier transport properties in the homologous series of 2,3,6,7,10,11-hexaalkoxytriphenylene. *Mol. Cryst. Liq. Cryst.* **2007**, *476*, 31–41.

27. B. R. Wegewijs, L. D. A. Siebbeles, N. Boden, R. J. Bushby, B. Movaghar, O. R. Lozman, Q. Liu, A. Pecchia, and L. A. Mason. Charge-carrier mobilities in binary mixtures of discotic triphenylene derivatives as a function of temperature. *Phys. Rev. B* **2002**, *65*, 245112-1–245112-8.

28. M. Funahashi and J. Hanna. Mesomorphic behavior and charge carrier transport in terthiophene derivatives. *Mol. Cryst, Liq. Cryst.* **2004**, *410*, 1–12.

29. M. Funahashi, F. Zhang, N. Tamaoki, and J. Hanna. Ambipolar transport in the smectic E phase of 2-propyl-5′-hexynylterthiophene derivative over a wide temperature range. *Chemphyschem*. **2008**, *9*, 1465–1473.

30. M. Funahashi and N. Tamaoki. Electronic conduction in the chiral nematic phase of an oligothiophene derivative. *Chemphyschem* **2006**, *7*, 1193.

31. K. Tokunaga, Y. Takayashiki, H. Iino, and J. Hanna. Electronic conduction in nematic phase of small molecules. *Phys. Rev. B* **2009**, *79*, 033201.

32. M. Funahashi and N. Tamaoki. Electronic conduction in the chiral nematic phase of an oligothiophene derivative. *Chemphyschem* **2006**, *7*, 1193–1197.

33. H. Maeda, M. Funahashi, and J. Hanna. Effect of domain boundary on carrier transport of calamitic liquid crystalline photoconductive materials. *Mol. Cryst. Liq. Cryst*. **2000**, *346*, 193.

34. H. Maeda. M. Funahashi, and J. Hanna. Electrical properties of domain boundaries in photoconductive smectic mesophases and crystal phases. *Mol. Cryst. Liq. Cryst*. **2001**, *366*, 369–376.

35. H. Zhang and J. Hanna. High $\mu\tau$-product in a smectic liquid crystalline photoconductor of a 2-phenylnaphthalene derivative. *Appl. Phys. Lett*. **2004**, *86*, 5251–5253.

36. A. Ohno, A. Haruyama, K. Kurotaki, and J. Hanna. Charge-carrier transport in smectic mesophases of biphenyls. *J. Appl. Phys*. **2007**, *102*, 083711.

37. V. Duzhko, A. Semyonov, R. J. Twieg, and K. D. Singer. Correlated polaron transport in a quasi-one-dimensional liquid crystal. *Phys. Rev. B* **2006**, *73*, 064201.

38. M. Funahashi and J. Hanna. Fast ambipolar carrier transport in smectic phases of phenylnaphthalene liquid crystal. *Appl. Phys. Lett*. **1997**, *71*, 602;   Fast hole transport in a new calamitic liquid crystal of 2-(4′-heptyloxyphenyl)-6-dodecylthiobenzothiazole. *Phys. Rev. Lett*. **1997**, *78*, 2184;   First electronic conduction with high hole mobility in smectic A phase of a calamitic liquid crystal. *Mol. Cryst. Liq. Cryst*. **1997**, *304*, 429.

39. H. Maeda, M. Funahashi and J. Hanna. Influence of domain boundaries on the carrier transport characteristics of the liquid crystalline semiconductors. *Mol. Cryst. Liq. Cryst*. **2000**, *346*, 183.

40. T. Holstein. Studies of polaron motion. Part I. The molecular-crystal model. *Ann. Phys. (N.Y.)* **1959**, *8*, 325;   Studies of polaron motion: Part II. The "small" polaron *Ann. Phys (N.Y.)*. **1959**, *8*, 343.

41. J. J. Sakurai. *Modern Quantum Mechanics (revised edition)*. Addison Wesley, **1994**.

42. Y. Toyozawa. *Optical Processes in Solids*. Cambridge University Press, **2003**.

43. R. Kubo and Toyozawa. Applicaiton of the method of generating function to radiative and non-radiative transitions of a trapped electron in a crystal. *Prog. Theor. Phys*. **1955**, *13*, 160.

44. R. A. Marcus and N. Sutin. Electron transfers in chemistry and biology. *Biochim. Biophys. Acta* **1985**, *811*, 265.

45. R. A. Marcus and N. Sutin. The relation between the barriers for thermal and optical electron transfer reactions in solution. *Comments Inorg. Chem*. **1986**, *5*, 119.

46. M. D. Newton and N. Sutin. Electron-transfer reactions in condensed phases. *Annu. Rev. Phys. Chem*. **1984**, *35*, 437.

47. D. Emin. Phonon-assisted transition rates I. Optical-phonon-assisted hopping in solids. *Adv. Phys*. **1975**, *24*, 305.

48. R. A. Marcus and N. Sutin. Electron transfers in chemistry and biology. *Biochim. Biophys. Acta. Rev. Bioenerg*. **1985**, *811*, 265.

49. K. F. Freed and J. Jortner. Multiphonon processes in the nonradiative decay of large molecules. *J. Chem. Phys*. **1970**, *52*, 6272.

50. J. L. Brédas, D. Beljonne, V. Coropceanu, and J. Cornil. Charge-transfer and energy-transfer processes in π-conjugated oligomers and polymers: A molecular picture. *Chem Rev*. **2004**, *104*, 4971.

51. D. A. Silva Filho, E. Kim, and J. L. Bredas. Transport properties in the rubrene crystal: Electronic coupling and vibrational reorganization energy. *Adv. Mater*. **2005**, *17*, 1072.

52. V. Coropceanu, J. Cornil, D. A. Silva Filho, Y. Olivier, R. Silbey, and J. C. Bredas. Charge transport in organic semiconductors. *Chem. Rev*. **2007**, *107*, 926.

53. G. R. Hutchison, M. A. Ratner, and T. J. Marks. Hopping transport in conductive heterocyclic oligomers. Reorganization energies and substituent effects. *J. Am Chem. Soc*. **2005**, *127*, 2339.

54. X. Amashukeli, J. R. Winkler, H. B. Gray, Nadine E. Gruhn, and D. L. Lichtenberger. Electron-transfer reorganization energies of isolated organic molecules. *J. Phys. Chem*. **2002**, *106*, 7953.

55. J. C. Sancho-Garcia. Assessment of density-functional models for organic molecular semiconductors: The role of Hartree-Fock exchange in charge-transfer processes. *Chem. Phys*. **2007**, *331*, 321.

56. M. D. Newton. Quantum chemical probes of electron-transfer kinetics: The nature of donor-acceptor interactions. *Chem. Rev*. **1991**, *91*, 767.

57. G. T. Velde, F. M. Bickelhaupt, E. J. Baerends, C. Fonseca Guerra, S. J. A. Van Gisvergen, J. G. Snijders, and T. Ziegler. Chemistry with ADF. *J. Comp. Chem*. **2001**, *22*, 931.

58. M. J. Frisch et al. *Gaussian 03*, Gaussian, Inc., Wallingford, CT, **2004**.

59. F. C. Grozema, R. Telesca, J. G. Snijders, and L. D. A. Siebbeles. Tuning of the excited state properties of phenylenevinylene oligomers: a time-dependent density functional theory study. *J. Chem. Phys*. **2003**, *118*, 9441.

60. K. Senthilkumar, F. C. Grozema, F. M. Bickelhaupt, and L. D. A. Siebbeless. Charge transport in columnar stacked triphenylenes: Effects of conformational fluctuations on charge transfer integrals and site energies. *J. Chem. Phys*. **2003**, *119*, 9809.

61. V. Lemaur, D. A. Silva Filho, V. Coropceanu, M. Lehmann, Y. Geerts, J. Piris. M. G. Debjie, A. M. Van de Craats, Kittusamy Senthinkumar, L. D. A. Siebbeles, J. M. Waarman, J. L. Bredas, and J. Cornil. Charge transport properties in discotic liquid crystals: A quantum-chemical insight into structure-property relationships. *J. Am. Chem. Soc*. **2004**, *126*, 3271.

62. J. Kirkpatrick. An approximate method for calculating transfer integrals based on the ZINDO Hamiltonian. *Int. J. Quantum Chem*. **2008**, *108*, 51.

63. X. Feng, V. Marcon, W. Pisula, M. R. Hansen, J. Kirkpatrick, F. Grozema, D. Andrienko, K. Kremer, and K. Müllen. Towards high charge-carrier mobilities by rational design of the shape and periphery of discotics. *Nat. Mater*. **2009**, *8*, 421.

64. M. M. Mikolajczyk, P. Toman, and W. Bartkowiak. Theoretical study of influence of the structural disorder on the charge carrier mobility in triphenylene stacks. *Chem. Phys. Lett*. **2010**, *485*, 253.

65. S. Chandrasekhar. *Liquid Crystals*. Cambridge University Press, **1993**.

66. P. G. de Gennes and J. Prost. *The Physics of Liquid Crystals*. Oxford University Press, USA, **1995**.

67. D. Demus, J. Goodby, G. W. Gray, H. W. Spiess, and V. Vill, eds. *Handbook of Liquid Crystals: Vol 1. Fundamentals*. New York: Wiley-VCH, **1998**.

68. K Tokunaga, H. Iino, and J. Hanna. Charge carrier transport properties in liquid crystalline 2-phenylbenzothiazole derivatives. *Mol. Cryst Liq. Cryst*. **2009**, *510*, 241.

69. H. Bässler. Charge transport in disordered organic photoconductors. *Phys. Stat. Sol (b)* **1993**, *175*, 15.

70. Yu. N. Gartstein and E. M. Conwell. High-field hopping mobility in disordered molecular solids: A Monte Carlo study of off-diagonal disorder effects. *J. Chem. Phys*. **1994**, *100*, 9175.

71. A. Dieckmann, H. Bässler, and P. M. Borsenberger. An assessment of the role of dipoles on the density-of-states function of disordered molecular solids. *J. Chem. Phys*. **1993**, *99*, 8136.

72. S. V. Novikov and A. V. Vannikov. Distribution of the electrostatic potential in a lattice of randomly oriented dipoles. *JETP* **1994**, *79*, 482.

73. A. Hirao and H. Nishizawa. Effect of dipoles on carrier drift and diffusion of molecularly doped polymers. *Phys. Rev. B* **1997**, *56*, R2904.

74. M. Funahashi and J. Hanna. Probe of molecular ordering in photoconductive smectic mesophases by transient photocurrent measurement. *Mol. Cryst. Liq. Cryst*. **2001**, *368*, **303**.

75. A. Miller and E. Abrahams. Impurity conduction at low concentrations. *Phys. Rev*. **1960**, *120*, 745.

76. A. Ohno and J. Hanna. Simulated carrier transport in smectic mesophase and its comparison with experimental result. *Appl. Phys. Lett*. **2003**, *82*, 751.

77. A. Ohno, K. Kurotaki, A. Haruyama, M. Funahashi, and J. Hanna. Modeling of electronic charge transport in smectic liquid crystals. *Proc. SPIE* **2003**, *4991*, 274–281.

78. S. V. Novikov, D. H. Dunlap, V. M. Kenkre, P. E. Parris, and A. V. Vannicov. Essential role of correlations in governing charge transport in disordered organic materials. *Phys. Rev. Lett*. **1998**, *81*, 4472.

79. I. Bleyl, C. Erdelen, H.-W. Schmidt, and D. Haarer. One-dimensional hopping transport in a columnar discotic liquid-crystalline glass. *Philos. Mag. B* **1999**, *79*, 463.

80. H. Cordes, K. Kohary, P. Thomas, S. Yamasaki, F. Hensel, and J.-H. Wendorrff. One-dimensional hopping transport in disordered organic solids. I. Analytic calculations. *Phys. Rev. B* **2001**, *63*, 094201; K. Kohary, K. Kohary, H. Cordes, S. D. Barnovskii, P. Thomas, S. Yamasaki, F. Hensel, and J.-H. Wendorff. One-dimensional hopping transport in disordered organic solids. II. Monte Carlo simulations. *Phys. Rev. B* **2001**, *63*, 094202.

81. D. H. Dunlap, P. E. Parris, and V. M. Kenkre. Charge-dipole model for the universal field dependence of mobilities in molecularly doped polymers. *Phys. Rev. Lett*. **1996**, *77*, 542.

82. M. Funahashi and J. Hanna. Fast ambipolar carrier transport in smectic phases of phenylnaphthalene liquid crystal. *Appl. Phys. Lett*. **1997**, *71*, 602.

83. V. I. Arkipov, E. V. Emelianova and H. Bässler. Equilibrium carrier mobility in disordered hopping systems. *Philos. Mag. B* **2001**, *81*, 985.

84. Bijan Movaghar, M. Grünewald, B. Ries, and H. Bässler. Diffusion and relaxation of energy in disordered organic and inorganic materials. *Phys Rev B* **1986**, *33*, 5545.

85. M. Funahashi and J. Hanna. High carrier mobility up to $0.1 cm^2/Vs$ and a wide mesomorphic temperature range of alkynyl-substituted terthiophene and quaterthiophene derivatives. *Mol. Cryst. Liq.Cryst*. **2005**, *436*, 1179.

CHAPTER 4

Self-Organized Discotic Liquid Crystals as Novel Organic Semiconductors

MANOJ MATHEWS and QUAN LI

Liquid Crystal Institute, Kent State University, Kent, Ohio

4.1. INTRODUCTION

Organic semiconductors are being intensely investigated and incorporated in device applications such as thin-film transistors, light-emitting diodes, solar cells, and sensors [1, 2]. Organic molecules offer the advantage of low-cost synthesis and easy manufacture of large-area thin films by solution processing for the fabrication of a new generation of low-cost, lightweight, and flexible devices that would be inaccessible by conventional methods using inorganic semiconducting materials. The efficiency of these devices is directly related to the mobility of the charges achievable in the conducting layer fabricated by solution-processing techniques such as spin-coating, casting, or printing at ambient conditions exploitable on an industrial scale [3]. A number of different molecular and structural features (chemical purity, solubility, degree of crystallinity, energy band gap, absorption, emission, charge generation, and transport, etc.) have to be combined for optimizing device performance. On the basis of morphology of the materials in the active layers, they can be broadly classified into crystalline, amorphous, and liquid crystalline semiconductors. The highest charge-carrier mobilities (>2 $cm^2V^{-1}s^{-1}$) in organic systems have been measured in organic single crystals of pentacene and rubrene [4, 5]. However, it is difficult to grow single-crystalline thin films. Charge mobilities in the polycrystalline films of polymeric semiconducting materials such as oligothiophenes, poly(phenylenevinylene), poly(thiophene), and poly(fluorine) compounds have reached values exceeding that of amorphous silicon (>1.0 $cm^2V^{-1}s^{-1}$) [6, 7]. Thermal sublimation in vacuum is a common method for depositing polymeric

Self-Organized Organic Semiconductors: From Materials to Device Applications, First Edition.
Edited by Quan Li.

thin films. However, the vacuum deposition technique is expensive. Solution processing would be key to realizing low-cost, large-area devices on flexible substrates. Considerable progress has been made over the past several years in developing solution-processible polymers [8, 9]. Amorphous molecular semiconductors based on π-conjugated polymers are soluble in organic solvents, allowing convenient processing from solution. In general, because of high structural disorder and lack of band transport they show poor mobilities on the order of 10^{-3} cm^{2}V^{-1}s^{-1} or less [10]. On the other hand, liquid crystalline semiconductors exhibit reasonably high charge-carrier mobility ($\sim$1.0 cm^{2}V^{-1}s^{-1}) and allow melt/solution processing [11, 12]. The self-organizing nature of the liquid crystalline phase also facilitates the easy formation of monodomain films and self-healing of defects and offers the possibility of material alignment in a controlled way over a large area [13, 14].

Liquid crystalline (LC) phases or mesophases are thermodynamically stable states of matter with an intermediate degree of order between crystals and liquids. LC phases can be induced in some shape-anisotropic molecules between their crystalline and liquid phases by the action of temperature and/or solvent. Liquid crystals or mesogens (LC phase-forming materials) can be broadly classified into thermotropic or lyotropic depending on whether the phase transitions are induced by change in temperature alone (thermotropic) or by the influence of both temperature and solvent (lyotropic) [15–17]. Molecules exhibiting thermotropic LC phases are usually composed of rigid anisometric core (for example, rod or disc shaped) and flexible alkyl side chains. Stimulus-responsive LC phases of calamitic (rod shaped) mesogens are at the forefront in the development of optoelectronic devices such as LC displays (LCDs) and continue to attract innovative research in view of other possible biomedical applications [18, 19]. On the other hand, LC molecules with a delocalized π-electron system can self-organize spontaneously into highly ordered columnar and smectic mesophase structures that are capable of electronic conduction in one and two dimensions, respectively. Significant attention has been paid to this class of new materials because of their potential applications in organic semiconductor-based devices. The presence of flexible alkyl side chains promotes the solubility of the LC molecules in most organic solvents and allows solution processing of the materials. Solution processing followed by thermal annealing is a facile and cheap method for the fabrication of devices. Furthermore, liquid crystallinity imparts unique mechanical properties and provides the molecules enough motional freedom to self-heal possible defects along the conducting channels. Other characteristics that make LC semiconductors attractive are that they are inexpensive and relatively easy to synthesize and purify and there are a large variety of possible molecular structures to optimize the physical properties through synthetic variation.

This chapter highlights the progress made in the field of discotic liquid crystalline semiconductors. Discotic LCs are particularly attractive for various device applications owing to their unique properties for self-assembly into ordered columnar structures with high charge-carrier mobility. A brief discussion of the electronic aspects of organic semiconductors is presented for understanding of the essential molecular engineering criteria. An overview of structural and meso-morphic properties of discotic molecules with emphasis on their semiconducting properties including charge transport and conductivity in terms of novel systems, control of alignment, and applications is provided through presentation of selected examples. Examples in which the materials have been characterized for charge transport properties and/or in device applications are considered. No discussion of calamitic and polymeric LC semiconductors is presented, and for these impor-tant topics we refer the readers to other chapters of this book and some recent review articles [11, 12].

4.2. SEMICONDUCTING PROPERTIES OF DISCOTIC LIQUID CRYSTALS

4.2.1. Discotic Liquid Crystalline Phases

Discotic liquid crystals (DLCs) consist of disc-shaped aromatic cores that are circularly surrounded by flexible alkyl chains. Phase segregation arising from these chemically and conformationally different molecular parts drives the self-organization of the molecules into hierarchically ordered mesophase structures. The discovery of DLCs dates back to the report of columnar mesophases in hexa-substituted esters of benzene **1** by Chandrasekhar et al. in 1977 [20]. Sub-sequently, numerous disc-shaped molecules with different central aromatic cores and peripheral substitutions showing various mesophases have been reported. More than 50 aromatic cores capable of forming discotic mesophases have been identified [21]. Figure 4-1 shows the molecular structures of the most promi-nent discotic cores. Mesophases formed by DLCs can be broadly classified into nematic (N), columnar (Col), and lamello-columnar (Col$_L$) phases. The least ordered N mesophase is characterized by the presence of only orientational order wherein the mesogens align with their principal axes along a common direc-tion. Most discotic molecules show the tendency to self-organize one on top of the other into columns. A subsequent two-dimensional lattice arrangement of different columns results in columnar mesophases. Columnar phases exhibit rich polymesomorphism and can be further classified depending upon their phase symmetry and varying degrees of molecular order within the columns [13]. For example, upright columns forming a hexagonal columnar phase (Col$_h$) and tilted columns resulting in a rectangular columnar phase (Col$_r$) are the two most com-mon types of columnar phases. In some cases, the columns are liquid-like, that is, there is no ordering of discs within or between the columns (disordered heexag-onal columnar phase, Col$_{hd}$), while in others they are arranged in an ordered fashion (ordered heexagonal columnar phase, Col$_{ho}$). More ordered phases such as columnar oblique (Col$_{ob}$), columnar plastic (Col$_p$), and columnar helical (H) mesophases are not as common. Figure 4-2 shows the most common phase struc-tures of the discotic molecules. These different mesophases can be identified experimentally by the combined use of polarizing optical microscopy (for textural observation), differential scanning calorimetry (for phase transition temperatures),

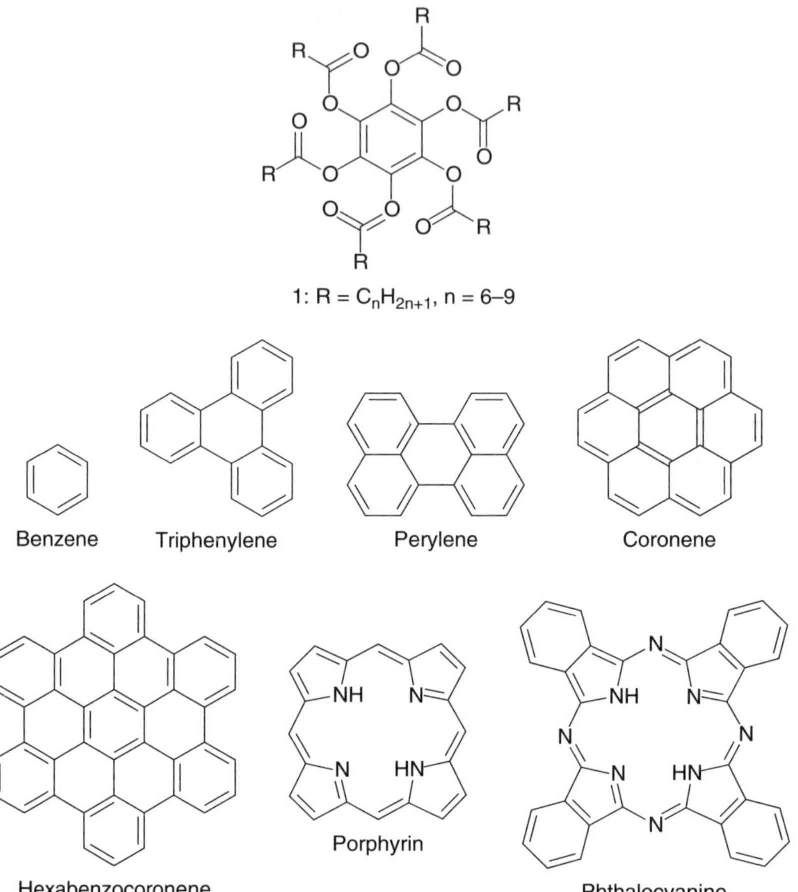

Figure 4-1 Molecular structure of hexa-ester benzene **1** and selected examples of aromatic cores of the most widely studied discotic molecules.

and X-ray diffraction studies (for molecular ordering in the phase). It is worth mentioning here that not only the disclike aromatic compounds but also molecules with unconventional shapes such as linear π-conjugated oligomers, rings, cones, and bowls have been shown to self-assemble into columnar structures by using supramolecular material design concepts [22].

4.2.2. Charge Transport in Discotic Columnar Phases

The essential electronic feature that all organic semiconductors share in common is the π-conjugated system, that is, single and double or single and triple bonds alternate throughout the molecule. Each conjugated carbon atom in the sp^2 hybridization (systems with alternating single and double bonds) of a molecule forms 3 σ-bonds from the overlap of hybridized $2s$, $2p_x$ and $2p_y$ valence atomic

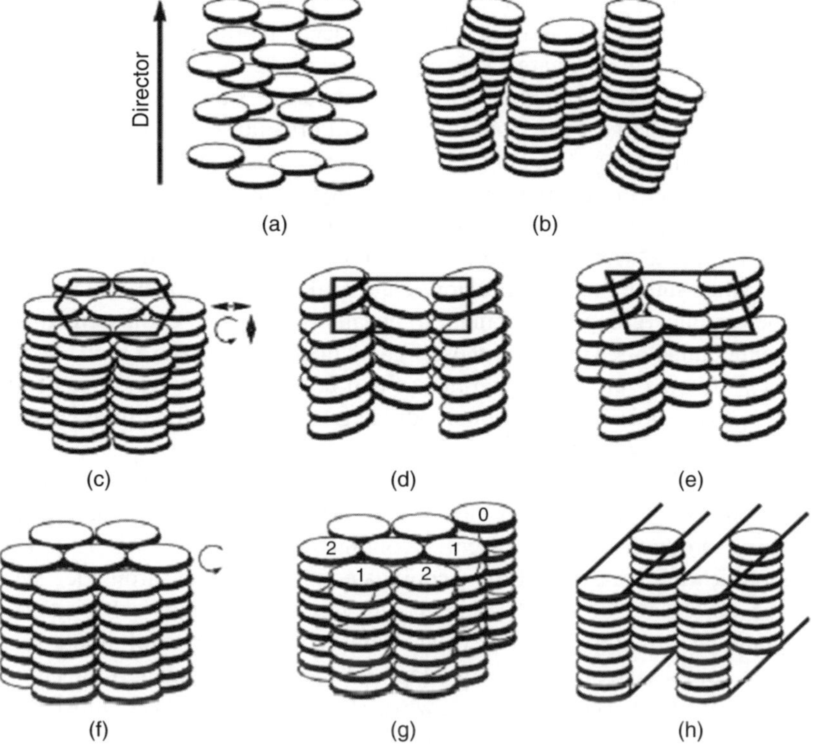

Figure 4-2 Schematic drawings of the most common mesophases of disc-shaped molecules: (a) discotic nematic, (b) columnar nematic, (c) hexagonal columnar, (d) rectangular columnar, (e) columnar oblique, (f) columnar plastic, (g) columnar helical and (h) columnar lamellar phase. Reprinted with permission from Ref. 21.

orbitals. The fourth orbital $2p_z$ lies perpendicular to the σ-bond plane and laterally overlaps to form the π-bonds (Fig. 4-3a). A σ-bond is strong because of the involvement of s-orbitals, and the energy difference between the occupied bonding orbitals (σ) and the unoccupied antibonding orbitals (σ^*) is quite large. However, a π-bond is weaker than a σ-bond, and it has a much smaller energy difference between the highest occupied molecular π-orbital (HOMO) and the lowest unoccupied molecular π^*-orbital (LUMO), and the corresponding band gap energy determines the semiconducting properties of a molecule (Fig. 4-3b). In a conjugated system, π-orbital wave functions of adjacent carbon atoms overlap and the electrons occupying such orbitals become relatively delocalized to form energy band gaps in the semiconducting range (1.5–3 eV). There are several differences in physical properties of organic and inorganic semiconductors due to the extent of variation in orbital overlap along the conducting pathway. In an inorganic semiconductor with a three-dimensional crystal lattice, the individual LUMOs and HOMOs form a conduction band (CB) and a valence band (VB) throughout the material. On the other hand, most organic semiconducting materials have weak intermolecular orbital overlap and do not form a CB and a VB. Charge carriers are localized in organic materials, and transport proceeds by hopping between neighboring sites rather than within a band. Therefore charge-carrier mobility in most organic semiconductors is generally low (typically 10^{-5}–0.1 $cm^2V^{-1}s^{-1}$) compared to inorganic semiconductors. Photon absorption in an organic semiconductor results in the generation of an exciton (bound electron-hole pair) as opposed to the creation of free charge carriers in an inorganic semiconductor. Furthermore, an organic semiconductor tends to have rather small exciton diffusion lengths (the distance over which excitons travel before undergoing recombination) usually of 3–40 nm compared with inorganic semiconductors. However, most organic materials possess a high absorption coefficient ($\alpha > 10^5$ cm^{-1}), which means that layer thickness can be kept thin to preserve exciton diffusion in a semiconducting layer and yet be highly absorptive for photoconducting applications.

Discotic molecules derived from large π-conjugated aromatic cores have attracted much attention as a novel type of organic semiconductor. Semiconducting properties in the discotic columnar mesophases can be attributed to the following reasons. First, conjugated π-electronic aromatic systems form the core parts of the discotic molecules. Second, their self-organization into columnar structures (with typical inter core distances of about 3.5 Å) facilitates long-range π-orbital overlap between the adjacent molecules for the formation of charge carrier pathways (Fig. 4-3c). However, most discotic molecules have a very low intrinsic carrier concentration because of the fairly large energy band gap (2–4 eV) and behave as insulators in the pure state. Charge carriers must be injected into the molecules by chemical doping, pulse radiolysis, or photoirradiation or from an electrode surface at high fields before electronic conduction can occur. Conduction in the columnar mesophase is highly anisotropic, that is, the charges move preferentially along the conducting aromatic cores of the columns. Flexible alkyl chains linked to the rigid core act as an insulating hydrocarbon matrix and decrease the probability for intercolumnar tunneling of the charge carriers. Therefore, discotic columns can act as one-dimensionally conducting molecular wires. Depending on the ionization potential (HOMO) or electron affinity (LUMO), these discotic molecules prefer to transport either positive holes or negative electrons, and, accordingly, the materials can be classified as p-type or n-type semiconductors. Based on the extent of aromatic core conjugation and the degree of order in the columnar stacking, charge-carrier mobility in discotic columns can generally range from 10^{-3} to 1.0 $cm^2V^{-1}s^{-1}$. The charge transport in liquid crystalline semiconductors with low mobility values (<0.1 $cm^2V^{-1}s^{-1}$) is described via a one-dimensional hopping process [23, 24]. A bandlike charge transport that involves the formation of conduction bands across several molecules is proposed to explain the high carrier mobilities (>0.1 $cm^2V^{-1}s^{-1}$) [25]. The exact mechanism of charge transport, however, is not completely understood and is still the subject of debate in the literature.

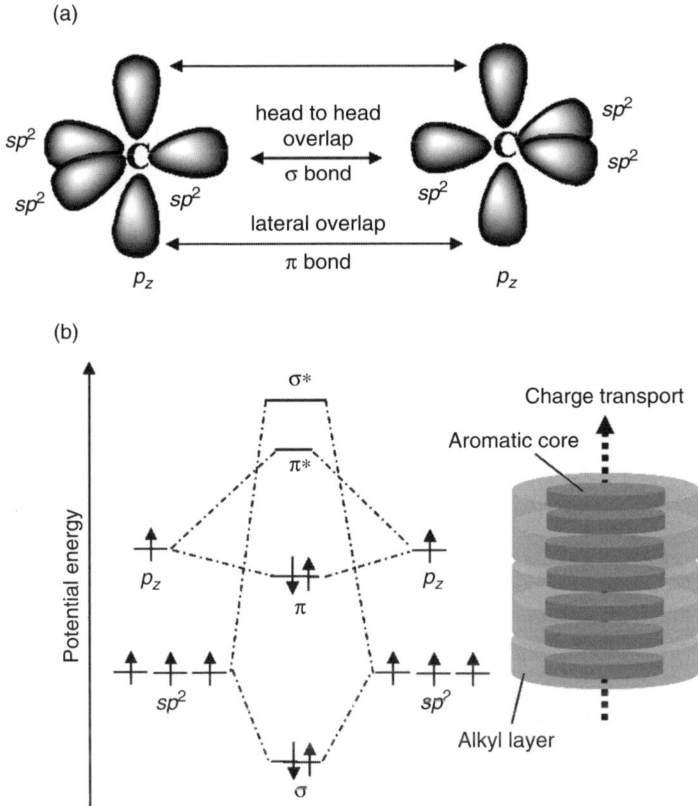

Figure 4-3 (a) Schematic illustration of σ-bond and π-bond formation between two sp^2 hybridized carbons. (b) An orbital energy-level diagram for the C–C bond formation between two sp^2 hybridized carbons. (c) Schematic representation of the organization of discotic molecules into one-dimensionally conducting columns.

4.2.3. Charge Mobility Measurement Techniques

Charge-carrier mobility in discotic columnar materials can be determined with the use of several different methods, such as pulse radiolysis-time-resolved microwave conductivity (PR-TRMC), time of flight (TOF), space charge-limited current (SCLC), and field-effect transistor (FET) techniques. Simple schematic representations of PR-TRMC and TOF techniques are shown in Figure 4-4 and Figure 4-5, respectively. In the PR-TRMC method, charge carriers are created throughout the bulk material by using nanosecond-duration pulses of ionizing radiation from a Van de Craats accelerator [26]. Generation of mobile charge carriers increases the conductivity of the sample. Microwaves are then used to monitor this change in conductivity of the medium. This method requires no electrode deposition onto the sample and therefore offers the practical advantage of excluding the interference of the electrode. Moreover, the mobility

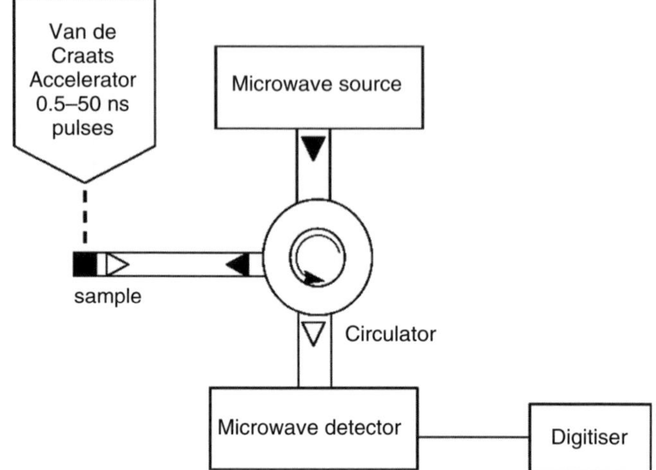

Figure 4-4 Schematic illustration of the PR-TRMC experimental setup, redrawn from Ref. 26.

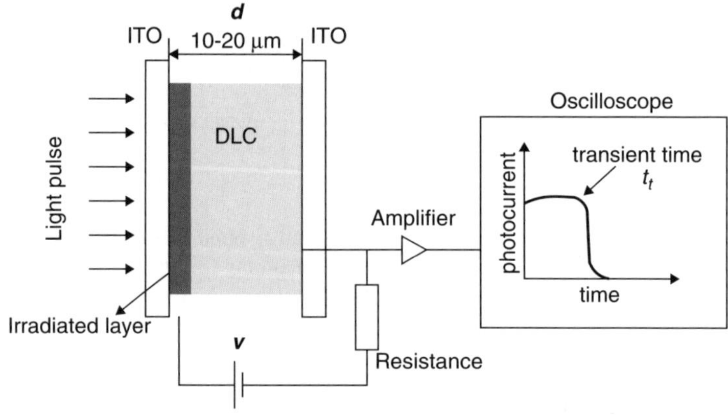

Figure 4-5 Schematic representation of the TOF experiment. The mobility μ depends on the applied voltage V and transit time t_t according to the equation $\mu = v/E = d^2/Vt_t$, where v is the drift velocity, d is the film thickness, and E is the applied electric field.

of intrinsic charge carriers can be determined even for nonaligned multidomain samples. It is believed that the PR-TRMC mobility is close to the maximum value that could be obtained for a sample because of the measurement of the mobility over a short range (a few molecules) and on a nanosecond timescale. PR-TRMC measurements yield the sum of positive and negative charge-carrier mobilities, $\sum \mu_{1D} = \mu(+) + \mu(-)$, and do not allow the separate determination of positive and negative carriers. With this technique, mobility values as high as $1.1\ \mathrm{cm^2V^{-1}s^{-1}}$ have been measured in a discotic hexabenzocoronene derivative [27]. Comparison of PR-TRMC data on different discotic systems has shown a strong correlation between the core size and the charge mobility. It is generally observed that the mobility values tend to increase with increasing size of the conjugated core [28].

In the TOF technique, charge carriers are generated in the sample via a short pulse of laser light (by the absorption of photons of appropriate energy greater than or equal to the band gap) and made to drift under an applied electric field, enabling one to study the mobility by measuring their transient time over a known interelectrode distance [29]. The sign of the applied voltage determines whether electron or hole transport is monitored. Most of the discotic molecules only support or strongly favor one type of charge carrier (positive or negative) and are most often classified as a hole- or electron-transporting material. It is common in the literature to refer to the hole-transporting materials as p-type and the electron-transporting materials as n-type. Measurement of the TOF mobility in a discotic material requires an extremely high degree of self-organization of the molecules within the interelectrode gap, because any imperfection in the columnar arrangement can have a negative influence on charge carrier drift and the mobility measured will represent a lower limit of the value. Although TOF mobilities up to 0.2 $cm^2V^{-1}s^{-1}$ have been reported for some cases, most DLCs exhibit values on the order of 10^{-3} $cm^2V^{-1}s^{-1}$ [11]. The charge mobilities determined by the TOF method are often found to be lower than the corresponding PR-TRMC values obtained for a sample. This is because TOF mobility is measured over relatively thick samples (usually tens of microns) on a millisecond or microsecond timescale and therefore is more susceptible to defects and macroscopic grain boundaries that impede charge-carrier transport. Investigations have also revealed that the mobility is discontinuous at crystalline, liquid crystalline, and isotropic boundaries and tends to decrease at each phase transition because of the decreasing molecular order [30].

Charge mobility measured by a TOF technique results from extremely low carrier densities generated via a short pulse of light irradiation. However, devices such as OLEDs and OFETs operate at much higher densities of injected charges, which lead to both partial trap filling and space charge effects. Therefore, SCLC techniques, for example, current-voltage (I-V) measurements on Schottky diode-like structures or FETs, that are close to the device operational conditions are used to determine the effective mobility of materials [31, 32]. The SCLC technique is sensitive to the structural defects of the macroscopic alignment and to the charge injection effectiveness at the electrodes.

4.3. DISCOTIC LIQUID CRYSTALS WITH HIGH CHARGE-CARRIER MOBILITY

Semiconducting properties of DLCs can be tailored by varying the nature of the central core and the peripheral substituents. To achieve high charge-carrier mobility in DLC materials, it is important to have a high degree of molecular order within the columnar mesophase either by structural design or by physical processing. One of the most successful molecular design strategies to achieve high charge-carrier mobility has been to increase the DLC core size to enhance the π-orbital overlap within the columnar structure. Decreasing the interdisc distance within the conducting columns by various noncovalent interactions such as charge transfer complexes, metal coordination, and hydrogen bonding have also been actively explored in order to enhance the π-π stacking of the molecules. Discotic cores rely on alkyl chains on their periphery for inducing liquid crystallinity and solubility. In addition, by varying the chain length and its chemical nature, a certain degree of control over mesophase order and alignment can be achieved. For practical device applications, it is also necessary to design DLCs with good thermal and oxidation stability and tunable electronic properties such as HOMO and LUMO energy levels. HOMO and LUMO energy levels of a molecule not only determine its p-type or n-type semiconducting properties but also affect the hole or electron injection barrier at the interface of the metal electrode in devices. These intrinsic electronic properties can often be tuned by electron-donating or electron-withdrawing substitutions on the discotic core.

Since pure DLCs are intrinsically insulating, charge carriers (electrons or holes) must be introduced into the system to make it (semi)conducting. Optimization of charge-carrier mobility of these materials is crucial to their potential device application. The idea of using discotic liquid crystals as quasi-one-dimensional conductors was first discussed by Simon and coworkers in 1982 [33]. Early studies on the charge transport behavior of columnar discotics involved chemical doping of charge carriers using strong oxidants. For example, when electron-rich hexahexyloxytriphenylene derivative **2c** (Fig. 4-6) was doped with 1 mol% of electron acceptor $AlCl_3$, the conductivity σ increased from an undoped value of less than 10^{-9} S m^{-1} to about 10^{-3} S m^{-1}, with the conductivity along the columns ($\sigma_\parallel$) being about three orders of magnitude greater than that in the perpendicular direction ($\sigma_\perp$) [34, 35]. However, liquid crystallinity is highly dependent on the purity of the materials, and doping often leads to a change in properties. Alternatively, in the early 1990s researchers were successful in employing ionizing radiation/photoirradiation as a means to generate charge carriers in chemically pure (nondoped) DLC molecules and measured the mobility by PR-TRMC or TOF technique. Porphyrin and phthalocyanine derivatives were the first examples of DLCs to be studied for the conductivity by a PR-TRMC method [36, 37]. In 1993, Adam et al. measured photoinduced charge-carrier mobility of 1×10^{-3} cm^{2}V^{-1}s^{-1} in the columnar phase of hexapentyloxytriphenylene (**2b**) by the TOF technique [29]. This study led to valuable insights into the charge transport in discotic liquid crystals for the first time by identification of the nature of charge carriers (whether holes or electrons) and the observation of nondispersive transport along the column axis as well as the temperature and field independence of the mobility of the charge carriers. Since then, the major focus of research on DLCs has shifted from simple structure-mesophase property relationship studies to better understand charge transport and to develop new high-mobility systems for various practical applications.

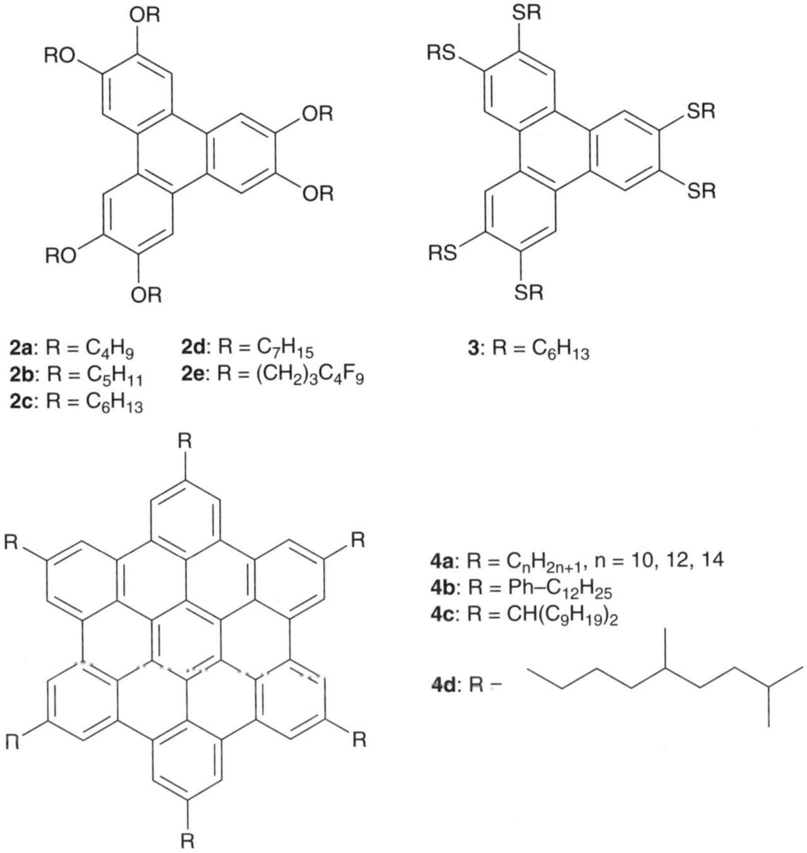

2a: R = C_4H_9 **2d**: R = C_7H_{15} **3**: R = C_6H_{13}
2b: R = C_5H_{11} **2e**: R = $(CH_2)_3C_4F_9$
2c: R = C_6H_{13}

4a: R = C_nH_{2n+1}, n = 10, 12, 14
4b: R = Ph–$C_{12}H_{25}$
4c: R = $CH(C_9H_{19})_2$

4d: R =

Figure 4-6   Molecular structures of DLCs **2–4**.

The majority of DLCs known today are derived from electron-rich aromatic cores and are known to be predominantly hole carriers. Representative examples include derivatives of triphenylenes, hexa-*peri*-benzocoronenes, porphyrins, and phthalocyanines. It is now well known that the extent of π-overlap between the core units is the most important parameter for the charge transport process in DLCs. Therefore, a molecule with a larger π-conjugated core is expected to show higher charge-carrier mobility. In addition, the charge transport properties are strongly correlated with the order present in the mesophases. This phenomenon is clearly evident from the systematic charge mobility studies done on triphenylene-based DLCs (**2-3**). For example, carrier mobilities of the order of 10^{-3} $cm^2V^{-1}s^{-1}$ for holes were measured in the Col_h phase of triphenylene DLC **2b** [29]. Approximately an order of magnitude increase in the hole transport was observed in the more ordered plastic columnar phase of a lower homolog **2a** [38]. Even higher hole mobility value (1×10^{-1} $cm^2V^{-1}s^{-1}$) was found in the three-dimensionally ordered helical columnar phase of triphenylene

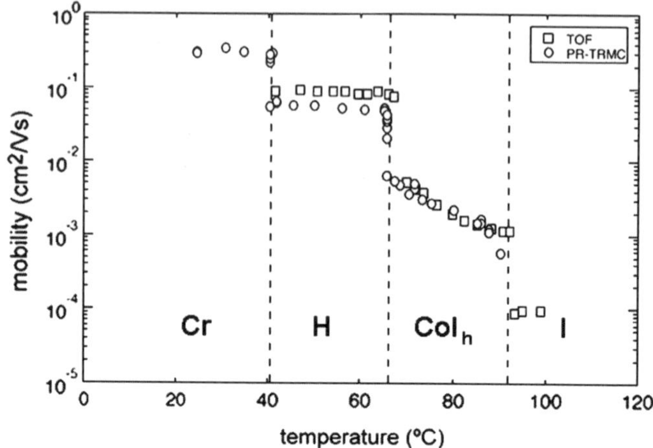

Figure 4-7   The charge-carrier mobility in different phases of **3** measured by TOF (μ_h; squares) and sum of one-dimensional intracolumnar mobility $\Sigma\mu_{1D}$ (open circles) estimated by PR-TRMC experiments while cooling the sample from its isotropic phase. Reproduced with permission from Ref. 30.

DLC **3** [39]. Short-side chain derivatives of triphenylenes, that is, **2a–b**, allow for better interaction of the cores, resulting in higher mobilities in comparison with the long-side chain analog compounds **2c–d** [40]. DLC compounds for which the alkyl chains are coupled to the core via sulfur (compound **3**) show, in general, considerably higher charge mobility values than those for which the chains are coupled either directly or via oxygen. This increase in charge-carrier mobility is attributed to the higher mesophase order induced by the sulfur atoms. Compound **3** exhibits phase transitions from isotropic to Col_h phase, from Col_h to the more ordered columnar helical (H) phase, and from H to the crystal phase below 40°C. A very low mobility of 1×10^{-4} cm^{2}V^{-1}s^{-1} is measured in the isotropic liquid phase because of the absence of any columnar order. Charge-carrier mobility is found to increase from 5×10^{-3} cm^{2}V^{-1}s^{-1} in the Col_h to 0.1 cm^{2}V^{-1}s^{-1} in the helical columnar phase (Fig. 4-7). Because of trapping at grain boundaries, charge-carrier mobility in the crystalline phase of **3** could not be determined by TOF technique. The sum of the one-dimensional charge-carrier mobilities, $\sum\mu_{1D}$, of about 0.4 cm^2 V^{-1}s^{-1} has been derived from the PR-TRMC technique for **3** in its crystalline phase [30]. This increase in short-range carrier mobility on transition from mesophase to crystal phase is a common feature of a number of different discotic materials and is attributed to the greater short-range structural order within the crystal packing of the molecules. By combining the results of both techniques, a change in mobility by more than three orders of magnitude in going from the crystalline phase to the isotropic liquid of **3** was obtained.

Initial studies reported that the mobility of electrons in discotic triphenylene derivatives was much smaller than that of the holes, indicating the latter to be the major charge carrier. Recently, ambipolar charge transport (electron as well as hole charge transport) in the columnar phases of **2a** and **3** was found by improving the sample purity [41, 42]. For both compounds, the carrier mobility for electrons were shown to be as fast as the hole mobility reported previously. Hydrogen bonding interactions are known to enhance mesophase order and charge transport in columnar phases. Paraschiv et al. recently reported high charge mobility of 0.2 $cm^2V^{-1}s^{-1}$ in a hydrogen-bonded 1,3,5-benzenetrisamide derivative with three pendant hexaalkoxytriphenylene groups [43]. This was attributed to the highly ordered columnar hexagonal plastic phase that was induced by the intermolecular hydrogen bonding of the 1,3,5-benzenetrisamide groups. Kato and coworkers succeeded in threefold enhancement in the hole mobility of **2b–d** by suppressing molecular fluctuations in the columnar phase by physical gelation of molecules with hydrogen-bonded fibers [44]. Strong modifications in the order within the columnar phase of discotic triphenylenes are also achieved by variation of the lateral chain or by direct polar substitutions on the aromatic core [45, 46]. Very recent studies revealed that doping of gold nanoparticles into the columnar phase of triphenylene DLCs increases conductivity significantly [47, 48].

Müllen and coworkers developed a new class of DLC materials derived from polycyclic aromatic hydrocarbon (PAH) cores such as hexa-*peri*-benzocoronenes (HBCs) **4** having high charge-carrier mobility and a large mesophase range that in some cases extends even to room temperature [49]. Charge-carrier mobility in the range of 0.2–0.4 $cm^2V^{-1}s^{-1}$ in the columnar mesophases and values exceeding 1.1 $cm^2V^{-1}s^{-1}$ in the crystal phase of HBC derivatives (**4a**) have been found by the PR-TRMC method [27]. By the TOF technique, long-range hole mobilities on the order of 10^{-3} $cm^2V^{-1}s^{-1}$ have been found at room temperature in compound **4c** [50]. The high charge-carrier mobility found in discotic HBCs can be attributed to their unique, highly ordered columnar phase structures leading to large intermolecular π-orbital overlap between the aromatic cores. Several HBC derivatives have so far been successfully employed as active conducting layers in molecular devices such as OFETs and solar cells [51]. HBC is a planar aromatic molecule consisting of 13 fused six-membered rings. Studies on larger discotic molecules derived from peripherally alkyl-substituted PAH cores consisting of 42, 60, 78, 96, and 132 carbon atoms revealed that there is no substantial increase in charge-carrier mobility with increasing core size [51]. This observation contrasted with the expectation that increasing core size increases the charge mobility [28]. HBCs and their larger derivatives mentioned above possess sixfold (D_{6h}) symmetry and stack with an average twist of 30° in their discotic phases. Molecules of threefold symmetry (e.g., triangularly shaped PAHs) with a helical packing structure and 60° twist were theoretically predicted to form optimal local arrangement for higher charge transport [52]. On the basis of this theoretical prediction, Müllen's research group synthesized triangle-shaped PAHs (**5a–b**) (Fig. 4-8) having the same number of carbons as fullerene and reported improved photovoltaic performance in comparison to the analog HBC DLCs [53]. Nuckolls and coworkers reported the stabilization of columnar mesophase in a contorted HBC derivative **6** (Fig. 4-8) [54]. Relatively high carrier mobilities and current

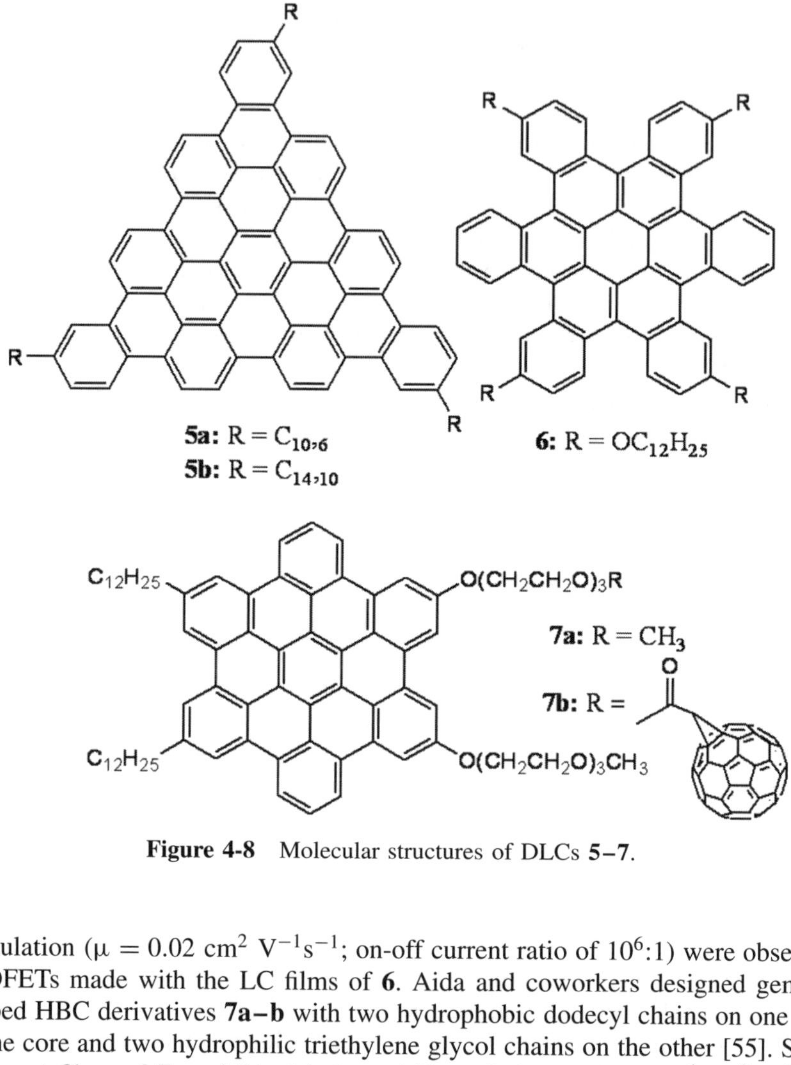

5a: $R = C_{10,6}$
5b: $R = C_{14,10}$

6: $R = OC_{12}H_{25}$

$C_{12}H_{25}$

$O(CH_2CH_2O)_3R$

7a: $R = CH_3$

7b: $R =$

$C_{12}H_{25}$

$O(CH_2CH_2O)_3CH_3$

Figure 4-8   Molecular structures of DLCs **5–7**.

modulation ($\mu = 0.02$ cm^2 V^{-1}s^{-1}; on-off current ratio of 10^6:1) were observed in OFETs made with the LC films of **6**. Aida and coworkers designed gemini-shaped HBC derivatives **7a–b** with two hydrophobic dodecyl chains on one side of the core and two hydrophilic triethylene glycol chains on the other [55]. Solution cast films of **7a** exhibited hole mobility of about 1.0×10^{-4} cm^{2}V^{-1}s^{-1}. On the other hand, C$_{60}$-appended HBC **7b** self-assembled into a coaxial nanotube with an ambipolar field effect mobility ($\mu_e = 1.1 \times 10^{-5}$ cm^{2}V^{-1}s^{-1} and $\mu_h = 9.7 \times 10^{-7}$ cm^{2}V^{-1}s^{-1}). More interestingly, mixing 10 mol% of **7b** in **7a** resulted in coassembled nanotube structures with intratubular hole mobility of 2.0 cm^{2}V^{-1}s^{-1}, as determined by the flash-photolysis time-resolved conductivity (FP-TRMC) technique.

Early research on DLCs showed that the π-electron-rich macrocycles such as porphyrins and phthalocyanines substituted by alkyl chains at the periphery tend to self-assemble to form columnar mesophases [33, 56]. Subsequent studies revealed the role of chain length and metal ion on phase transition temperatures, mesophase range, and carrier mobilities. Initial charge-carrier

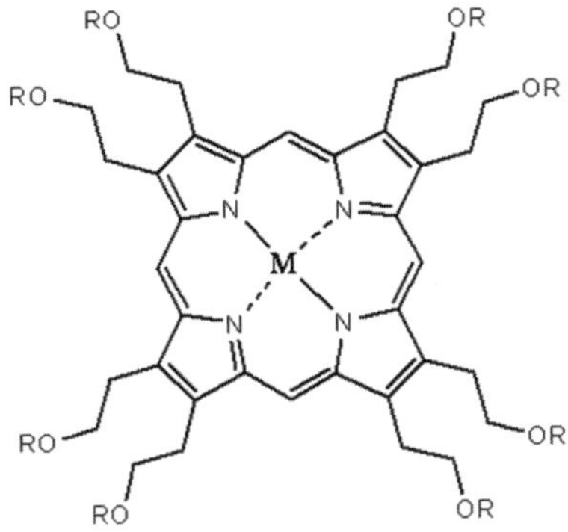

8a: R = C_9H_{19}, M = 2 x H
8b: R = C_9H_{19}, M = Zn

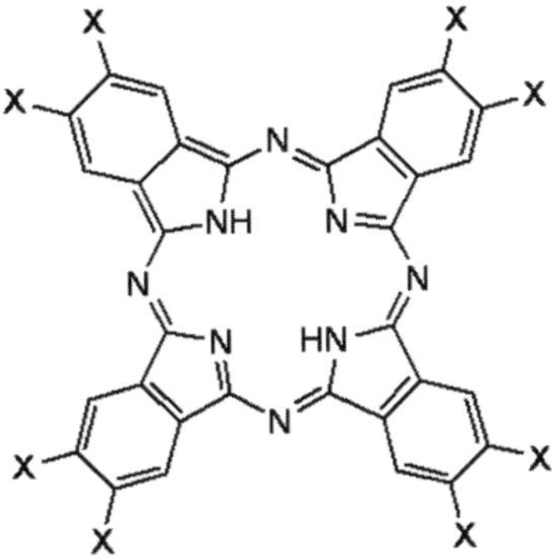

9a: X = $OC_{12}H_{25}$
9b: X = $SO_2C_{12}H_{25}$

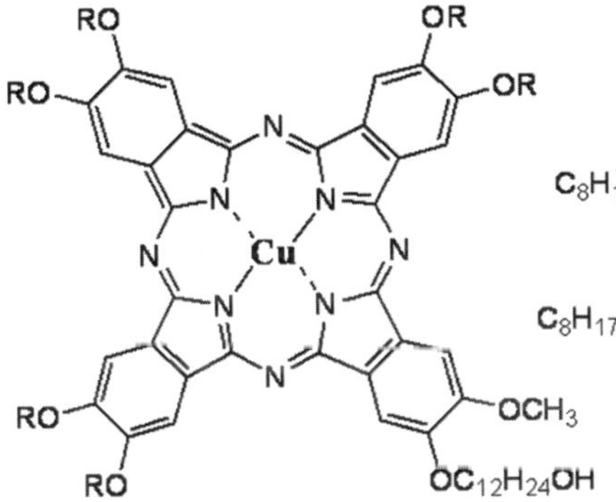

10: R =
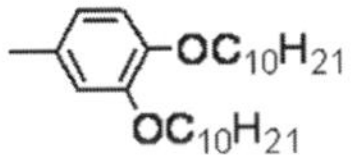

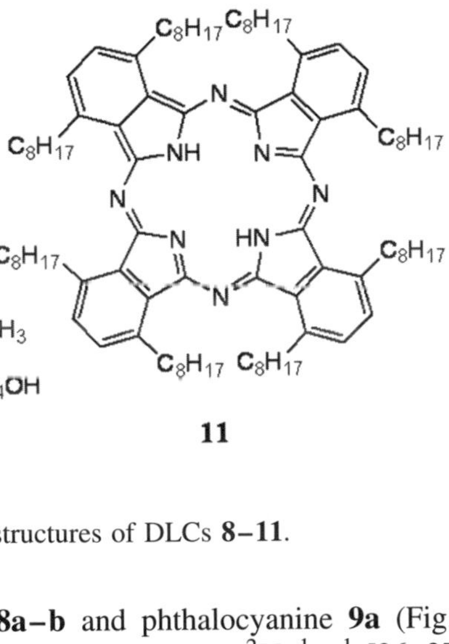

11

Figure 4-9 Molecular structures of DLCs **8–11**.

mobility measurements on porphyrin **8a–b** and phthalocyanine **9a** (Fig. 4-9) by PR-TRMC revealed $\sum \mu_{1D}$ values in excess of 0.1 $cm^2V^{-1}s^{-1}$ [36, 37]. By using the TOF technique, Ohta and coworkers found ambipolar charge transport (negative charge mobility of 2.4×10^{-3} $cm^2V^{-1}s^{-1}$ and a positive mobility of 2.2×10^{-3} $cm^2V^{-1}s^{-1}$) in the Col_h of a Cu-phthalocyanine compound **10** [57]. This was surprising at the time, since it was generally believed that the negative carrier mobilities should be much lower than the positive carrier mobilities in electron-rich organic semiconductors, including conventional vacuum-deposited films of Cu-phthalocyanines. Bushby and colleagues recently reported high hole mobility (~0.2 $cm^2V^{-1}s^{-1}$) in the Col_r phase of an octaalkyl phthalocyanine **11** [58]. Even higher mobility values (0.7 $cm^2V^{-1}s^{-1}$) were reported in the columnar phase of phthalocyanine metallomesogens [59].

As part of our ongoing research on self-organizing charge transport materials for photovoltaic device applications, we designed and synthesized some porphyrin-based DLCs **(12a–d)** (Fig. 4-10) that combine easy synthesis,

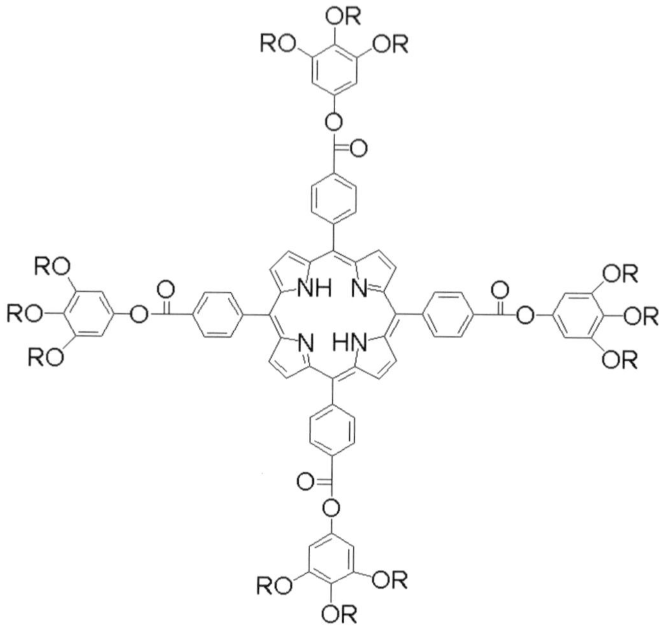

12a R=C$_{12}$H$_{25}$; **12b** R = C$_{14}$H$_{29}$

12c R = C$_{16}$H$_{33}$; **12d** R =(CH$_2$)$_{10}$C$_4$F$_9$

Figure 4-10   Molecular structures of DLCs **12a–d**.

high chemical purity, and unique self-aligning properties together with wide temperature range for their columnar LC phase, which extends even to room temperature [60]. More interestingly, they were found to form defect-free large-area monodomain films spontaneously with homeotropic alignment [61]. This feature is quite beneficial for potential photovoltaic applications. Encouraged by these results, we solution processed bilayer- and bulk-heterojunction cells based on **12a–b**. The power conversion efficiency achieved under ambient conditions appears to be the highest of any reported solar cells using columnar DLC materials so far [62].

The vast majority of the known discotic mesogens are relatively rich in electrons, and thus are better at transporting holes (donor materials) than electrons. Electron-transporting materials (acceptor materials) with high mobility are essential for the improvement of charge separation in solar cells and for applications in bipolar transistors and OLEDs. Sandwiching a layer of *n*-type semiconducting material between the cathode and the emission layer enhances the performance of LEDs by facilitating the electron injection and by relegating holes to the emission layer. Therefore more electron-deficient discotic molecules are being synthesized and characterized as *n*-type materials than ever before. They can be obtained either by substitution of electron-withdrawing peripheral groups onto a well-known p-type discotic core or by designing new electron-deficient hetero-cyclic aromatic cores. Substitution of electron-withdrawing groups lowers both the LUMO and HOMO levels in π-conjugated systems and enhances the electron transport. The electron transport in the columnar mesophase was first evidenced in an electron-deficient thioether-substituted tricycloquinoxaline derivative **13a** (Fig. 4-11) [63]. Doping of compound **13a** with 6 mol% of potassium metal, an

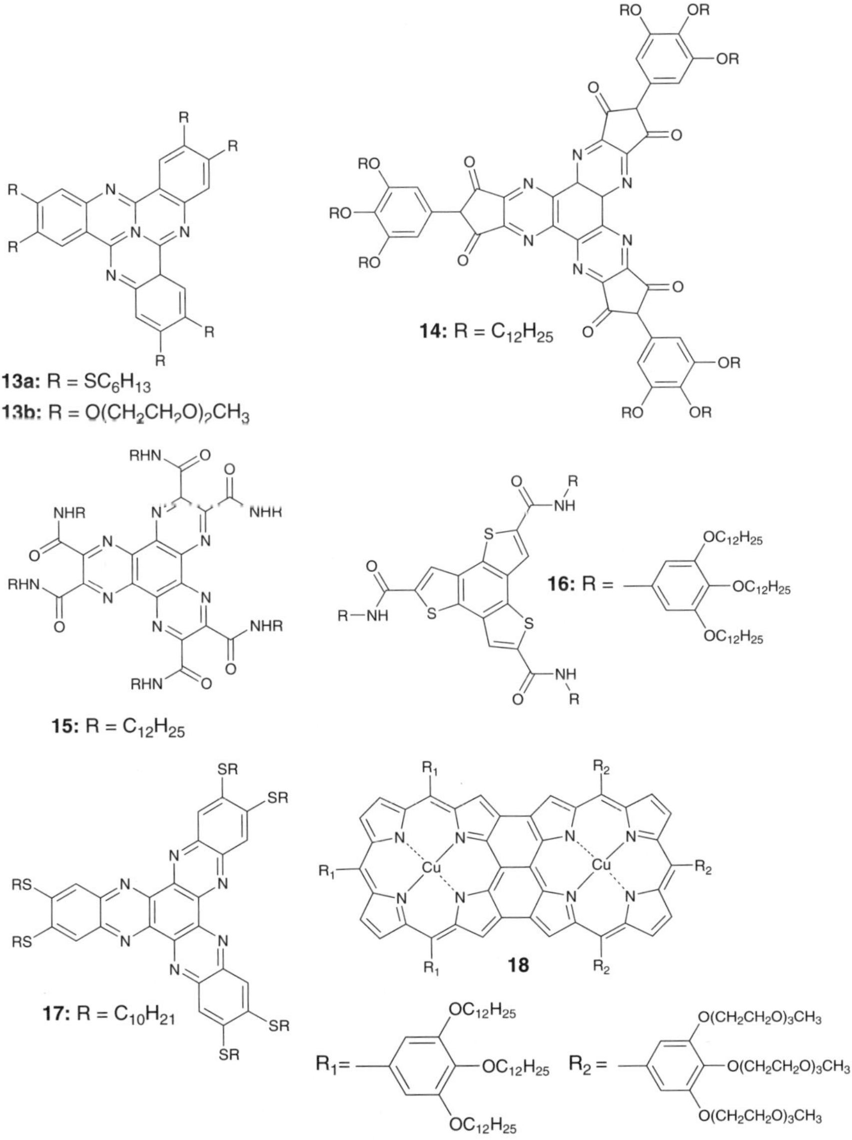

Figure 4-11   Molecular structures of DLCs **13–18**.

electron donor, resulted in conductivity ($\sigma_\parallel$) of about 2.9×10^{-5} S m^{-1} in the columnar phase. It was later found that replacing the hexylthio side chains in **13a** with ethyleneoxy chains (**13b**) enhances the K$^+$ solubility, leading to higher doping levels and higher conductivities ($\sigma_\parallel = 1.1 \times 10^{-3}$ S m^{-1}). Photoconductivity studies on **13b** using the TOF technique revealed transient photocurrents for both electrons ($\sim 10^{-4}$ cm^{2}V^{-1}s^{-1}) and holes [64]. Hexaazatriphenylene (**14** and **15**) is another example of an electron-deficient aromatic heterocyclic core exhibiting a columnar mesophase. Meijer and coworkers demonstrated applicability of DLC **14** as an electron-acceptor material by measuring photoinduced electron transfer processes in blend films with a well-known donor material poly(3-hexylthiophene) [65]. In another study, Gearba et al. reported $\sum \mu_{1D}$ mobilities up to 2.0×10^{-2} cm^{2}V^{-1}s^{-1} in the columnar mesophase of hydrogen-bonded hexaazatriphenylene **15** [66]. The interdisc distance of 3.18 Å reported for **15** is the smallest value ever found in the columnar phase, which is attributed to the presence of intermolecular hydrogen bonding. Very recently, Demenev et al. reported electron mobility of 2.0×10^{-3} cm^{2}V^{-1}s^{-1} in two hydrogen-bonded thermotropic hexagonal columnar mesophases of a benzotristhiophene derivative **16** [67]. Interestingly, the charge mobility was found to be temperature independent even across the phase transitions between the two mesophases of **16** because of the presence of an intracolumnar hydrogen bonding. Lehmann et al. reported DLCs based on electron-deficient hexaazatrinaphthylene core (for example, **17**) [68]. The electron-deficient character of **17** was demonstrated by cyclic voltammetry. By PR-TRMC measurements, charge carrier mobilities of 0.9 and 0.3 cm^{2}V^{-1}s^{-1}, respectively, for crystalline and columnar LC phases of **17** were found. Aida and coworkers recently synthesized a fused metalloporphyrin dimer **18** bearing hydrophobic (alkyl) and hydrophilic (triethylene glycol) chains in the periphery and reported its self-organization into a room-temperature columnar phase. High electron mobilities were found for the LC film of **18** (0.27 cm^{2}V^{-1}s^{-1} at 16°C) by FP-TRMC and the transient optical spectroscopy (TOS) technique [69]. Very recently, Geerts and colleagues described a series of electron-deficient discotic mesogens based on phthalocyanine-bearing peripheral alkylsulfonyl substituents (for example, **9b**). Electrochemical studies revealed the first reduction potential value of about -0.14 V versus the saturated calomel electrode (SCE) for **9b**, indicating the potential application of this material as an air-stable n-type semiconductor [70].

Notable among the high-performance n-type DLC materials are the perylene derivatives functionalized with electron-withdrawing imide substituents such as compounds **19–22** (Fig. 4-12). Photoconductivity and n-type semiconducting properties of these materials have found applications in prototype devices such as solar cells, OFETs, and OLEDs. Struijk et al. first reported the charge transport properties in perylene-based DLCs [71]. They measured charge mobilities of 0.1 cm^{2}V^{-1}s^{-1} and 0.2 cm^{2}V^{-1}s^{-1} in the liquid crystalline and crystalline phases, respectively, for N,N'-alkylperylene tetracarboxyldiimide **19d** by the PR-TRMC method. TOF measurements of mobility on a similar compound **19a**, however, yielded very similar mobility values for both electrons and holes [72].

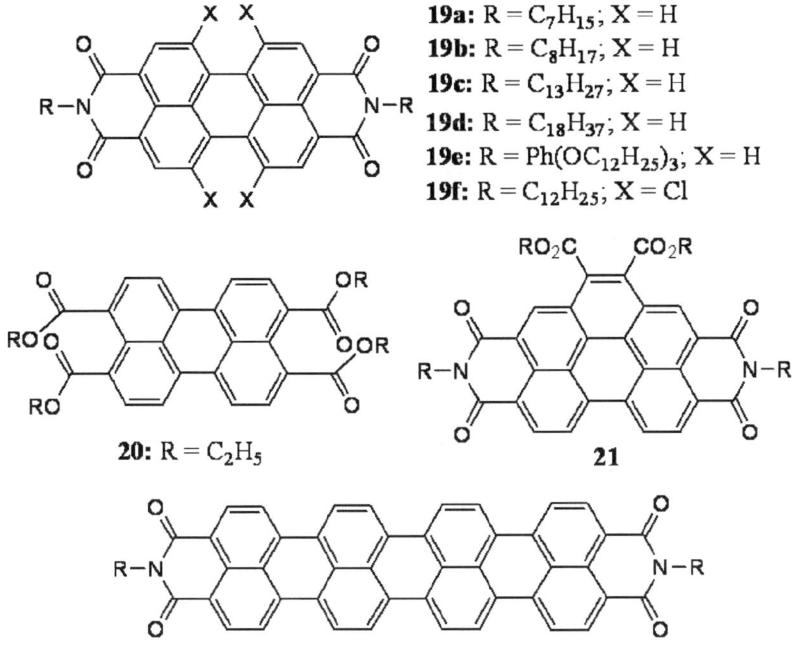

19a: $R = C_7H_{15}$; $X = H$
19b: $R = C_8H_{17}$; $X = H$
19c: $R = C_{13}H_{27}$; $X = H$
19d: $R = C_{18}H_{37}$; $X = H$
19e: $R = Ph(OC_{12}H_{25})_3$; $X = H$
19f: $R = C_{12}H_{25}$; $X = Cl$

20: $R = C_2H_5$

21

22: $R = CH(C_7H_{15})_2$

Figure 4-12   Molecular structures of DLCs **19–22**.

Field-effect mobility up to 0.6 $cm^2V^{-1}s^{-1}$ and current on/off ratios $>10^5$ were obtained for thin-film transistors based on perylene bisimide **19b**, when operated under nitrogen atmosphere [73]. Charge-carrier mobility values as high as 1.1×10^{-2} $cm^2V^{-1}s^{-1}$ have been found in the room-temperature Col_h phase of perylene bisimide **19e** [74]. The latter work showed a strong dependence of mobility values on film morphology, which in turn was sensitive to processing conditions such as melting temperature, cooling rate, and shear applied during sample preparation. Functionalization of perylene bisimide cores at bay positions was considered by researchers in order to induce mesomorphism and improve charge carrier properties. Accordingly, enhanced charge transport properties were reported for tetrachloro-substituted perylene bisimide **19f** in comparison to a nonchlorinated analog compound [75].

Destruel and coworkers reported fluorescence in the columnar mesophase of perylene tetraester compound **20** [76]. They demonstrated red light emission by combining electron-deficient **20** with electron-rich triphenylene **2a** (as a hole transport material) in a bilayer OLED device. Columnar LC diimide-diester compounds (**21**) are known to support electron transport in their mesophases because of the presence of strong electron-withdrawing substituents. Combining **21** with LC benzoperylene triester compounds in bilayer OLED devices was reported to yield red color emission [77]. Higher homolog of perylene diimides, namely, terrylene and quaterrylene tetracarboxyldiimides, also exhibit columnar mesophase with attractive semiconducting properties [78]. For example, recently a "swallow-tailed" quaterrylene tetracarboxyldiimide **22** was utilized as an active conducting layer in an ambipolar thin-film transistor [79]. However, postproduction thermal annealing of the device resulted solely in electron transport.

A few other cores like coronene, decacyclene, and rufigallol are also known to exhibit electron transport in their columnar mesophases when appropriately substituted with electron-withdrawing substituents. A coronene derivative **23** (Fig. 4-13) is reported to show a room-temperature discotic mesophase having intracolumnar charge-carrier mobility of $\sim$0.2 cm^{2}V^{-1}s^{-1} [80]. Hashimoto and coworkers found electron-accepting properties in discotic 1,7,13-heptanoyldecacyclene **24**. Photo-induced electron transfer was observed when **24** was combined with a p-type semiconducting polymer [81]. Kumar and colleagues reported stabilization of a room-temperature Col$_h$ phase in a series of electron-deficient rufigallol derivatives **25** [82]. More recently, his group reported wide range stabilization of the Col$_h$ phase in DLC oligomers **26**, wherein an electron-rich triphenylene moiety was combined with an electron-deficient anthraquinone moiety to obtain donor-acceptor-donor triads [83]. Although charge transport behavior of these compounds has not yet been reported, such a study can be highly useful for potential application in devices like solar cells where a pure single compound with the capability to form separate charge conduction pathways for electrons and holes is highly desirable. In another interesting study, Bunning et al. reported electrons as majority carriers in a triblock copolymer with a main chain polymeric triphenylene liquid crystal capped at either end with poly(ethylene oxide) chains [84].

Highly fluorescent discotic columnar phases are beneficial for OLED applications. However, the majority of the DLCs discussed so far (except for some perylene derivatives) are only weakly fluorescent in their mesophases. Recently, pyrene-based DLCs have been synthesized in order to induce strong fluorescence in the columnar mesophase [85, 86]. Shimizu and coworkers reported ambipolar charge transport in the Col$_h$ phase of a pyrene derivative **27** (Fig. 4-14) with charge mobilities on the order of 10^{-3} cm^{2}V^{-1}s^{-1} [86]. Interestingly, when the Col$_h$ phase of **27** was supercooled to room temperature, a rigid glassy state was formed (with structural features of the preceding Col$_h$ phase) without disrupting the charge transport.

In recent years, several other π-conjugated systems based on triazatruxenes [87, 88], corannulenes [89], ovalenes [90], tris(N-salicylideneanilines) [91], and dibenzophenazine [92] have been identified as promising DLC cores. Talarico et al. found charge mobilities of 0.03 cm^{2}V^{-1}s^{-1} in the Col$_h$ phase of a discotic triazatruxene derivative **28** and 0.09 cm^{2}V^{-1}s^{-1} in its crystal phase [87]. More recently, Zhao et al. reported higher charge mobility (0.8 cm^{2}V^{-1}s^{-1}) in the crystal phase of a triazatruxene derivative with an extended π-system where three triphenylene units were fused together with a triazatruxene core [88]. Charge transport properties of other DLCs mentioned above are not yet reported.

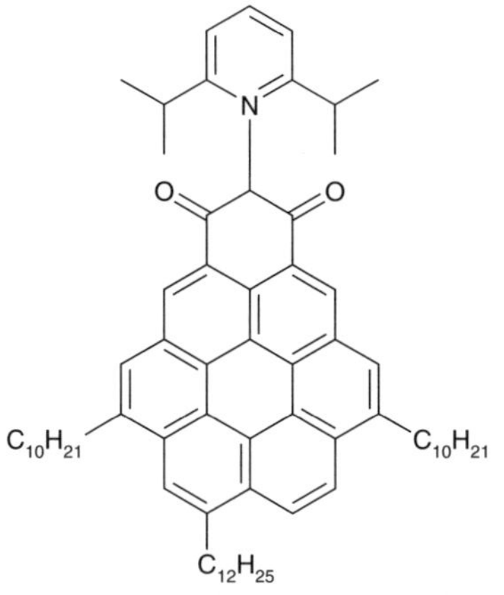

25: R = C_nH_{2n+1}; n = 4–12

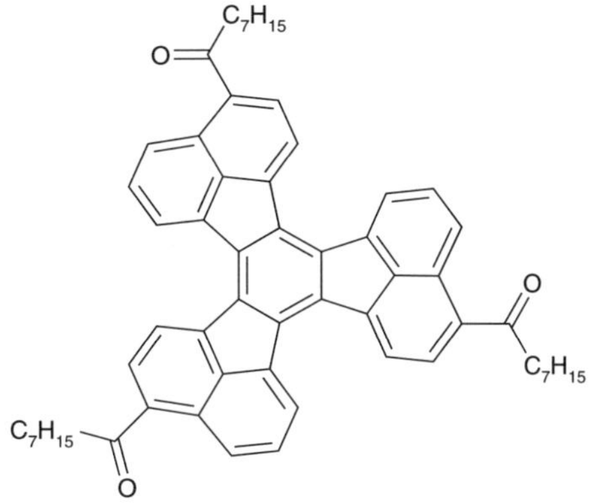

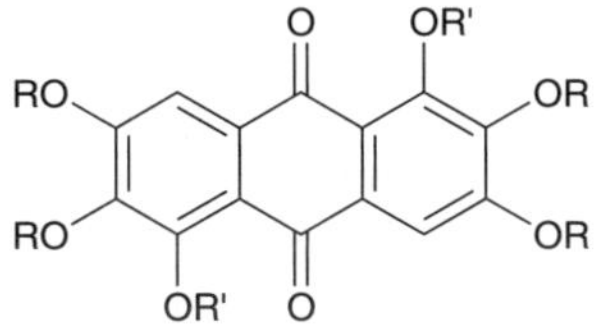

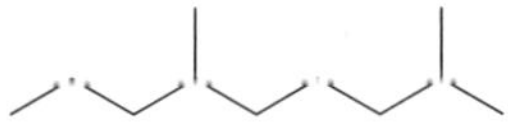

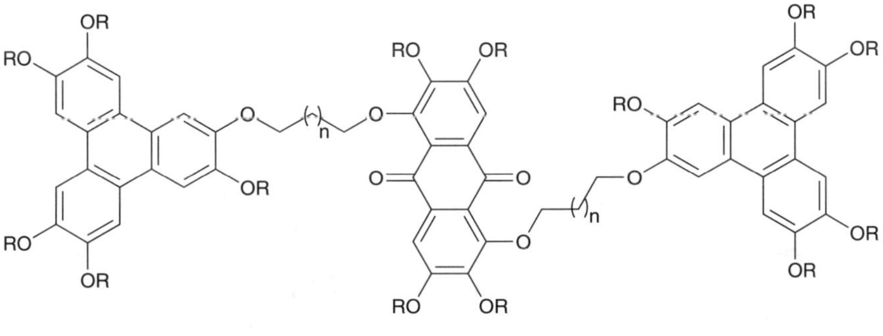

26: R = OC_6H_{13}; n = 8, 10

Figure 4-13 Molecular structures of DLCs **23–26**.

4.4. PROCESSING OF DISCOTIC MATERIALS INTO ACTIVE SEMICONDUCTING LAYERS

The efficiency of devices based on semiconducting columnar DLCs depends not only on the availability of materials with high charge-carrier mobilities but also

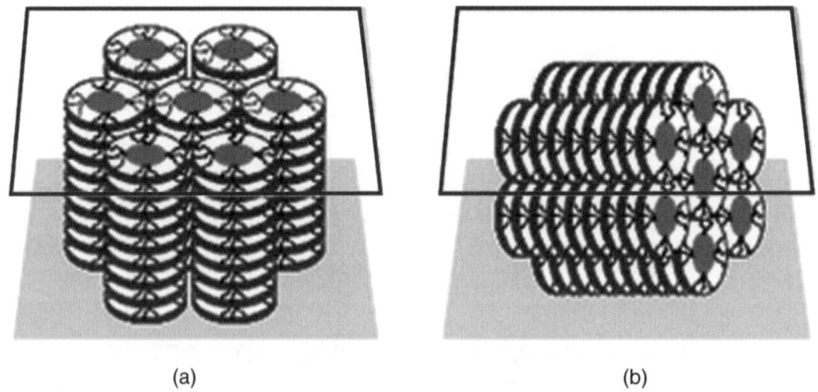

Figure 4-14 Molecular structures of DLCs **27** and **28**.

on the processabilty of these into highly ordered monodomain thin-film structures. Local defects at domain boundaries in unoriented layers can trap charge carriers and significantly decrease device performance. Furthermore, DLCs are quasi-one-dimensional semiconductors, that is, charges and excitons travel much faster along the columnar stacks than between columns. This implies that the columns must be appropriately aligned in a direction that the current is desired to flow. DLCs can align either perpendicular to the substrate surface, referred to as homeotropic alignment, or parallel to the substrate surface, referred to as planar alignment, i.e., homogeneous alignment. This is illustrated in Figure 4-15. A planar alignment (or "edge-on" orientation) of columns on a substrate is required for OFET applications, while a homeotropic alignment (also known as "face-on" orientation) is preferred for solar cells and OLED applications. In devices the conducting columns have to bridge the gap between anode and cathode all within a defect-free and long-range ordered thick (a few micrometers) or thin (a few tens of nanometers) film. However, conventional techniques used for alignment

(a)   (b)

Figure 4-15 Schematic representations of homeotropic (a) and planar (b) orientations of the hexagonal columnar discotic phase. (A full color version of this figure appears in the color plate section.)

of calamitic LC phases are not useful for the alignment control of highly viscous columnar LC phases [93]. Although the most promising way is to design columnar discotics that self-organize spontaneously with planar or homeotropic alignment onto a substrate, general molecular design principles that allow this controllability on the molecular orientation have not been established. Therefore, the self-organizing ability of discotic molecules is often combined with homeotropic or planar alignment control techniques for device fabrication.

4.4.1. Homeotropic Alignment

The influence of different parameters such as molecular structure, film thickness, and surface interactions for obtaining homeotropic alignment of various columnar phases has been the subject of intense study during the past several years. In general, it is observed that a "face-on" alignment of the columnar mesophase can be obtained when the material is thermally annealed between two substrates. Representative examples of triphenylene, phthalocyanine, porphyrin, and hexabenzocoronene DLCs have been shown to adapt such an alignment in their columnar mesophases spontaneously in relatively thick films (a few micrometers) when the material is confined between two substrates and slowly cooled from their isotropic state. In most cases, the uniform alignment requires a slow cooling rate. Moreover, high viscosity of the columnar mesophase often promotes multidomains and prevents the perfect homeotropic alignment over a large area, which is required for device applications. As a molecular design strategy, attempts have been made to decrease the isotropic melt viscosity of mesogens either by incorporating hetero atoms in the flexible side chains or by introducing sterically hindering groups on to the mesogenic core. Accordingly, phthalocyanine molecules bearing alkoxy groups at or within their core periphery revealed the tendency to spontaneously align "face on" in their columnar mesophase [94]. Mullen and coworkers reported spontaneous homeotropic alignment in branched long alkyl chain-substituted (with or without ether linkages) HBC derivatives on indium tin oxide (ITO) substrates by melt processing in the columnar mesophase [95]. A systematic study of triphenylene DLCs on various substrates revealed that the homeotropic alignment is thermodynamically favored in the columnar phases, while the alignment changes to planar when kept in the glassy or crystalline state [96]. Some other studies focused on modifying the surface affinity of molecules by changing the chemical nature of the side chains. Shimizu and coworkers found that substitution of perfluoroalkyl group into the peripheral chains of triphenylene mesogens (for example, **2e**) promoted homeotropic alignment in the hexagonal columnar (Col_h) phase on polyimide; ITO-, and cetyltrimethylammonium bromide-coated glass substrates [97]. The influence of number as well as nature of substrates on the alignment of a discotic phthalocyanine molecule **29** (Fig. 4-16) was investigated by Geerts and coworkers [98]. They found that thermal annealing of **29** in the Col LC phase induced homeotropic alignment irrespective of the surface, while a planar alignment was always preferred in thin films on a single substrate. Their study also revealed that the alignment in the case of **29** was governed by the

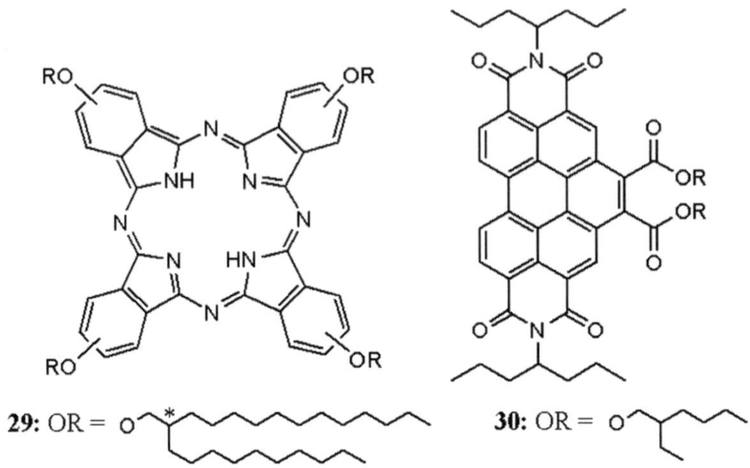

29: OR =   **30:** OR =

Figure 4-16   Molecular structures of DLCs **29** and **30**.

confinement induced by the two solid substrates rather than the nature/polarity of the substrates. Recently, homeotropic alignment of columns by application of an electric field was reported in a corannulene-based DLC molecule [89]. This observation is noteworthy, considering that most DLCs known so far are not responsive to the applied electric field in their columnar mesophases.

In recent years, we have investigated several porphyrin-based DLCs (**12a–d**) for their alignment behavior and found that the control of orientational order of columns can be achieved by varying the film thickness, thermal annealing process, and mechanical shearing. Our experiments revealed that the surface modification is not necessary for obtaining uniform alignment in the columnar mesophases of **12a–d**. The preferential orientation of these materials between two substrates was found to be homeotropic. Thinner cells ($<$10 μm) promoted more uniform homeotropic alignment. Slow cooling at a rate of $\sim$2°C/min from the isotropic phase of **12a–c** promotes selective nucleation and slow growth of homeotropic domains, while abrupt cooling results in multidomain structures. The homeotropic anchoring in the columnar LC films of **12a–c** was demonstrated by the lack of birefringence (Fig. 4-17) as well as by the typical hexagonal pattern obtained by X-ray scattering studies (Fig. 4-18a). The observation of charac- teristic isogyre interference (commonly refered to as a "Maltese cross") in the conoscopic study further supported the homeotropic alignment in the colum- nar LC phase (Fig. 4-18b). Samples deposited on a single substrate displayed multiple homeotropic and randomly oriented planar domains. However, shear- ing and annealing the sample film resulted in highly ordered planar alignment along the shear direction. We also studied alignment properties of partially per- fluoroalkylated discotic porphyrins (for example, **12d**) [99]. Porphyrin **12d** can be easily capillary filled into thin cells from its isotropic phase. In films thinner than 10 μm, it showed an exceptionally strong tendency toward spontaneous

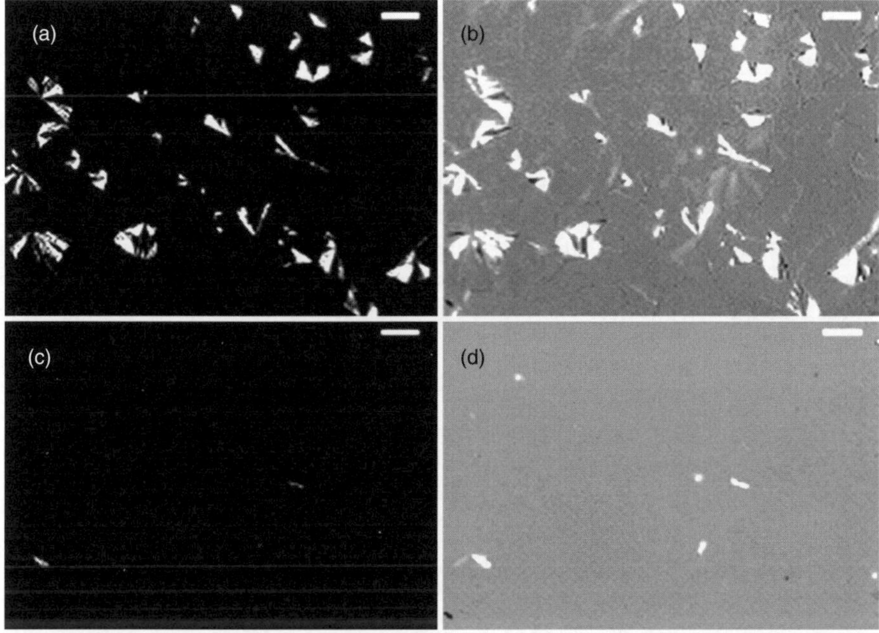

Figure 4-17   Spontaneous transition from planar to homeotropic orientation observed for compound **12c** in a 1.8-μm cell. The planar domains (a, b) obtained upon fast cooling slowly transform to the homeotropic state (c, d) during annealing at 118.0°C for 2 h. Optical images were taken at 90° (a, c) and 70° (b, d) angles between the polarizers. The scale bar corresponds to 50 μm. Reproduced with permission from Ref. 60. (A full color version of this figure appears in the color plate section.)

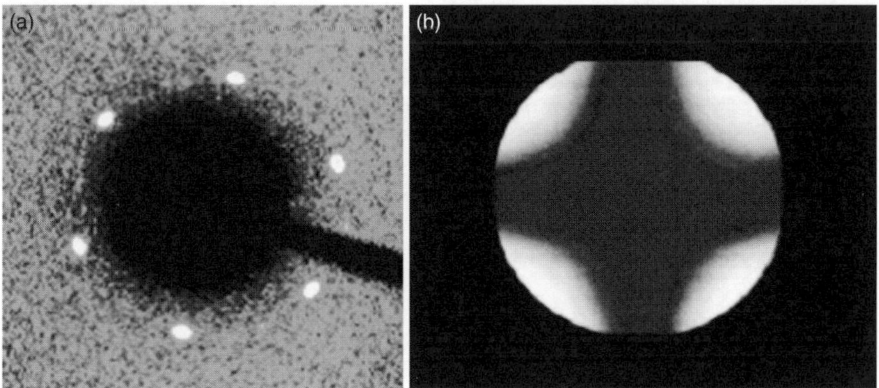

Figure 4-18   (a) X-ray diffraction pattern of hexagonal columnar structures obtained from homeotropic monodomain of sample **12c** in a 8.0-μm cell at 75.0°C. (b) Conoscopic image of **12c** at 75.0°C in a 1.8-μm cell. Reproduced with permission from Ref. 60. (A full color version of this figure appears in the color plate section.)

homeotropic alignment while cooling at 2°C/min. The completely dark texture and the symmetric cross in the conoscopic image indicated the homeotropic alignment. Synchrotron X-ray diffraction obtained from an optically dark homeotropic monodomain thin film of **12d** confirmed the homeotropic alignment. Interestingly, it was much easier to obtain a large-area defect-free homeotropically aligned thin film of **12d** than its nonfluorinated analog compound **12b** (Fig. 4-19). Another appealing aspect of these porphyrin DLCs is their capability to retain stable homeotropic alignment through room temperature.

Fabrication of devices such as solar cells and OLEDs requires sequential deposition of the active conducting layers on one electrode before the deposition of the second electrode. Therefore, it is desirable to achieve homeotropic

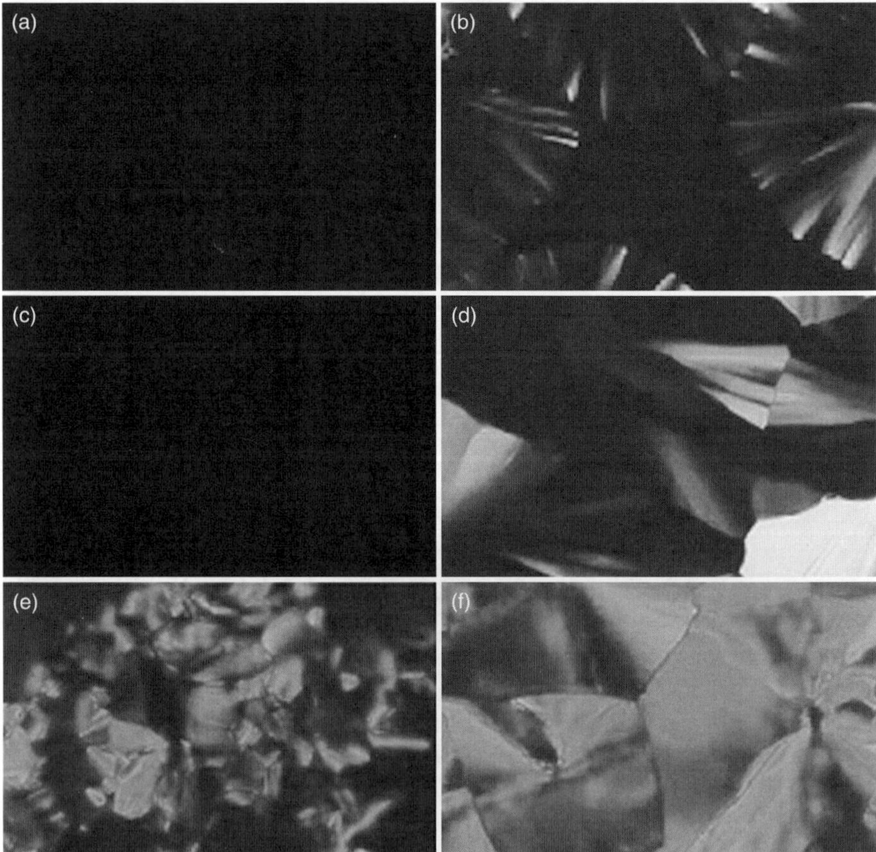

Figure 4-19 Cross-polarized optical textures of **12d** (a: 5 μm thick; c: 9 μm thick; e: 20 μm thick) and its corresponding nonfluorinated counterpart **12b** (b: 5 μm thick; d: 9 μm thick; f: 20 μm thick) at room temperature. The dark areas represent homeotropic alignment, and the bright domains appear when the porphyrin planes are oblique to the substrates. (A full color version of this figure appears in the color plate section.)

alignment of the materials in open films (i.e., one interface of the film being in contact with the air) before the deposition of the second electrode. However, most of the discotic mesogens which show homeotropic alignment between two surfaces, do not exhibit such an alignment on a single surface. In some cases, kinetically controlled thermal annealing has been shown to be useful to obtain homeotropic alignment in relatively thick ($\geq$300 nm) open supported columnar LC films [100]. However, planar alignment of columnar LC films was favored during controlled growth of thinner films ($\leq$100 nm). Organic molecules generally have strong optical absorption coefficients, so thin films (50–100 nm) are required for characterization of their optoelectronic properties. Some recent studies have demonstrated homeotropic alignment of columnar mesophase on thinner open films depending on the nature of the columnar mesophase and the surface treatment. For example, Gearba et al. reported the formation of homeotropically aligned open films of a hexagonal columnar phase on poly(tetrafluoroethylene) (PTFE)-coated substrates [101]. Their study on discotic phthalocyanine **29** revealed that such an alignment is possible only in the Col_h mesophase of the material and not in the underlying Col_r phase. In another study, Charlet et al. reported "face-on" orientation of **30** in thin open films ($\geq$50 nm) on ITO substrates that were subjected to either UV ozone or nitrogen plasma surface treatment [102]. Steinhart and coworkers achieved homeotropically aligned nanowires of a triphenylene discotic compound within ordered porous alumina templates [103]. Very recently, Geerts and coworkers demonstrated the use of a polymer sacrificial layer to induce homeotropic alignment in the columnar LC films. In their method, "face-on" alignment is imposed by thermal annealing of the confined LC film with a top sacrificial polymer layer that is later removed by washing with a selective solvent (Fig. 4-20). As a proof-of-principle experiment, they demonstrated homeotropic alignment of discotic columns in open films (50–300 nm) for a phthalocyanine mesogen **29** [104].

4.4.2. Planar Alignment

The most important application envisioned for discotic columns with planar alignment, i.e., homogeneous alignment, is in the field of OFETs. A well-ordered and uniaxially aligned thin layer of discotic columns is a prerequisite for high charge-carrier transport and thus for device performance. Various approaches including application of PTFE alignment layers, zone casting, zone crystallization, Langmuir–Blodgett (LB) films, self-assembled monolayers, pulsed infrared irradiation, and magnetic field have been developed to achieve planar alignment. Among these methods, the use of a PTFE alignment layer and zone casting offers the advantage of sample processing from solution. Zimmermann and coworkers first demonstrated the "edge-on" orientation of a triphenylene DLC in its hexagonal plastic columnar phase by spin coating the material onto a PTFE alignment layer [105]. PTFE-coated surfaces were later shown to be an excellent substrate for aligning several HBC derivatives. Friend and coworkers reported planar alignment of HBC **4d** on rubbed-PTFE layer and obtained field-effect

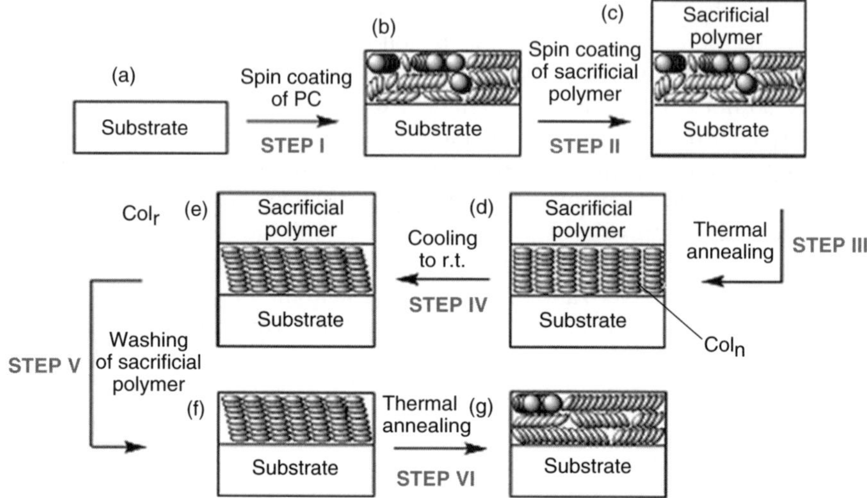

Figure 4-20 Schematic representation of various processing steps involved in obtaining homeotropically aligned thin films of phthalocyanine derivative **29** by the application of a sacrificial polymer layer. The last step demonstrates the scenario without the use of a sacrificial layer. Reprinted with permission from Ref. 104.

mobility as high as 0.5×10^{-3} cm^{2}V^{-1}s^{-1} even without the need for an annealing step in the mesophase [106]. Zone casting and zone crystallization techniques are very similar in principle and differ only in processing conditions. While the former method involves solution processing, the latter depends on melt processing of the sample. In the zone-casting method, a solution of the material is deposited through a nozzle onto a moving glass support so that the temperature/concentration gradient guides the molecules to uniaxially aligned columns parallel to the moving substrate (see Fig. 4-21). The film morphology depends on

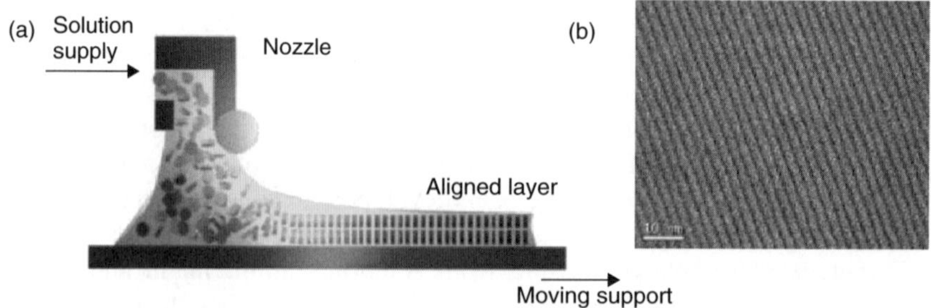

Figure 4-21 (a) Schematic presentation of the zone-casting technique. (b) Large-area image from HRTM of a zone-cast **4a** ($n = 12$) displaying homogeneous film formation with single columnar features lying in the zone-casting direction. Reproduced with permission from Ref. 107. Copyright 2005, Wiley-VCH.

processing conditions such as evaporation temperature, polarity of the solvent, concentration, solvent flow, substrate velocity, as well as the distance between the nozzle and the substrate. This is now a well-known method to process various materials such as low-molecular-weight semiconductors, discotic molecules, and block copolymers. Müllen and coworkers reported higher FET device performance (on-off ratio of 10^4 and field-effect mobility of $\sim5 \times 10^{-3}$ cm^{2}V^{-1}s^{-1}) with zone-cast films of discotic HBC **4a** ($n = 12$) [107]. In another study, Tracz et al. demonstrated planar alignment in the Col$_r$ phase of phthalocyanine **29** by using this technique [108]. In the zone crystallization method, the sample in its isotropic phase is moved in a defined velocity from a hot substrate to a cold substrate, allowing the material to crystallize along the temperature gradient as an aligned film [109]. Discotic HBC derivatives with branched long alkyl substituents were successfully processed into macroscopically ordered domains with "edge-on" orientation by using the above technique [110]. In another approach, application of a strong magnetic field (20 T) was used to produce large-area monodomain films of discotic HBC molecules. On the basis of this method, solution-processed FETs had been constructed with charge-carrier mobilities up to 10^{-3} cm^{2}V^{-1}s^{-1}, which were significantly higher than the mobilities obtained for the unaligned material [111]. Creating a Langmuir–Blodgett film is another frequently applied method for fabricating ultrathin ordered layers of discotic molecules. In this technique, the ability of amphiphilic molecules to align "edge on" at the air-water interface (the Langmuir film) is effectively transferred to a solid substrate by vertical dipping/raising of the substrate while maintaining a constant surface pressure [112]. DLCs of triphenylenes, phthalocyanines, and HBCs that are partially functionalized with hydrophilic side chains are reported to form well-ordered LB films when spread from solution at the air-water interface [113–115]. Armstrong and coworkers fabricated thin-film transistors based on a discotic phthalocyanine molecule on silicon substrates by the LB technique [116]. In another emerging technique, self-assembled monolayers (SAMs) of discotic molecules with either planar or homeotropic orientation of headgroups are constructed in order to induce the same orientation into the bulk sample. An "edge-on" orientation of the SAMs obtained from thiol-functionalized triphenylene derivatives on a gold surface has been reported [117]. Shimizu and coworkers reported selective switching between planar and homeotropic alignments in the Col$_h$ phase of the triphenylene mesogen **2c** by the excitation of C–C stretching vibration of the molecule using a combination of linearly and circularly polarized infrared irradiation [118]. Recently, this technique was extended to achieve alignment control of plastic columnar and helical columnar mesophases of other triphenylene derivatives [119]. Räder et al. reported a soft-landing methodology that exploits matrix-assisted laser desorption/ionization (MALDI) mass spectrometry to produce ordered structures of organic macromolecules at surfaces [120]. In this technique, the molecules are first converted to the gas phase, accelerated by strong electric fields, purified according to their mass-to-charge ratio, and then decelerated and softly deposited at surfaces. Soft-landing deposition of HBC molecules resulted in planar alignment on surfaces. Although this technique has not yet been used to align any columnar mesophase, the versatility of the method promises such studies in the near future.

4.5. APPLICATIONS OF SEMICONDUCTING DISCOTIC LIQUID CRYSTALS

Discotic nematic liquid crystals have found their first commercial application in the development of an optical compensator film for improving the view angle of LCDs [121]. Also, semiconducting columnar DLCs are very promising materials for applications in various molecular electronic devices where the phenomena of photoconductivity, charge transport, and charge recombination are utilized. With an increasing fundamental understanding of the charge transport in bulk and in thin films of DLC materials, the interest in these materials for potential applications is rapidly growing. In this section we discuss the progress that has been made so far toward DLC-based prototype devices such as solar cells, OLEDs, and OFETs.

4.5.1. Solar Cells

Solar cells (also known as photovoltaic cells) allow the conversion of light energy into electrical energy. A conventional photovoltaic (PV) cell consists of a thin film of semiconductor sandwiched between two electrodes with different work functions. In such a device, a built-in electric field is generated by the equalization of the Fermi energies of the two electrodes by flow of electrons from the low-work-function electrode to the high-work-function electrode. In principle, this built-in electric field can pull the photo-induced charge carriers (electrons and holes) to their corresponding electrodes, thereby generating a current and a voltage. In contrast to the generation of free carriers in inorganic semiconducting materials, light absorption in organic semiconducting materials leads to the generation of tightly bound (binding energy of 0.1–0.4 eV) charge carriers in the form of excitons. In most cases, an additional driving force is required to dissociate photo-generated excitons into the separate charges, limiting the operation of such single-layer organic photovoltaic (OPV) cells. Excitons can be split into separate charges at the electrode interfaces or by charge transfer interaction between donor and acceptor molecules. In the case of heterojunction devices, two organic semiconducting materials with different electron affinities (LUMO) or ionization potentials (HOMO) are combined together to provide sufficient chemical potential energy to overcome the intrinsic exciton-binding energy (Fig. 4-22). Also, charge separation is shown to be much more efficient at the donor-acceptor interface than at the electrode interface. Once an exciton reaches the donor-acceptor interface, the electron promoted in the LUMO level of the excited donor can be transferred to the low-lying LUMO level of the acceptor, with the hole remaining on the HOMO level of the donor, thereby forming a polaron pair. To produce current, the dissociated holes and electrons have to move to their respective electrodes under the influence of built-in electric potential created by connecting two

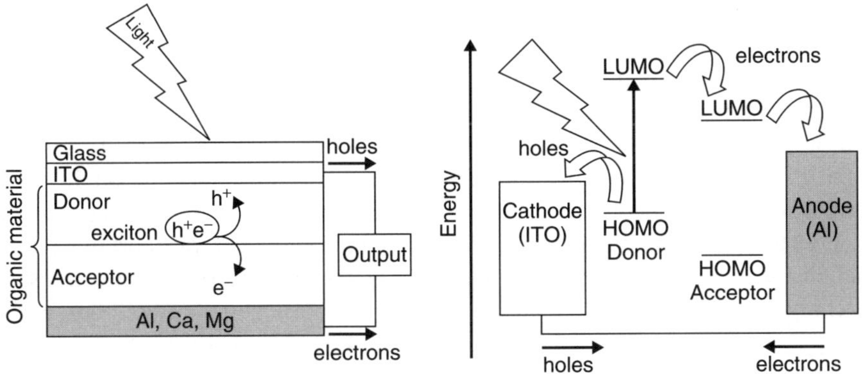

Figure 4-22 Schematic diagram of an organic heterojunction solar cell (left) and the corresponding energy-level diagram showing charge separation and transport (right).

electrodes. Therefore, organic materials having a long exciton diffusion length and high carrier mobility are desirable for improving energy conversion efficiency. In solar cells, the UV-VIS absorption profile of the materials is also very important, as it directly relates to the quantity of photons that the device can potentially capture. Equally important are the relative energy levels of the donor and acceptor materials. The energy gap between HOMO of the donor and LUMO of the acceptor determine the potential output (open-circuit voltage) of the device. Therefore, the performance of a solar cell device critically depends on several factors such as absorption of incident light, exciton diffusion, charge separation, and charge collection at the electrodes [122].

The concept of a heterojunction cell structure for organic solar cells was first introduced by Tang in 1986 [123]. By making a bilayer device using two different dyes, about 1% power conversion efficiency was obtained, which was an order of magnitude greater than single-material OPV cells known at that time. The morphology of a donor-acceptor interface in a heterojunction cell was soon recognized as an important parameter to determine the device performance. In a bilayer device, the donor-acceptor interface is planar. To increase the donor-acceptor interface area, Yu et al. in 1995 introduced the concept of bulk heterojunction (BHJ) cells in which the donor and acceptor materials are blended together in a film to form an interpenetrating network of donors and acceptors such that the charge carriers can have unrestricted conduction pathways to their respective electrodes [124]. Recently, power conversion efficiencies of about 5% have been reported in BHJ devices having mixtures of fullerene derivative PCBM (6,6-phenyl-C_{61}-butyric acid methyl ester) and an electron donor material P3HT [poly(3-hexylthiophene-2,5-diyl)] [125, 126].

Discotic columns as an active semiconducting layer in OPV cells have a number of advantages over other organic materials. Apart from the ease of solution processibility into large-area thin films on flexible substrates, the highly ordered columnar mesophase of DLCs offers large exciton diffusion length and high charge-carrier mobilities, which are both key parameters for the device efficiency. Moreover, the large π-conjugated aromatic cores of DLCs can be particularly attractive because of their absorption capabilities of the entire solar spectrum. Calculations based on the absorption coefficient of organic materials reveal that a film thickness of 50–100 nm is often required to absorb most of the incident light. Experiments have shown that an exciton can diffuse only around 5–40 nm in most organic semiconductors before recombination. On the other hand, highly ordered homeotropically aligned DLC films support efficient transport of excitons and charge carriers in relatively thick films (up to a few 100 nm). Slow transport of charges can result in recombination during the journey to the electrodes. In this context, the high intrinsic charge-carrier mobilities reported for several discotic materials further highlight their potential for solar cell applications.

Among the most promising columnar discotics that have been investigated so far for solar cell applications include derivatives of porphyrin, phthalocyanine, HBC, triphenylene, and perylene cores. The first report on a photovoltaic effect in a discotic LC film was published in 1990 by Gregg et al., who sandwiched a porphyrin DLC layer between identical ITO-coated glass electrodes [127]. Several years later, Petritsch et al. demonstrated photovoltaic performance in a bilayer heterojunction cell comprising a discotic phthalocyanine derivative as an electron donor and a perylene derivative as an electron-acceptor material, albeit with poor external quantum efficiencies (EQE $\sim$0.5%) [128]. Schmidt-Mende et al. reported a significant improvement in photovoltaic performance by combining discotic HBC **4b** (electron donor) with an n-type perylene tetracarboxdiimide molecule in a BHJ device [129]. They achieved a power conversion efficiency around 2% under monochromatic illumination (490 nm). This performance was mainly attributed to the nanophase segregation of perylene and HBC systems in the blend films to establish separate percolation pathways for electron and hole transport. However, photovoltaic efficiency of their device was limited under standard solar illumination conditions because of weak absorption in the red part of the spectrum.

Recently, we have fabricated bilayer- and bulk-heterojunction solar cells by solution processing of discotic porphyrin donors **12a** or **12b** with C_{60} or PCBM acceptors [62]. Discotic porphyrins are good candidates for photovoltaic applications because (i) they are highly absorptive over the wavelengths of the solar spectrum, (ii) they exhibit spontaneous homeotropic alignment in their columnar mesophase, and (iii) their energy levels are matched well with the electron acceptors and anode materials to facilitate charge separation and transfer. The fact that nature employs porphyrin-based molecules such as chlorophylls for photosynthesis provided further incentive to study these materials. Both bilayer- and bulk-heterojunction devices were fabricated on ITO glass coated with a 30-nm-thick poly(3,4-ethylenedioxythiophene)/poly(styrenesulfonate) (PEDOT-PSS) conductive polymer mixture. Next on top of the PEDOT-PSS was deposited the active layer(s). In the bilayer solar cell, an active layer was composed of a porphyrin layer (80 nm thick, spin coated from its chlorobenzene solution) and a C_{60} layer (30 nm thick, thermally deposited under vacuum). In the bulk-heterojunction device, a blend thin film of the porphyrin (**12a** or **12b**) and PCBM (1:1 *w/w* and 230–250 nm thick spin coated from a chlorobenzene solution) was used as the active layer. This active photoconducting layer was responsible for light absorption, exciton generation/dissociation, and charge transport. On top of the active layer was deposited the 15-nm-thick Ca/Al electrode (150 nm thick). Devices were characterized under illumination of AM 1.5 G, 100 mW/cm^2. The pristine bilayer device exhibited an open circuit voltage (V_{OC}) of 0.495 V, a short-circuit current density (J_{SC}) of 0.400 mA/cm^2, and power conversion efficiency (PCE) of 0.070%. Postproduction thermal annealing of the device resulted in more than 100% increase in both the J_{SC} (0.400–0.870 mA/cm^2) and PCE (0.070–0.141%) (Fig. 4-23). This difference in performance can be attributed to the improved intracolumnar packing of the porphyrin layer upon annealing favoring the charge-carrier transport and exciton diffusion length, which are both key parameters for device efficiency. Compared to the bilayer PV cells, BHJ devices showed better performance. After thermal annealing, BHJ cells exhibited a J_{SC} of 3.990 mA/cm^2 and a PCE of 0.712%. The preliminary photovoltaic performance is inspiring, since a homeotropic alignment is difficult to achieve in such a blend film. We anticipate dramatic improvements in efficiency by tailoring molecular structure, optimized cell structure, and engineering if we can maintain the homeotropically aligned architecture in bulk, double, or multiheterojunction solar cells.

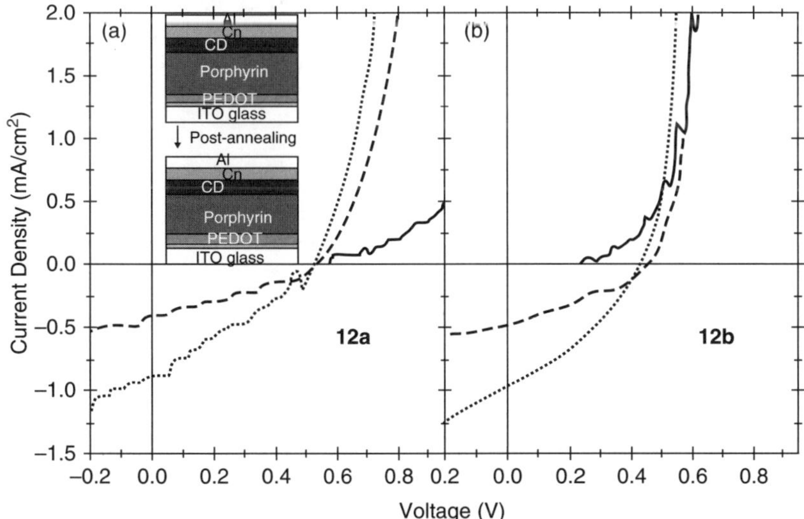

Figure 4-23 *J-V* characteristics of the bilayer solar cells based on (a) **12a** and (b) **12b** under dark (solid line), and before (dash line) and after (dot line) annealing under the illumination of AM 1.5 G and 100 mW/cm^2. The inset (a) schematically shows possible ordering of active porphyrin layer before and after annealing. Reproduced with permission from Ref. 62. (A full color version of this figure appears in the color plate section.)

In BHJ solar cells, the nanoscale phase segregation of donor (D) and acceptor (A) molecules within the blend films has a significant influence on device performance. However, it is often difficult to optimize the degree of phase separation and domain size in blend D/A systems because of several governing factors such as the choice of solvent, speed of evaporation, and solubility and miscibility of the donor and acceptor, for example. One way to obtain bicontinuous phase separation to ensure a large interfacial area between donor and acceptor is to design covalently or noncovalently linked D/A dyads [130]. Although D/A linked polymeric systems (the "double-cable" polymers) are well known for OPV device applications [131, 132], it is only recently that researchers have started to focus on such hybridized columnar discotic systems. Müllen and coworkers described HBC-pyrene and HBC-anthraquinone-based molecular systems and their bulk self-assembly into noncovalent columnar versions of the so-called "double cable" systems [133, 134]. Very recently, Aida and colleagues reported photovoltaic response in a fullerene (C_{60})-appended HBC molecule **7b** [55]. Although OPV cells employing discotic molecules are yet to match the performance of polymer-based devices, the progress made so far has been very encouraging.

4.5.2. Organic Light-Emitting Diodes

Organic light-emitting diodes (OLEDs) generate light by electroluminescence. The basic structure of an OLED is shown in Figure 4-24. It generally consists of a thin film of stacked organic semiconductors composed of a hole transport layer, an emission layer, and an electron transport layer sandwiched between two electrodes. An OLED device configuration is the reverse of a solar cell. The low-work-function electrode (for example, Al or Mg) injects electrons into the LUMO of the electron-transporting material. A thin film of the transparent

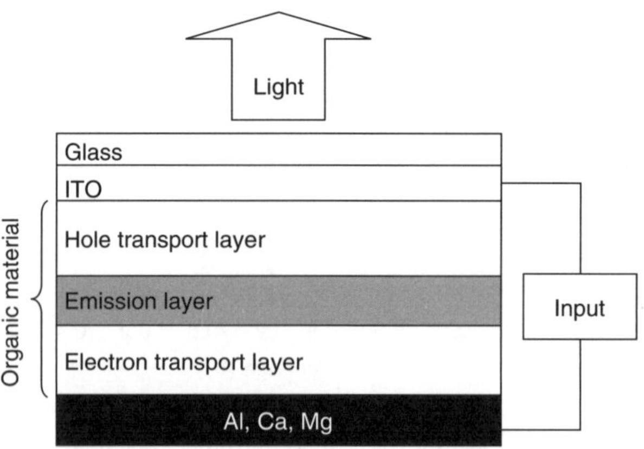

Figure 4-24   Schematic diagram of an OLED device. Electrons introduced at the metal electrode (cathode) recombine with holes introduced at the ITO electrode (anode) to emit light.

ITO deposited onto the substrate usually serves as the anode and injects holes into the HOMO of the hole-transporting material. With application of a voltage across the OLED, electrons and holes drift through their respective transporting layers and recombine near the interface (emission layer) to emit light. In many cases, an electron-transporting layer and the light-emitting layer are combined into a single layer. Progress achieved on the performance of OLEDs regarding brightness, efficiency, and stability have resulted in their commercial applications in various portable display devices [135].

The most important material properties that are required for OLED applications are efficient charge transport and luminescence. Initial interest in DLC materials for OLED application stemmed from their excellent charge transport properties evidenced in the columnar mesophases. Wendorff and coworkers first demonstrated the potential use of columnar discotics for applications in single-layer OLEDs, while almost simultaneously Bacher et al. reported bilayer OLEDs with low-molar-mass and polymeric discotic triphenylene compounds [136–138]. Destruel and coworkers reported red light emission by combining highly fluorescent and electron-deficient perylene derivative **20** with an electron-rich triphenylene **2a** (as hole transport material) in a bilayer OLED device [76]. They later achieved red, green, and blue color emissions by combining electron-deficient triphenylene and perylene discotic derivatives (as an electron transport layer) with a non-liquid crystalline hole-transporting layer in multilayer OLED devices [139]. With the increasing focus of several research groups on design of highly fluorescent columnar mesophases, performances of discotic-based OLEDs are expected to improve significantly in the coming years.

4.5.3. Organic Field Effect Transistors

In field-effect transistor (FET) devices, current flowing along a semiconductor path called the *channel* is controlled by an electric field. Electrons can move along the channel by continually putting electrons in one end (the *source electrode*) and removing them at the other (the *drain electrode*). A large variation in the current flow from the source to the drain can be achieved by the application of voltage through a control electrode called the *gate*. In an FET device, the same metal electrode acts as the source or drain, depending on the sign of the gate voltage. A typical OFET device consists of a thin organic semiconducting film, ideally forming ohmic contacts (negligible resistance) to the drain and source electrodes (Au is commonly used, and deposition *in vacuo* is the typical technique for fabrication). An OFET can be fabricated in either top- or bottom-contact geometry (Fig. 4-25). In "top-contact" geometry, the drain and source electrodes are deposited on top of the semiconducting layer, while the electrodes are positioned under the active layer in "bottom-contact" geometry. Highly doped silicon substrate is often used as the gate electrode. In addition, a thin layer of gate dielectric (SiO_2) acts as an insulating layer, preventing current conduction between the gate and source/drain electrodes. If there is no gate voltage applied, the organic semiconductor, which is intrinsically undoped, will not show any

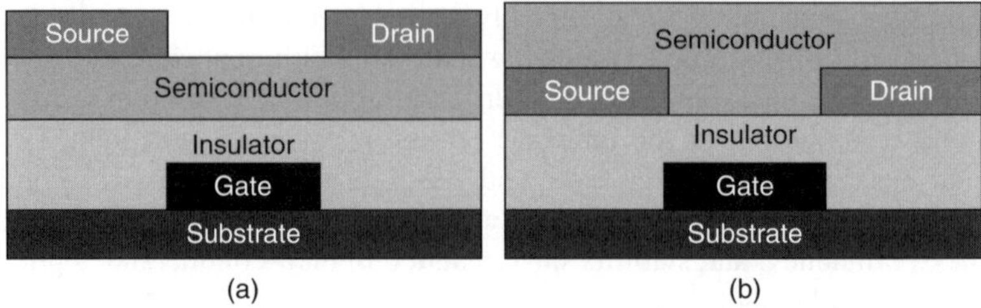

Figure 4-25   OFET device configurations: (a) Top-contact device, with source and drain electrodes deposited onto the organic semiconducting layer. (b) Bottom-contact device, with organic semiconductor deposited onto prefabricated source and drain electrodes.

charge carriers. A voltage applied to the gate electrode causes the formation of a thin layer of accumulated charge at the semiconductor-dielectric interface, and results in current flow between source and drain electrodes. As mentioned above, organic semiconductors can function as either p-type or n-type. Most organic semiconductors are p-type, that is, the application of a negative gate voltage results in the formation of an accumulated layer of positive hole carriers. If the Fermi level of the source/drain metal is close to the HOMO level of the organic semiconductor, then positive charges can be extracted by electrodes with the application of a voltage between the source and drain. When a positive voltage is applied to the gate, negative charges are induced in an n-type transistor. If the Fermi level of source/drain metal is close to the LUMO level of the organic semiconductors, then negative charges can be injected and extracted by the electrodes by applying a voltage between the source and drain. In some organic semiconductors, both electrons and holes can be injected and transported, an effect known as ambipolar behavior. Although OFETs are not suitable for use in applications requiring very high switching speeds (due to low charge mobility of organic semiconductors), they can be competitive candidates for novel thin-film transistor applications in active matrix displays, sensors, smart cards, and radiofrequency identification tags (RFIDs) [140].

Dramatic improvement in the performance of OFETs has been achieved in recent years. Organic FETs based on polycrystalline films of oligothiophenes or pentacenes prepared by thermal evaporation have reached hole mobilities exceeding that of amorphous silicon, with values larger than $1.0\ \mathrm{cm^2V^{-1}s^{-1}}$. Also, high hole mobilities (up to $0.10\ \mathrm{cm^2V^{-1}s^{-1}}$) have been obtained for solution-processed OFETs of poly(3-hexylthiophene) compound. Notable among high-performance DLCs for OFET applications are the derivatives of HBCs, phthalocyanines, and perylenes [32]. Solution-processed OFETs of HBC **4d** are reported to yield field-effect mobilities of $0.5-1.0 \times 10^{-3}\ \mathrm{cm^2V^{-1}s^{-1}}$ with a turn-on voltage of -5 V to -10 V and on/off ratios of more than 10^4 [106]. An order of magnitude increase in FET mobility had been reported with zone-cast films of HBC **4a** ($n = 12$) [107]. Nuckolls and coworkers reported OFETs (top contact with Au as source and drain electrode) based on nonplanar

HBC **6** with high charge-carrier mobilities (0.02 $cm^2V^{-1}s^{-1}$), high on/off ratios (10^6), and low turn-on voltages (-3 V) [54]. Very recently they have achieved further improvement in OFET device performance by integrating HBC **6** with single-walled carbon nanotube (SWNT) electrodes [141]. Armstrong and coworkers fabricated bottom-contact transistors based on a discotic Cu-phthalocyanine derivative on silicon substrates with the LB alignment technique, and their devices showed field mobility of 10^{-2} $cm^2V^{-1}s^{-1}$ [116].

Perylene diimide derivatives are one of the most important n-type semiconducting mesogenic materials with high charge-carrier mobility. High field-effect mobilities have been demonstrated in n-channel thin-film transistors using vapor-deposited perylene diimide derivatives. In 2002, Malenfant et al. reported n-channel OFETs based on **19b** with saturation electron mobilities ranging from 0.3 to 0.6 $cm^2V^{-1}s^{-1}$ and on-off ratios greater than 10^5 in a bottom-contact configuration [73]. Chesterfield et al. reported saturation electron mobilities as high as 1.7 $cm^2V^{-1}s^{-1}$ and current on-off ratios of 10^7 in an OFET device based on **19b** by controlling film growth conditions to attain well-organized thin films of the material [142]. Recently, Tatemichi et al. achieved an even higher field-effect mobility (2.1 $cm^2V^{-1}s^{-1}$) with perylene diimide **19c** by annealing at an adequate temperature (140°C) after device fabrication [143]. One grave drawback of semiconducting perylene diimide derivatives is the lack of stability of devices under ambient conditions in air. Substitution of fluoro alkyl chains in the imide positions has been shown to improve the air stability of perylene diimide organic semiconductors [144].

Single-channel OFETs with simultaneous or selective transport of electrons and/or holes (ambipolar charge transport) are highly desirable owing to their application in complementary-like circuits. Recently, swallow-tailed quaterrylene tetracarboxdiimide **22**, which forms a discotic columnar mesophase, was solution processed to form an active layer in an ambipolar OFET device [79]. Saturated electron and hole mobilities were found to be 1.5×10^{-3} $cm^2V^{-1}s^{-1}$ and 1.0×10^{-3} $cm^2V^{-1}s^{-1}$, respectively. However, postproduction thermal annealing of these devices resulted solely in electron transport. Vapor annealing of the resulting unipolar transistors helped to switch back to ambipolar behavior, though with significantly reduced charge-carrier mobilities. To realize the advantages of LC organic semiconductors in practical applications, the FETs (both n-channel and p-channel) should at least provide field-effect mobilities above 1 cm^2 and on-off ratios in the range of 10^6 and operate under ambient conditions, preferably in a bottom-contact configuration.

4.6. CONCLUSION

In this chapter, the semiconducting properties of discotic liquid crystals and their potential for various applications have been discussed. It is clear from the discussions here that the research activities on these materials have been primarily focused on synthesis and evaluation of new materials. Detailed studies have revealed the subtle dependence of mesomorphic and charge transport properties on minor changes of the molecular architectures. It is now widely accepted that the DLCs offer several advantages over other organic semiconductors; they are solution processable and self-organizing and possess moderately high charge-carrier mobilities. Integrating them into existing, well-studied organic semiconducting devices such as solar cells, light-emitting diodes, and field-effect transistors allows for the simplification of the device fabrication processes in a highly cost-effective way. Although encouraging progress has been made in recent years, the efficiency of these devices is not yet sufficient for their commercial implementation. The need to improve device efficiency and stability requires state-of-the art material design and device fabrication under more rigorously defined conditions. Therefore, a great deal of progress in the next decade can be expected with respect to (i) materials, (ii) device performance, stability, and versatility, (iii) lower manufacturing cost, and (iv) the use of environmentally friendly materials and processes.

ACKNOWLEDGMENTS

The preparation of this chapter benefited from the support to Quan Li by the Ohio Board of Regents under its Research Challenge program, the Department of Energy (DOE DE-SC0001412), the Air Force Office of Scientific Research (AFOSR FA9550-09-1-0193 and FA9550-09-1-0254), and the National Science Foundation (NSF IIP 0750379).

REFERENCES

1. S. R. Forrest. The path to ubiquitous and low-cost organic electronic appliances on plastic. *Nature* **2004**, *428*, 911–918.

2. *Organic Electronics: Materials, Manufacturing and Applications*, H. Klauk, Ed., Wiley-VCH, Weinheim, **2006**.

3. *Printed Organic and Molecular Electronics*, D. R. Gamota, P. Braziz, K. Kalyanasundaram and J. Zhang, Eds., Kluwer Accademic, Boston, **2004**.

4. O. D. Jurchescu, J. Baas, and T. T. M. Palstra. Effect of impurities on the mobility of single crystal pentacene. *Appl. Phys. Lett.* **2004**, *84*, 3061–3063.

5. V. Podzorov, S. E. Sysoev, E. Loginova, V. M. Pudalov, and M. E. Gershenson. Single-crystal organic field effect transistors with the hole mobility ∼8 cm^2/Vs. *Appl. Phys. Lett.* **2003**, *83*, 3504–3506.

6. *Semiconducting Polymers: Chemistry, Physics and Engineering*, G. Hadziioannou and G. G. Malliavas, Eds., Wiley-VCH, Weinheim, **2007**.

7. J. Roncali, P. Leriche, and A. Cravino. From one-to three-dimensional organic semiconductors: In search of the organic silicon. *Adv. Mater.* **2007**, *19*, 2045–2060.

8. H. Pan, Y. Li, Y. Wu, P. Li, B. S. Ong, S. Zhu, and G. Xu. Low-temperature, solution-processed, high-mobility polymer semiconductors for thin-film transistors. *J. Am. Chem. Soc.* **2007**, *129*, 4112–4113.

9. H. Yan, Z. Chen, Y. Zheng, C. Newman, J. R. Quinn, F. Dotz, M. Kastler, and A. Facchetti. A high-mobility electron-transporting polymer for printed transistors. *Nature* **2009**, *457*, 679–686.

10. S. Tiwari and N. C. Greenham. Charge mobility measurement techniques in organic semiconductors. *Opt. Quant. Electron.* **2009**, *41*, 69–89.

11. W. Pisula, M. Zorn, J. Y. Chang, K. Mullen, and R. Zental. Liquid crystalline ordering and charge transport in semiconducting materials. *Macromol. Rapid Commun.* **2009**, *30*, 1179–1202.

12. M. Funahashi. Development of liquid-crystalline semiconductors with high carrier mobilities and their application to thin-film transistors. *Polym. J.* **2009**, *4*, 459–469.

13. S. Laschat, A. Baro, N. Steinke, F. Giesselmann, C. Hagele, G. Scalia, R. Judele, E. Kapatsina, S. Sauer, and A. Schreivogel. Discotic liquid crystals: from tailor-made synthesis to plastic electronics. *Angew. Chem. Int. Ed.* **2007**, *46*, 4832–4887.

14. S. Sergeyev, W. Pisula, and Y. H. Geerts. Discotic liquid crystals: A new generation of organic semiconductors. *Chem. Soc. Rev.* **2007**, *36*, 1902–1929.

15. *Handbook of Liquid Crystals*, Vol. 2A, D. Demus, J. Goodby, G. W. Gray, H.-W. Spiess, and V. Vill, Eds., Wiley-VCH, Weinheim, **1998**.

16. *Liquid Crystals*, S. Chandrasekhar, Ed., Cambridge Univ. Press, Cambridge, **1992**.

17. *Introduction to Liquid Crystals: Chemistry and Physics*, P. J. Collings and M. Hird, Eds., Taylor & Francis Ltd, London, **1997**.

18. *Liquid Gold: The Story of Liquid Crystal Displays and the Creation of an Industry*, J. A. Castellano, Ed., World Scientific, **2005**.

19. *Liquid Crystals: Frontiers in Biomedical Applications*, S. J. Woltman, G. D. Jay, and G. P. Crawford, Eds., World Scientific, Singapore, **2007**.

20. S. Chandrasekhar, B. K. Sadashiva, and K. A. Suresh. Liquid crystals of disc-like molecules. *Pramana* **1977**, *9*, 471–480.

21. S. Kumar, Self-organization of disc-like molecules: chemical aspects. *Chem. Soc. Rev.* **2006**, *35*, 83–109.

22. T. Kato, T. Yasuda, Y. Kamikawa, and M. Yoshio. Self-assembly of functional columnar liquid crystals. *Chem. Commun.* **2009**, 729–739.

23. N. Boden, R. J. Bushby, J. Clements, B. Movaghar, K. J. Donovan, and T. Kreouzis. Mechanism of charge transport in discotic liquid crystals. *Phys. Rev. B* **1995**, *52*, 13274–13279.

24. K. Kohary, H. Cordes, S. D. Baranovskii, P. Thomas, S. Yamasaki, F. Hensel, and J.-H. Wendorff. One-dimensional hopping transport in disordered organic solids. II. Monte Carlo simulations. *Phys. Rev. B* **2001**, *63*, 094202.

25. L. J. Lever, R. W. Kelsall, and R. J. Bushby. Band transport for discotic liquid crystals. *Phys. Rev. B* **2005**, *72*, 035130.

26. J. M. Warman and A. M. Van de Craats. Charge mobility in discotic materials studied by PR-TRMC. *Mol. Cryst. Liq. Cryst.* **2003**, *396*, 41–72.

27. A. M. Van de Craats, J. M. Warman, A. Fechtenkötter, J. D. Brand, M. A. Harbison, and K. Müllen. Record charge carrier mobility in a room-temperature discotic liquid-crystalline derivative of hexabenzocoronene. *Adv. Mater.* **1999**, *11*, 1469–1472.

28. A. M. Van de Craats and J. M. Warman. The core-size effect on the mobility of charge in discotic liquid crystalline materials. *Adv. Mater.* **2001**, *13*, 130–133.

29. D. Adam, F. Closs, T. Frey, D. Funhoff, D. Haarer, H. Ringsdorf, P. Schuhmacher, and K. Siemensmeyer. Transient photoconductivity in a discotic liquid crystal. *Phys. Rev. Lett.* **1993**, *70*, 457–460.

30. A. M. Van de Craats, J. M. Warman, M. P. De Haas, D. Adam, J. Simmerer, D. Haarer, and P. Schuhmacher. The mobility of charge carriers in all four phases of the columnar discotic material hexakis (hexylthio) triphenylene: Combined TOF and PR-TRMC results. *Adv. Mater.* **1996**, *8*, 823–826.

31. J. Reynaert, V. I. Arkhipov, G. Borghs, and P. Heremans. Current-voltage characteristics of a tetracene crystal: Space charge or injection limited conductivity. *Appl. Phys. Lett.* **2004**, *85*, 603–605.

32. Y. Shimizu, K. Oikawa, K. Nakayama, and D. Guillon. Mesophase semiconductors in field effect transistors. *J. Mater. Chem.* **2007**, *17*, 4223–4229.

33. C. Piechocki, J. Simon, A. Skoulios, D. Guillon, and P. Weber. Discotic mesophases obtained from substituted metallophthalocyanines. Toward liquid crystalline one-dimensional conductors. *J. Am. Chem. Soc.* **1982**, *104*, 5245–5247.

34. N. Boden, R. J. Bushby, J. Clements, M. V. Jesudason, P. F. Knowles, and G. Williams. One-dimensional electronic conductivity in discotic liquid crystals. *Chem. Phys. Lett.*, **1988**, *152*, 94–99.

35. N. Boden, R. Borner, D. R. Brown, R. J. Bushby, and J. Clements. ESR studies of radical cations produced on doping discotic liquid crystals with lewis acids. *Liq. Cryst.* **1992**, *11*, 325–334.

36. P. G. Schoten, J. M. Gorman, M. P. deHaas, M. A. Fox, and H. L. Pan. Charge migration in supramolecular stacks of peripherally substituted porphyrins. *Nature* **1991**, *353*, 736–737.

37. P. G. Schoten, J. M. Gorman, M. P. deHaas, J. F. van der Pol, and J. W. Zwikker. Radiation-induced conductivity in polymerized and nonpolymerized columnar aggregates of phthalocyanine. *J. Am. Chem. Soc.* **1992**, *114*, 9028–9034.

38. J. Simmerer, B. Glusen, W. Paulus, A. Kettner, P. Schuhmacher, D. Adam, K.-H. Etzbach, K. Siemensmeyer, J. H. Wendorff, H. Ringsdorf, and D. Haarer. Transient photoconductivity in a discotic hexagonal plastic crystal. *Adv. Mater.* **1996**, *8*, 815–819.

39. D. Adam, P. Schumacher, J. Simmerer, L. Haussling, K. Siemensmeyer, K. H. Etzbachi, H. Ringsdorf, and D. Haarer. Fast photoconduction in the highly ordered columnar phase of a discotic liquid crystal. *Nature* **1994**, *371*, 141–143.

40. H. Bengs, F. Closs, T. Frey, D. Funhoff, H. Ringsdorf, and K. Siemensmeyer. Highly photoconductive discotic liquid crystals: Structure–property relations in the homologous series of hexa-alkoxytriphenylenes. *Liq. Cryst.* **1993**, *15*, 565–574.

41. H. Iino, J. Hanna, C. Jager, and D. Haarer. Fast electron transport in discotic columnar phase of triphenylene derivative, hexabutyloxytriphenylene. *Mol. Cryst. Liq. Cryst.* **2005**, *436*, 217–224.

42. H. Iino, Y. Takayashiki, J. Hanna, R. J. Bushby, and D. Haarer. High electron mobility of 0.1 $cm^{-1}V^{-1}s^{-1}$ in the highly ordered columnar phase of hexahexylthiotriphenylene. *Appl. Phys. Lett.* **2005**, *87*, 192105.

43. I. Paraschiv, K. de Lange, M. Giesbers, B. van Lagen, F. C. Grozema, R. D. Abellon, L. D. A. Siebbeles, E. J. R. Sudholter, H. Zuilhof, and A. T. M. Marcelis. Hydrogen-bond stabilized columnar discotic benzenetrisamides with pendant triphenylene groups. *J. Mater. Chem.* **2008**, *18*, 5475–5481.

44. N. Mizoshita, H. Monobe, M. Inoue, M. Ukon, T. Watanabe, Y. Shimizu, K. Hanabusa, and T. Kato. The positive effect on hole transport behaviour in anisotropic gels consisting of discotic liquid crystals and hydrogen-bonded fibres. *Chem. Commun.* **2002**, 428–429.

45. J. A. Rego, S. Kumar, and H. Ringsdorf. Synthesis and characterization of fluorescent, low-symmetry triphenylene discotic liquid crystals: Tailoring of mesomorphic and optical properties. *Chem. Mater.* **1996**, *8*, 1402–1409.

46. A. Kettner and J. H. Wendorff. Modifications of the mesophase formation of discotic triphenylene compounds by substituents. *Liq. Cryst.* **1999**, *26*, 483–487.

47. S. Kumar, S. K. Pal, P. S. Kumar, and V. Lakshminarayanan. Novel conducting nanocomposites: synthesis of triphenylene-covered gold nanoparticles and their insertion into a columnar matrix. *Soft Matter* **2007**, *3*, 896–900.

48. L. A. Holt, R. J. Bushby, S. D. Evans, A. Burgess, and G. Seeley. A 10^6-fold enhancement in the conductivity of a discotic liquid crystal doped with only 1% (w / w) gold nanoparticles. *J. Appl. Phys.* **2008**, *103*, 063712.

49. M. Kastler, F. Laquai, K. Müllen, and G. Wegner. Room-temperature nondispersive hole transport in a discotic liquid crystal. *Appl. Phys. Lett.* **2006**, *89*, 252103.

50. J. Wu, W. Pisula, and K. Müllen. Graphenes as potential material for electronics. *Chem. Rev.* **2007**, *107*, 718–747.

51. M. G. Debije, J. Piris, M. P. de Haas, J. M. Warman, Z. Tomovic, C. D. Simpson, M. D. Watson, and K. Müllen. The optical and charge transport properties of discotic materials with large aromatic hydrocarbon cores. *J. Am. Chem. Soc.* **2004**, *126*, 4641–4645.

52. X. Feng, V. Marcon, W. Pisula, M. R. Hansen, J. Kirkpatrick, F. Grozema, D. Andrienko, K. Kremer, and K. Müllen. Towards high charge-carrier mobilities by rational design of the shape and periphery of discotics. *Nature Mater.* **2009**, *8*, 421–426.

53. X. Feng, M. Liu, W. Pisula, M. Takase, J. Li, and K. Müllen. Supramolecular organization and photovoltaics of triangle-shaped discotic graphenes with swallow-tailed alkyl substituents. *Adv. Mater.* **2008**, *20*, 2684–2689.

54. S. Xiao, M. Myers, Q. Miao, S. Sanaur, K. Pang, M. L. Steigerwald, and C. Nuckolls. Molecular wires from contorted aromatic compounds. *Angew. Chem. Int. Ed.* **2005**, *44*, 7390–7394.

55. Y. Yamamoto, G. Zhang, W. Jin, T. Fukushima, N. Ishii, A. Saeki, S. Seki, S. Tagawa, T. Minari, K. Tsukagoshi, and T. Aida. Ambipolar-transporting coaxial nanotubes with a tailored molecular graphene–fullerene heterojunction. *PNAS* **2009**, *106*, 21051–21056.

56. J. W. Goodby, P. S. Robinson, B.-K. Teo, and P. E. Cladi. The discotic phase of uroporphyrin octa-*N*-dodecyl ester. *Mol. Cryst. Liq. Cryst.* **1980**, *56*, 303–309.

57. H. Fujikake, T. Murashige, M. Sugibayashi, and K. Ohta. Time-of-flight analysis of charge mobility in a Cu-phthalocyanine-based discotic liquid crystal semiconductor. *Appl. Phys. Lett.* **2004**, *85*, 3474.

58. H. Iino, J. Hanna, R. J. Bushby, B. Movaghar, B. J. Whitaker, and M. J. Cook. Very high time-of-flight mobility in the columnar phases of a discotic liquid crystal. *Appl. Phys. Lett.* **2005**, *87*, 132102.

59. K. Ban, K. Nishizawa, K. Ohta, A. M. Craats, J. M. Warman, I. Yamamoto, and H. Shirai. Discotic liquid crystals of transition metal complexes 29: mesomorphism and charge transport properties of alkylthio-substituted phthalocyanine rare-earth metal sandwich complexes. *J. Mater. Chem.* **2001**, *11*, 321–331.

60. S. W. Kang, Q. Li, B. D. Chapman, R. Pindak, J. O. Cross, L. Li, M. Nakata, and S. Kumar. Microfocus x-ray diffraction study of the columnar phase of porphyrin-based mesogens. *Chem. Mater.* **2007**, *19*, 5657–5663.

61. L. Li, S. W. Kang, J. Harden, Q. Sun, X. Zhou, L. Dai, A. Jakli, S. Kumar, and Q. Li. Nature-inspired light-harvesting liquid crystalline porphyrins for organic photovoltaics. *Liq. Cryst.* **2008**, *35*, 233–239.

62. Q. Sun, L. Dai, X. Zhou, L. Li, and Q. Li. Bilayer-and bulk-heterojunction solar cells using liquid crystalline porphyrins as donors by solution processing. *Appl. Phys. Lett.* **2007**, *91*, 253505.

63. N. Boden, R. C. Borner, R. J. Bushby, and J. Clements. First observation of an *n*-doped quasi-one-dimensional electronically-conducting discotic liquid crystal. *J. Am. Chem. Soc.* **1994**, *116*, 10807–10808.

64. N. Boden, R. J. Bushby, K. Donovan, Q. Liu, Z. Lu, T. Kreouzis, and A. Wood. 2,3,7,8,12,13-Hexakis [2-(2-methoxyethoxy) ethoxy] tricycloquinazoline: a discogen which allows enhanced levels of *n*-doping. *Liq. Cryst.* **2001**, *28*, 1739–1748.

65. K. Pieterse, P. A. van Hal, R. Kleppinger, J. A. J. M. Vekemans, R. A. J. Janssen, and E. W. Meijer. An electron-deficient discotic liquid-crystalline material. *Chem. Mater.* **2001**, *13*, 2675–2679.

66. R. I. Gearba, M. Lahmann, J. Levin, D. A. Ivanov, M. H. J. Koch, J. Barbera, M. G. Debije, J. Piris, and Y. H. Geerts. Tailoring discotic mesophases: Columnar order enforced with hydrogen bonds. *Adv. Mater.* **2003**, *15*, 1614–1618.

67. A. Demenev, S. H. Eichhorn, T. Taerum, D. F. Perepichka, S. Patwardhan, F. C. Grozeme, L. D. A. Siebbeles, and R. Klenkler. Quasi temperature independent electron mobility in hexagonal columnar mesophases of an H-bonded benzotristhiophene derivative. *Chem. Mater.* **2010**, *22*, 1420–1428.

68. M. Lehmann, G. Kestemont, R. Gomez Aspe, C. Buess-Herman, M. H. J. Koch, M. G. Debije, J. Piris, M. P. de Haas, J. M. Warman, and M. D. Watson. High charge-carrier mobility in p-deficient discotic mesogens: Design and structure-property relationship. *Chem. Eur. J.* **2005**, *11*, 3349–3362.

69. T. Sakurai, K. Shi, H. Sato, K. Tashiro, A. Osuka, A. Saeki, S. Seki, S. Tagawa, S. Sasaki, H. Masunaga, K. Osaka, M. Takata, and T. Aida. Prominent electron transport property observed for triply fused metalloporphyrin dimer: Directed columnar liquid crystalline assembly by amphiphilic molecular design. *J. Am. Chem. Soc.* **2008**, *130*, 13812–13813.

70. B. Tylleman, G. Gbabode, C. Amato, C. Buess-Herman, V. Lemaur, J. Cornil, R. Go'mez Aspe, Y. H. Geerts, and S. Sergeyev. Metal-free phthalocyanines bearing eight alkylsulfonyl substituents: Design, synthesis, electronic structure, and mesomorphism of new electron-deficient mesogens. *Chem. Mater.* **2009**, *21*, 2789–2797.

71. C. W. Struijk, A. B. Sieval, J. E. J. Dakhorst, M. van Dijk, P. Kimkes, R. B. M. Koehorst, H. Donker, T. J. Schaafsma, S. J. Picken, and A. M. van de Craats. Liquid crystalline perylene diimides: Architecture and charge carrier mobilities. *J. Am. Chem. Soc.* **2000**, *122*, 11057–11066.

72. J. Y. Kim, I. J. Chung, Y. C. Kim, and J. W. Yu. Mobility of electrons and holes in a liquid crystalline perylene diimide thin film with time of flight technique. *Chem. Phys. Lett.* **2004**, *398*, 367–371.

73. P. R. L. Malenfant, C. D. Dimitrakopoulos, J. D. Gelorme, L. L. Kosbar, T. O. Graham, A. Curioni, and W. Andreoni. N-type organic thin-film transistor with high field-effect mobility based on a *N,N*-dialkyl-3,4,9,10-perylene tetracarboxylic diimide derivative. *Appl. Phys. Lett.* **2002**, *80*, 2517.

74. Z. An, J. Yu, S. C. Jones, S. Barlow, S. Yoo, B. Domercq, P. Prins, L. D. A. Siebbeles, B. Kippelen, and S. R. Marder. High electron mobility in room-temperature discotic liquid-crystalline perylene diimides. *Adv. Mater.* **2005**, *17*, 2580–2583.

75. Z. Chen, M. G. Debije, T. Debaerdemaeker, P. Osswald, and F. Wuerthner. Tetrachloro-substituted perylene bisimide dyes as promising *n*-type organic semiconductors: studies on structural, electrochemical and charge transport properties. *Chem Phys Chem*, **2004**, *5*, 137–139.

76. I. Seguy, P. Destruel, and H. Bock. An all-columnar bilayer light-emitting diode. *Synth. Met.* **2000**, *111*, 15–18.

77. S. Alibert-Fouet, S. Dardel, H. Bock, M. Oukachmih, S. Archambeau, I. Seguy, P. Jolinat, and P. Destruel. Electroluminescent diodes from complementary discotic benzoperylenes. *Chem. Phys. Chem.* **2003**, *4*, 983–985.

78. F. Nolde, W. Pisula, S. Muller, C. Kohl, and K. Mullen. Synthesis and self-organization of core-extended perylene tetracarboxdiimides with branched alkyl substituents. *Chem. Mater.* **2006**, *18*, 3715–3725.

79. H. N. Tsao, W. Pisula, Z. Liu, W. Osikowicz, W. R. Salaneck, and K. Müllen. From ambi-to unipolar behavior in discotic dye field-effect transistors. *Adv. Mater.* **2008**, *20*, 2715–2719.

80. U. Rohr, C. Kohl, K. Müllen, A. Craats, and J. Warman. Liquid crystalline coronene derivatives. *J. Mater. Chem.* **2001**, *11*, 1789–1799.

81. K. Hirota, K. Tajima, and K. Hashimoto. Physicochemical study of discotic liquid crystal decacyclene derivative and utilization in polymer photovoltaic devices. *Synth. Met.* **2007**, *157*, 290–296.

82. H. K. Bisoyi and S. Kumar. Room-temperature electron-deficient discotic liquid crystals: facile synthesis and mesophase characterization. *New J. Chem.* **2008**, *32*, 1974–1980.

83. S. K. Gupta, V. A. Raghunathan, and S. Kumar. Microwave-assisted facile synthesis of discotic liquid crystalline symmetrical donor–acceptor–donor triads. *New J. Chem.* **2009**, *33*, 112–118.

84. J. C. Bunning, K. J. Donovan, R. J. Bushby, O. R. Lozman, and Z. Lu. Electron photogeneration in a triblock co-polymer discotic liquid crystal. *Chem. Phys.* **2005**, *312*, 145–150.

85. A. Hayer, V. de Halleux, A. Kohler, A. El-Garoughy, E. W. Meijer, J. Barbera, J. Tant, J. Levin, M. Lehmann, and J. Gierschner, J. Cornil, and Y. H. Geerts. Highly fluorescent crystalline and liquid crystalline columnar phases of pyrene-based structures. *J. Phys. Chem. B*, **2006**, *110*, 7653–7659.

86. M. J. Sienkowska, H. Monobe, P. Kaszynski, and Y. Shimizu. Photoconductivity of liquid crystalline derivatives of pyrene and carbazole. *J. Mater. Chem.* **2007**, *17*, 1392–1398.

87. M. Talarico, R. Termine, E. M. Garcia-Frutos, A. Omenat, J. L. Serrano, B. Gomez-Lor, and A. Golemme. New electrode-friendly triindole columnar phases with high hole mobility. *Chem. Mater.* **2008**, *20*, 6589–6591.

88. B. Zhao, B. Liu, R. Q. Png, K. Zhang, K. A. Lim, J. Luo, J. Shao, P. K. H. Ho, C. Chi, and J. Wu. New discotic mesogens based on triphenyl-fused triazatruxenes: Synthesis, physical properties, and self-assembly. *Chem. Mater.* **2010**, *22*, 435–449.

89. D. Miyajima, K. Tashiro, F. Araoka, H. Takezoe, J. Kim, K. Kato, M. Takata, and T. Aida. Liquid crystalline corannulene responsive to electric field. *J. Am. Chem. Soc.* **2009**, *131*, 44–45.

90. S. S. Besbes, E. Grelet, and H. Bock. Soluble and liquid-crystalline ovalenes. *Angew. Chem. Int. Ed.* **2006**, *45*, 1783–1786.

91. C. V. Yelamaggad, A. S. Achalkumar, D. S. Shankar Rao, and S. K. Prasad. Self-assembly of C_{3h} and C_s symmetric keto-enamine forms of tris(N-salicylideneanilines) into columnar phases: A new family of discotic liquid crystals. *J. Am. Chem. Soc.* **2004**, *126*, 6506–6507.

92. C. Lavigueur, E. J. Foster, and V. E. Williams. Self-assembly of discotic mesogens in solution and in liquid crystalline phases: Effects of substituent position and hydrogen bonding. *J. Am. Chem. Soc.* **2008**, *130*, 11791–11800.

93. S. H. Eichhorn, A. Adavelli, H. S. Li, and N. Fox. Alignment of discotic liquid crystals. *Mol. Cryst. Liq. Cryst.* **2003**, *397*, 47–58.

94. K. Hatsusaka, K. Ohta, I. Yamamoto, and H. Shirai. Discotic liquid crystals of transition metal complexes, Part 30: spontaneous uniform homeotropic alignment of octakis (dialkoxyphenoxy) phthalocyaninatocopper (II) complexes. *J. Mater. Chem.* **2001**, *11*, 423–433.

95. W. Pisula, Z. Tomovic, B. El Hamaoui, M. D. Watson, T. Pakula, and K. Mullen. Control of the homeotropic order of discotic hexa-*peri*-hexabezocoronenes. *Adv. Funct. Mater.* **2005**, *15*, 893–904.

96. J. K. Vij, A. Kocot, and T. S. Perova. Order parameter, alignment and anchoring transition in discotic liquid crystals. *Mol. Cryst. Liq. Cryst.* **2003**, *397*, 231–244.

97. N. Terasawa, H. Monobe, K. Kiyohara, and Y. Shimizu. Strong tendency towards homeotropic alignment in a hexagonal columnar mesophase of fluoroalkylated triphenylenes. *Chem. Commun.* **2003**, 1678–1679.

98. V. De Cupere, J. Tant, P. Viville, R. Lazzaroni, W. Osikowicz, W. R. Salaneck, and Y. H. Geerts. Effect of interfaces on the alignment of a discotic liquid-crystalline phthalocyanine. *Langmuir* **2006**, *22*, 7798–7806.

99. X. Zhou, S. W. Kang, S. Kumar, and Q. Li. Self-assembly of discotic liquid crystal porphyrin into more controllable ordered nanostructure mediated by fluorophobic effect. *Liq. Cryst.* **2009**, *36*, 269–274.

100. E. Grelet and H. Bock. Control of the orientation of thin open supported columnar liquid crystal films by the kinetics of growth. *Europhys. Lett.* **2006**, *73*, 712–718.

101. R. I. Gearba, D. V. Anokhin, A. I. Bondar, W. Bras, M. Jahr, M. Lehmann, and D. A. Ivanov. Homeotropic alignment of columnar liquid crystals in open films by means of surface nanopatterning. *Adv. Mater.* **2007**, *19*, 815–820.

102. E. Charlet, E. Grelet, P. Brettes, H. Bock, H. Saadaoui, L. Cisse, P. Destruel, N. Gherardi, and I. Seguy. Ultrathin films of homeotropically aligned columnar liquid crystals on indium tin oxide electrodes. *Appl. Phys. Lett.* **2008**, *92*, 024107.

103. M. Steinhart, S. Zimmermann, P. Goring, A. K. Schaper, U. Gosele, C. Weder, and J. H. Wendorff. Liquid crystalline nanowires in porous alumina: Geometric confinement versus influence of pore walls. *Nano Lett.* **2005**, *5*, 429–434.

104. E. Pouzet, V. D. Cupere, C. Heintz, J. W. Andreasen, D. W. Breiby, M. M. Nielsen, P. Viville, R. Lazzaroni, G. Gbabode, and Y. H. Geerts. Homeotropic alignment of a discotic liquid crystal induced by a sacrificial layer. *J. Phys. Chem. C.* **2009**, *113*, 14398–14406.

105. S. Zimmermann, J. H. Wendorff, and C. Weder. Uniaxial orientation of columnar discotic liquid crystals. *Chem. Mater.* **2002**, *14*, 2218–2223.

106. A. M. van de Craats, N. Stutzmann, O. Bunk, M. M. Nielsen, M. Watson, K. Müllen, H. D. Chanzy, H. Sirringhaus and R. H. Friend. Meso-epitaxial solution-growth of self-organizing discotic liquid-crystalline semiconductors. *Adv. Mater.* **2003**, *15*, 495–499.

107. W. Pisula, A. Menon, M. Stepputat, I. Lieberwirth, U. Kolb, A. Tracz, H. Sirringhaus, T. Pakula, and K. Müllen. A zone-casting technique for device fabrication of field-effect transistors based on discotic hexa-peri-hexabenzocoronene. *Adv. Mater.* **2005**, *17*, 684–689.

108. A. Tracz, T. Makowski, S. Masirek, W. Pisula, and Y. H. Geerts. Macroscopically aligned films of discotic phthalocyanine by zone casting. *Nanotechnology*, **2007**, *18*, 485303.

109. C. Y. Liu and A. J. Bard. In-situ regrowth and purification by zone melting of organic single-crystal thin films yielding significantly enhanced optoelectronic properties. *Chem. Mater.* **2000**, *12*, 2353–2362.

110. W. Pisula, M. Kastler, D. Wasserfallen, T. Pakula, and K. Müllen. Exceptionally long range self assembly of hexa-peri-hexabenzocoronene with dove-tailed alkyl substituents. *J. Am. Chem. Soc.* **2004**, *126*, 8074–8075.

111. I. O. Shklyarevskiy, P. Jonkheijm, N. Stutzmann, D. Wasserberg, H. J. Wondergem, P. C. M. Christianen, A. P. H. J. Schenning, D. M. de Leeuw, Z. Tomovic, J. Wu, K. Mullen, and J. C. Maan. High anisotropy of the field-effect transistor mobility in magnetically aligned discotic liquid-crystalline semiconductors. *J. Am. Chem. Soc.* **2005**, *127*, 16233–16237.

112. T. Bjornholm, T. Hassenkam, and N. Reitzel. Supramolecular organization of highly conducting organic thin films by the Langmuir-Blodgett technique. *J. Mater. Chem.* **1999**, *9*, 1975–1990.

113. O. Y. Mindyuk and P. A. Heiney. Structural studies of langmuir films of disc-shaped molecules. *Adv. Mater.* **1999**, *11*, 341–344.

114. P. Smolenyak, R. Peterson, K. Nebesny, M. Torker, D. F. O'Brien, and N. R. Armstrong. Highly ordered thin films of octasubstituted phthalocyanines. *J. Am. Chem. Soc.* **1999**, *121*, 8628–8636.

115. N. Reitzel, T. Hassenkam, K. Balashev, T. R. Jensen, P. B. Howes, K. Kjaer, A. Fechtenkotter, N. Tchebotareva, S. Ito, and K. Mullen. Langmuir and Langmuir-Blodgett films of amphiphilic hexa-peri-hexabenzocoronene: New phase transitions and electronic properties controlled by pressure. *Chem. Eur. J.* **2001**, *7*, 4894–4901.

116. S. Cherian, C. Donley, D. Mathine, L. LaRussa, W. Xia, and N. Armstrong. Effects of field dependent mobility and contact barriers on liquid crystalline phthalocyanine organic transistors. *J. Appl. Phys.* **2004**, *96*, 5638–5643.

117. N. Boden, R. J. Bushby, P. S. Martin, S. D. Evans, R. W. Owens, and D. A. Smith. Triphenylene-based discotic liquid crystals as self-assembled monolayers. *Langmuir* **1999**, *15*, 3790–3797.

118. H. Monobe, K. Awazu, and Y. Shimizu. Alignment control of a columnar liquid crystal for a uniformly homeotropic domain using circularly polarized infrared irradiation. *Adv. Mater*. **2006**, *18*, 607–610.

119. H. Monobe, K. Awazu, and Y. Shimizu. Alignment change of hexahexylthiotriphenylene in the helical columnar phase by infrared laser irradiation. *Thin Solid Films* **2009**, *518*, 762–766.

120. H. J. Räder, A. Rouhanipour, A. M. Talarico, V. Palermo, P. Samorì, and K. Müllen. Processing of giant graphene molecules by soft-landing mass spectrometry. *Nature Mater*. **2006**, *5*, 276–280.

121. K. Kawata. Orientation control and fixation of discotic liquid crystal. *Chem. Rec*. **2002**, *2*, 59–80.

122. T. L. Benanti and D. Venkataraman. Organic solar cells: An over view focusing on active layer morphology. *Photosynth. Res*. **2006**, *87*, 73–81.

123. C. W. Tang. 2-Layer organic photovoltaic cell. *Appl. Phys. Lett*. **1986**, *48*, 183–185.

124. G. Yu, J. Gao, J. C. Hummelen, F. Wudl, and A. J. Heeger. Polymer photovoltaic cells: Enhanced efficiencies via a network of internal donor-acceptor heterojunctions. *Science* **1995**, *270*, 1789–1791.

125. M. Reyes-Reyes, K. Kim, and D. L. Carroll. High-efficiency photovoltaic devices based on annealed poly (3-hexylthiophene) and 1-(3-methoxycarbonyl)-propyl-1-phenyl-(6,6) C_{61} blends. *Appl. Phys. Lett*. **2005**, *87*, 083506.

126. W. Ma, C. Yang, X. Gong, K. Lee, and A. J. Heeger. Thermally stable, efficient polymer solar cells with nanoscale control of the interpenetrating network morphology. *Adv. Func. Mater*. **2005**, *15*, 1617–1622.

127. B. A. Gregg, M. A. Fox, and A. J. Bard. Photovoltaic effect in symmetrical cells of a liquid crystal porphyrin. *J. Phys. Chem*. **1990**, *94*, 1586–1598.

128. K. Petritsch, R. H. Friend, A. Lux, G. Rozenberg, S. C. Moratti, and A. B. Holmes. Liquid crystalline phthalocyanines in organic solar cells. *Synth. Met*. **1999**, *102*, 1776–1777.

129. L. Schmidt-Mende, A. Fechtenkotter, K. Mullen, E. Moons, R. H. Friend, and J. D. MacKenzie. Self-organized discotic liquid crystals for high-efficiency organic photovoltaics. *Science* **2001**, *293*, 1119–1122.

130. F. J. M. Hoeben, P. Jonkheijm, E. W. Meijer, and A. P. H. J. Schenning. About supramolecular assemblies of π-conjugated systems. *Chem. Rev*. **2005**, *105*, 1491–1546.

131. A. M. Ramos, M. T. Rispens, J. K. J. van Duren, J. C. Hummelen, and R. A. J. Janssen. Photoinduced electron transfer and photovoltaic devices of a conjugated polymer with pendant fullerenes. *J. Am. Chem. Soc*. **2001**, *123*, 6714–6715.

132. A. Cravino and N. S. Sariciftci. Double-cable polymers for fullerene based organic optoelectronic applications. *J. Mater. Chem*. **2002**, *12*, 1931–1943.

133. N. Tchebotareva, X. Yin, M. D. Watson, P. Samori, J. P. Rabe, and K. Müllen. Ordered architectures of a soluble hexa-peri-hexabenzocoronene-pyrene dyad: Thermotropic bulk properties and nanoscale phase segregation at surfaces. *J. Am. Chem. Soc*. **2003**, *125*, 9734–9739.

134. P. Samori, X. Yin, N. Tchebotareva, Z. Wang, T. Pakula, F. Jäckel, M. D. Watson, A. Venturini, K. Müllen, and J. P. Rabe. Self-assembly of electron donor-acceptor dyads into ordered architectures in two and three dimensions: Surface patterning and columnar double cables. *J. Am. Chem. Soc.* **2004**, *126*, 3567–3575.

135. C. R. McNeill and N. C. Greenham. Conjugated-polymer blends for optoelectronics. *Adv. Mater.* **2009**, *21*, 1–11.

136. T. Christ, B. Glusen, A. Greiner, A. Kellner, R. Sander, V. Stumpflen, V. Tsukruk, and J. H. Wendorff. Columnar discotics for light emitting diodes. *Adv. Mater.* **1997**, *9*, 48–51.

137. I. H. Stapff, V. Stümpflen, J. H. Wendorff, D. B. Spohn, and D. Möbius. Multi-layer light emitting diodes based on columnar discotics, *Liq. Cryst.* **1997**, *23*, 613–617.

138. A. Bacher, I. Bleyl, C. H. Erdelen, D. Haarer, W. Paulus, and H. W. Schmidt. Low molecular weight and polymeric triphenylenes as hole transport materials in organic two-layer LEDs. *Adv. Mater.* **1997**, *9*, 1031–1035.

139. T. Hassheider, S. A. Benning, H. S. Kitzerow, M. F. Achard, and H. Bock. Color-tuned electroluminescence from columnar liquid crystalline alkyl arenecarboxylates. *Angew. Chem. Int. Ed.* **2001**, *40*, 2060–2063.

140. H. Sirringhaus, Device physics of solution-processed organic field-effect transistors. *Adv. Mater.* **2005**, *17*, 2411–2425.

141. X. Guo, S. Xiao, M. Myers, Q. Miao, M. L. Steigerwald, and C. Nuckolls. Photoresponsive nanoscale columnar transistors. *PNAS* **2009**, *106*, 691–696.

142. R. J. Chesterfield, J. C. McKeen, C. R. Newman, P. C. Ewbank, D. A. da Silva Filho, J. L. Bredas, L. L. Miller, K. R. Mann, and C. D. Frisbie. Organic thin film transistors based on n-alkyl perylene diimides: Charge transport kinetics as a function of gate voltage and temperature. *J. Phys. Chem. B.* **2004**, *108*, 19281–19292.

143. S. Tatemichi, M. Ichikawa, T. Koyama and Y. Taniguchi. High mobility *n*-type thin-film transistors based on *N,N*-ditridecyl perylene diimide with thermal treatments. *Appl. Phys. Lett.* **2006**, *89*, 112108.

144. R. Schmidt, J. H. Oh, Y. S. Sun, M. Deppisch, A. M. Krause, K. Radacki, H. Braunschweig, M. Konemann, P. Erk, Z. Bao and F. Wurthner. High-performance air-stable *n*-channel organic thin film transistors based on halogenated perylene bisimide semiconductors. *J. Am. Chem. Soc.* **2009**, *131*, 6215–6228.

■■■■■ **CHAPTER 5**

Self-Organized Semiconducting Smectic Liquid Crystals

JI MA and QUAN LI

Liquid Crystal Institute, Kent State University, Kent, Ohio

5.1. INTRODUCTION

Liquid crystals (LCs) are mesophases between crystalline solids and isotropic liquids. They are an important class of organic materials that have been studied extensively in display and electro-optical devices because of their unique characteristics [1–3]. The molecular geometrical structures of LCs can be grouped by conventional types such as calamitic (rod shaped) and discotic (disclike) molecules and nonconventional configurations including bent-type (bowl shaped and banana shaped), sanidic (brick- or lathlike), and other types [4]. For a very long time, the electrical conductions in LCs were considered as an ionic effect [5–7]. However, it was not until 1993 that Haarer et al. reported that high charge-carrier mobilities were found in a discotic LC [8]. Several years later, photoconduction in smectic LCs was first established [9]. Compared with discotic LCs, the control of alignments and molecular arrangements of smectic LCs is easier because of the rodlike molecular structure. Good alignments and precise controls benefit the molecular self-organization of the calamitic LCs, providing less defects and grain boundaries. It was found that the charge-carrier mobilities are highly relevant to the molecular orientational order of the LC phases. Smectic LCs exhibit SmA, SmB, SmC...and SmL phases, the molecular orientations, arrangements, and distributions of which are different. This provides more choices and opportunities to exploit and develop materials and practical device applications.

In the following sections, we first introduce and discuss smectic LCs and their various molecular arrangements. Second, the techniques to measure charge-carrier mobility in the smectic phases or in the semiconducting devices are specified. These are helpful to understand the molecular arrangements of smectic liquid crystals and the research methods to characterize charge-carrier transports. Some of the smectic mesophases have been exploited as semiconductor materials or device applications. Through the summary of semiconducting smectic materials, we find that the molecular structures and arrangements have an intimate relationship with the charge-carrier mobilities. Related materials, charge-carrier transport properties, practical devices, and potential applications are presented and discussed. At the end of this chapter, the further research needed to advance the semiconducting field are proposed.

5.2. SMECTIC PHASES AND STRUCTURES

It is well known that LCs are between ordinary solid and liquid states. They are divided into two classes: *thermotropic* LCs, which have temperature-dependent phases, and *lyotropic* LCs, which have phases dependent on the ratio of solvent to mesogen [10–12]. *Lyotropic* LC is usually a solution, which makes it difficult to realize practical applications. *Thermotropic* LC is relatively stable in the defined temperature ranges. For instance, 4-*n*-pentyl-4′-cyano-biphenyl (5CB) exhibits a stable nematic mesophase between 22 and 35°C [13]. On the basis of molecular ordering, *thermotropic* LCs are classified into three phases: the nematic, smectic, and cholesteric phases [14]. For a rod-shaped molecule, the varieties and arrangements of its phases may be found depending on the temperature [15], as shown in Figure 5-1. When the temperature is low, the material is in the crystalline phase,

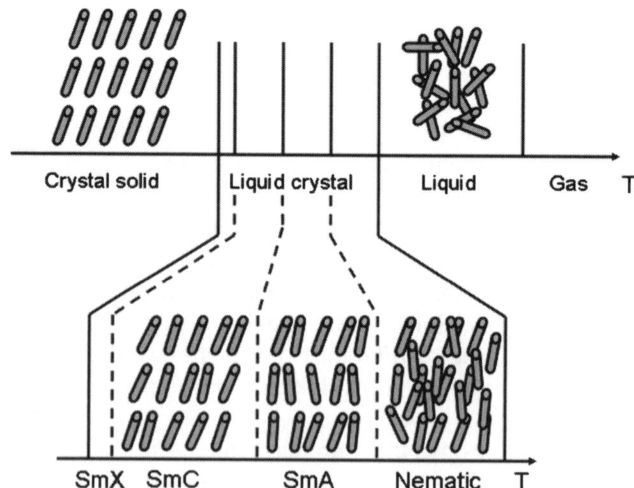

Figure 5-1 Schematic diagram of various phases and molecular arrangements of a rod-shaped molecular structure *thermotropic* liquid crystal (SmX: higher-ordered smectic phases).

which has positional and orientational orders. When the temperature is increased, the material may transform into the smectic-C (SmC) phase to form a layered structure, where molecules are tilted. When the temperature is further increased, the phase sequence may be the layered structure smectic-A (SmA) phase (not tilted), nematic (N) phase with orientational order but no position order, isotropic liquid (Iso) phase, and gaseous phase, respectively. Between the crystal phase and the nematic phase, several subgroups of smectic phases may exist.

In the following section, we discuss detailed correlations between charge-carrier mobility properties and molecular arrangements of the semiconducting smectic LCs. Therefore we first introduce the smectic phase subgroups. The word "smectic" comes from the Greek, meaning "soap", since the thick, slippery residual at the bottom of a soap dish is actually a type of smectic LC. Smectic phases exhibit more degrees in orientational order and positional order than nematic phases. Illustrations of smectic phase subgroups [14, 16] are summarized in Table 5-1 and Table 5-2, which detail an overview of each smectic phase subgroup with corresponding bond orientational order, positional order, molecular orientation, and molecular packing. To take the SmA phase as an example, the nematic-like director in each layer is perpendicular to the smectic plane and parallel with the layer normal. The SmC phase shows a molecular arrangement similar to that of the SmA phase, but with tilted molecular orientations in the layers. Some of the smectic phases show arrangements with

Table 5-1   Smectic Phase Subgroups and Structures.

Phase subgroup	Bond-orientational ordering	Positional ordering	Molecular orientation	Molecular packing	Phase type
SmA	Short range	Short range	Orthogonal	Random	Fluid
SmB	Long range	Short range	Orthogonal	Hexagonal	Hexatic
SmC	Short range	Short range	Tilted	Random	Fluid
SmD	None	None	None	Isotropic and cubic crystalline coexistence	Plastic Crystal
SmE	Long range	Long range	Orthogonal	Orthorhombic	Crystal
SmF	Long range	Short range	Tile to side of hexagon	Pseudo-hexagonal	Hexatic
SmG	Long range	Long range	Tile to side of hexagon	Pseudo-hexagonal	Crystal
SmH	Long range	Long range	Tilted to side b	Monoclinic	Crystal
SmI	Long range	Short range	Tilt to apex of hexagon	Pseudo-hexagonal	Hexatic
SmJ	Long range	Long range	Tilt to apex of hexagon	Pseudo-hexagonal	Crystal
SmK	Long range	Long range	Tilted to side a	Monoclinic	Crystal
SmL	Long range	Long range	Orthogonal	hexagonal	Crystal

Table 5-2   Smectic Phase Subgroups in Chiral Compounds.

Phase subgroup	Ordering	Molecular orientation	Molecular packing	Phase type
SmC*	Helical; no layer correlation; short-range in-plane correlation	Tilted	Random	LC
SmF*	Helical; no layer correlation; long-range in-plane correlation	Tilted to apex	Pseudo-hexagonal	LC
SmI*	Helical; no layer correlation; short-range in-plane correlation	Tilted to side	Pseudo-hexagonal	LC
SmG*	No helical structure; long-range layer correlation; long-range in-plane correlation	Tilted to side	Pseudo-hexagonal	Crystal
SmH*	No helical structure; long-range layer correlation; long-range in-plane correlation	Tilted to apex	Herringbone	Crystal
SmJ*	No helical structure; long-range layer correlation; long-range in-plane correlation	Tilted to apex	Pseudo-hexagonal	Crystal
SmK*	No helical structure; long-range layer correlation; long-range in-plane correlation	Tilted to side	Herringbone	Crystal

positional ordering within the layer, such as the SmB phase, which has short-range hexagonal molecular orientation but bond angles showing a long-range order. Long-range bond orientational-ordering and positional-ordering subgroups like SmE and SmG show a three-dimensional crystalline structure. The higher the order of the smectic subgroup is, the more solidlike is the molecular arrangement.

When examined under a cooling process, the variety of LC phases in rod-shaped molecules may appear in the phase sequence given below:

$$\text{Iso} \rightarrow \text{N} \rightarrow \text{SmA} \rightarrow \text{SmC} \rightarrow \text{SmC(Alt)} \rightarrow \text{SmB(Hex)} \rightarrow \text{SmI} \rightarrow \text{SmB(Cry)}$$

$$\rightarrow \text{SmF} \rightarrow \text{SmJ} \rightarrow \text{SmG} \rightarrow \text{SmE} \rightarrow \text{SmK} \rightarrow \text{SmH} \rightarrow \text{Crystal}$$

The lower the temperature is, the higher is the order of molecular arrangements. Further schematic illustrations of typical molecular arrangements in smectic phases [11] are shown in Figure 5-2. It is worth noting that charge transport property was observed in mesophases of SmA, SmB, SmC, SmE, SmF, SmG, and so on, based on the reports in the literature.

The smectic phase can be characterized and determined by polarizing optical microscopic (POM) texture and X-ray diffraction (XRD) patterns. For example, a smectic LC 4-4″-bis(dodecyloxy)-3-methyl-p-terphenyl (12O-MeTP-O12) with charge transport property showed SmC and SmG at different temperature ranges, as identified by the POM textures and XRD patterns as shown in Figure 5-3 [17].

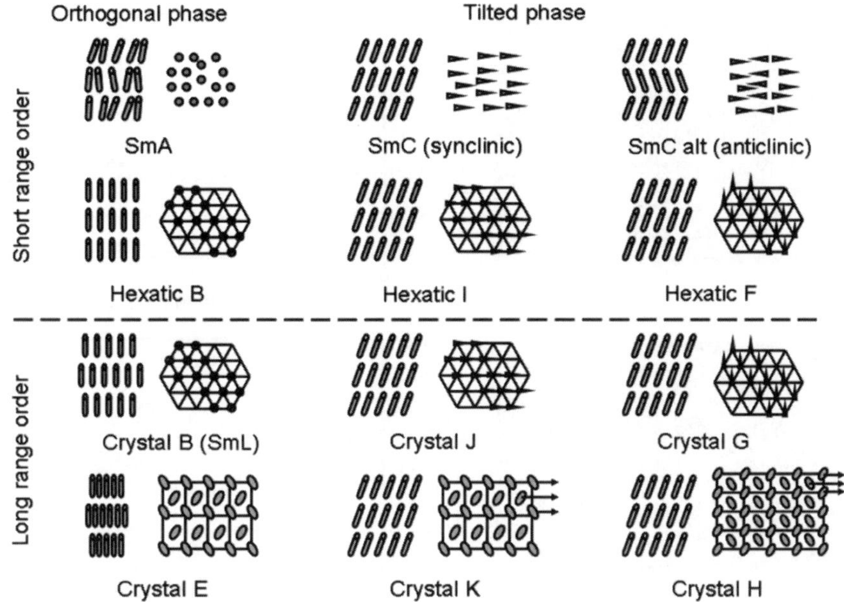

Figure 5-2 Molecular arrangements in the smectic phase. In each phase: layer stacking (left) and in-plane ordering (right).

The SmC phase showed a fanlike texture in POM and a peak spacing in XRD at the small angle region (33.5 Å). The calculated molecular length was 46 Å. Therefore, this indicated that the molecules were tilted against the smectic layer, which is the typical molecular structure and arrangement in the SmC phase, while the SmG phase showed a peak spacing with 31.4 Å at the small angle region and 4.72 Å and 4.11 Å at the wide angle region, indicating that the molecules were tilted against the smectic layer. Furthermore, it also revealed the hexagonal lattice structure in the smectic layer by judging the lattice constant ratio of $\sqrt{3}/2$. POM textures and XRD patterns provide a powerful method to distinguish different molecular structures and arrangements.

After presenting the concepts of smectic phase subgroups, phase transition sequences, and judgment methods, in the forthcoming sections we discuss the molecular design, phase types, properties, and device applications of smectic LCs as semiconductor materials.

5.3. CHARACTERIZATION TECHNIQUES

The carrier mobility of inorganic semiconductors is determined by the Hall effect and conductivity measurements, which has a large value of mobility. However, in organic materials, the resistance is so high and the mobility is low. In order to measure charge-carrier mobilities, other approaches have to be used. The

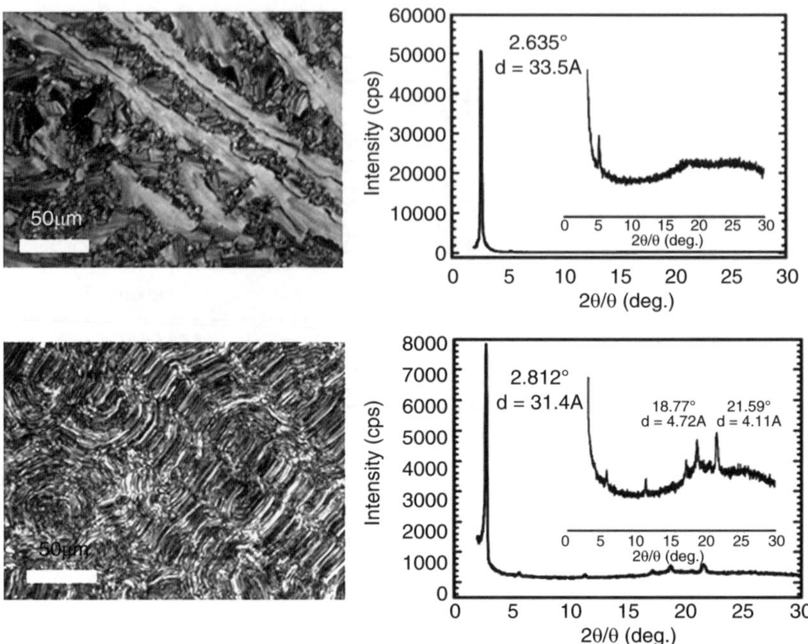

Figure 5-3 Polarizing optical microscopic textures and X-ray diffraction patterns of 12O-MeTP-O12 at SmC phase at 145°C (top) and SmG phase at 100°C (bottom). From Ref. 17 (Y. Takayashiki et al. Mol. Cryst. Liq. Cryst. 2008, 480, 295–301. Reprinted by permission of the Taylor & Francis Group).

experimental characterization techniques to detect and measure charge-carrier transport in organic materials, especially in LC materials, include the time-of-flight (TOF) technique, the pulse radiolysis time-resolved microwave conductivity (PR-TRMC) measurement, the space charge-limited current (SCLC) method and the field-effect transistor technique, etc.

5.3.1. Time-of-Flight Technique

Time-of-flight (TOF) measurement is a well-known technique and is widely used to measure the charge-carrier mobility in organic semiconducting materials. The main significance of this technique is that the electron and hole mobilities can be measured separately. The first TOF measurement was introduced to detect charge-carrier mobility of LC, or, more accurately, discotic LC in 1993 [8]. Since then, this method has often been exploited to study the charge-carrier transport of LCs. The most common use of TOF is to detect the transient photocurrent as a function of time. The typical TOF experimental setup is shown in Figure 5-4. A light pulse is exposed on the surface of the sample. A small amount of charge is generated near the transparent electrode, such as indium tin oxide (ITO). The photo-generated carriers can be transported to the opposite electrode through the material with thickness d when an external voltage V is applied. Once the

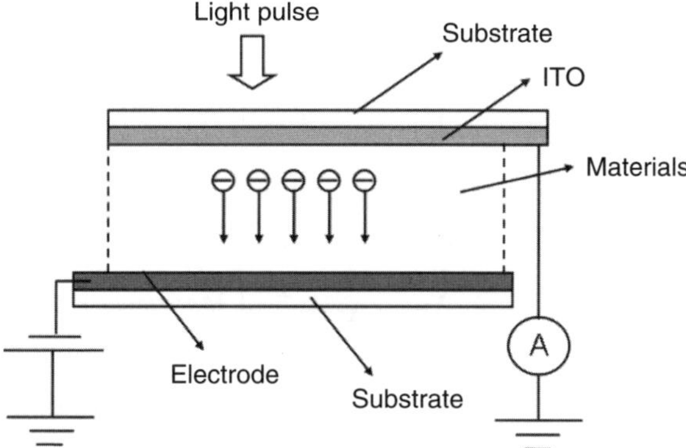

Figure 5-4 Typical TOF experimental setup.

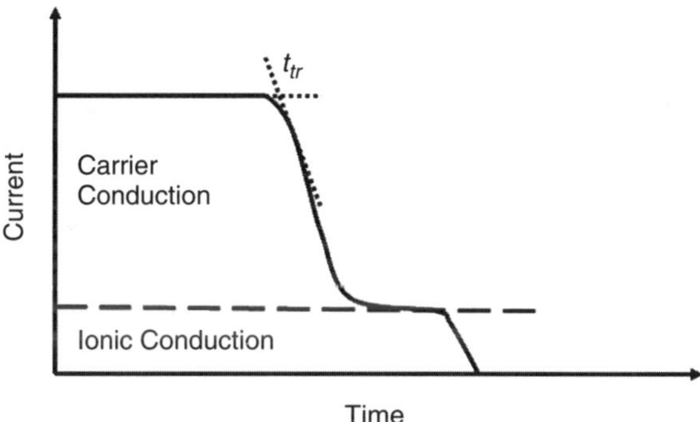

Figure 5-5 Schematic photocurrent curve with time measured by TOF method.

charge carriers reach the opposite electrode, the photocurrent drops, forming a characteristic feature in the oscilloscope.

A schematic curve of a typical nondispersive photocurrent with time, namely, the transient current response, is shown in Figure 5-5. The whole "spectrum" of the photocurrent transient features is detected where the electronic conduction and ionic conduction can be distinguished [18]. The charge-carrier mobility μ can be determined by the equation $\mu = d^2/t_{tr}V$, where t_{tr} is the transit time. Based on the polarity of applying bias voltage, both positive and negative charge-carrier mobilities can be measured. This measurement can detect the charge transport of different types of carriers in one single system. The temperature and polarity of the sample can be controlled easily.

5.3.2. Pulse Radiolysis Time-Resolved Microwave Conductibility Technique

The pulse radiolysis time-resolved microwave conductibility (PR-TRMC) method is a microwave absorption measurement to probe the charge-carrier mobility

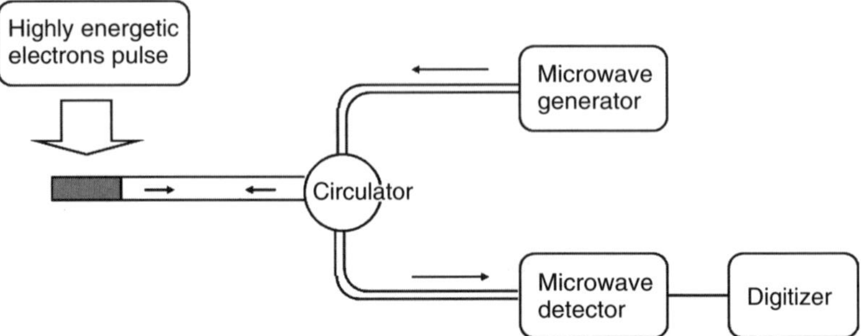

Figure 5-6   Experimental setup of PR-TRMC method.

[19–23]. In this method, the sample is contained in a microwave cell and is excited by a highly energetic electron (in the MeV range) pulse (0.5–50 ns) to generate a density of free carriers. Microwave is used to probe the conductivity of the carriers. The change in electrical conductivity $\Delta\sigma$ induced by the electron pulse is detected through the change of microwave power reflected by the sample. A brief schematic setup is shown in Figure 5-6. $\Delta\sigma$ is determined by the equation $\Delta\sigma = e \sum \mu_{\mathrm{TRMC}} N_{\mathrm{eop}}$, where $\sum \mu_{\mathrm{TRMC}}$ is the sum of the charge-carrier mobility including hole and electron carriers, N_{eop} is the concentration of charge-carrier electron-hole pairs at the end of the pulse, and $N_{\mathrm{eop}} = W_{\mathrm{p}} D / e E_{\mathrm{p}}$. Here E_{p} is the average energy in electron volts absorbed per ionization event, D is the dose (in units of $\mathrm{Gy} = \mathrm{Jkg}^{-1}$) and W_{p} is the probability that ion pairs survive to the end of the electron pulse. Hence the charge-carrier mobility, $\sum \mu_{\mathrm{TRMC}} = E_{\mathrm{p}}(\Delta\sigma/D)/W_{\mathrm{p}}$, is determined by the end-of-pulse conductivity per unit dose $\Delta\sigma/D$ (in units of Sm^2/J).

With this technique, no electrode contact is required for the sample, and it is not influenced by space charge effects. The charges are directly created in the bulk, and charges trapped by impurities or defects would not create a response. Compared with the previous TOF method, the PR-TRMC method is considered to provide intrinsic AC mobility for the bulk. The TOF method employs DC to probe the charge carriers in the macroscopic region between two electrodes. Some carriers are trapped by impurities, defects, and grain boundaries. Therefore the value obtained by the TOF method is generally lower than the value measured by the PR-TRMC method. However, the value obtained by the PR-TRMC method cannot distinguish the contribution from positive (hole) or negative (electron) carriers.

5.3.3. Space Charge-Limited Current Technique

The space charge-limited current (SCLC) technique is a charge-injection collection method based on the analysis of the relationship between the current density (J) and applied voltage (V) of a organic thin film sandwiched by the injecting electrodes [24]. It can be used to measure the mobility of a very thin film (< 1 μm), which is very difficult to measure by a TOF experiment. The current-voltage characteristic of the sample shows ohmic behavior at low applied voltage. If the applied voltage is high, the charge carriers are injected from one electrode. When the charge density is comparable to the charge density on the electrode, the characteristics of the current density J become space charge limited with a nearly quadratic dependence on the voltage V. The current is transport limited, and the electric characteristics can be decided by the Mott–Gurney equation: $J = (9/8)\varepsilon_0\varepsilon_r\mu(V^2/L^3)\Theta$ [25]. Here, ε_0 is the free-space permittivity, ε_r is the dielectric constant of the material, L is the film thickness, Θ is a factor of the ratio of the number of free carriers (not trapped) to the total number of carriers, and μ is the mobility independent of applied field. When $\Theta = 1$, it means trap free. This expression is derived for the materials in which the mobility is independent of the applied field. For organic materials, the charge mobility is often dependent on the electric field E, so it can be described by the disorder formalism, $\mu = \mu_0\exp(\gamma E^{1/2})$, where μ_0 is the charge mobility at zero electric field and γ is a constant. Then the expression of SCLC is approximated to $J = (9/8)\varepsilon_0\varepsilon_r\mu_0\exp[0.891\gamma(V/L)^{1/2}](V^2/L^3)$. The μ_0 and γ can be estimated by fitting the experimental $J-V$ curves from these equations. These parameters will determine the mobility μ. The SCLC technique is an indirect method to evaluate the mobility.

5.3.4. Current-Voltage Analysis Method

The common technique to probe the field-effect mobility (μ_{FET}) in organic semi-conductors is a current-voltage analysis method [26]. The field-effect mobility μ_{FET} is derived from the current-voltage curve measurement of the fabricated device. In the low-drain voltage (V_D) range, the current-voltage response curve is linear, while in the high-voltage range, the drain current I_D is saturated. In the linear region, $I_D = (WC_i\mu_{\text{FET}}/L)(V_G - V_{\text{th}} - V_D/2)V_D$. Here, W is the *channel* width, L is the *channel* length, C_i is the capacitance per unit area of the insulator, V_G is the gate voltage, and V_{th} is the threshold voltage. In the high-drain voltage region, the field-effect mobility of the device can be calculated by the equation $I_{\text{Dsat}} = (1/2)(W/L)\mu_{\text{FET}}C_i(V_G - V_{\text{th}})^2$, where I_{Dsat} is the saturated drain current [27]. The mobility μ_{FET} measured by this indirect method is the so-called "field-effect mobility."

5.3.5. Other Characterization Methods

Other characterization methods to detect charge-carrier mobility in organic materials include transient electroluminescence (EL) by applying a step voltage, Auston switch-based picoseconds photoconductivity technique, and the charge extraction by linearly increasing voltage (CELIV) technique, etc. [28]. It should be mentioned that more mobility measurement methods will arise in the future based on electric field generation, injecting generation, doping generation, and photo generation methods when more materials and devices are established.

5.4. CHARGE-CARRIER TRANSPORT IN SMECTIC LIQUID CRYSTALS

In the case of conventional organic materials (such as polymeric amorphous compounds), the charge-carrier mobility is on the order of 10^{-6} cm^2/ Vs, which is much smaller (about 6 orders of magnitude) than that in the single-crystal state. This is one of the disadvantages of a conventional organic system. However, the charge-carrier mobilities in LC materials are typically in the range of 10^{-4}–10^{-1} cm^2/Vs, or even 1 cm^2/Vs. Moreover, the features of LCs such as the ability to be processed via solution and the ability to dynamically self-repair the defects in the bulk make them earn more attention. Discotic and smectic LCs have been used as semiconducting materials recently. The first report of charge-carrier transport in the discotic hexapentyloxytriphenylene (HPT), was in 1993 [8]. The charge-carrier transport in the first rod-shaped smectic LC 2-(4′-heptyloxyphenyl)-6-dodecylthiobenzothiazole (7O-PBT-S12) was in 1997 [9]. The chemical structures and schematic molecular shapes of HPT and 7O-PBT-S12 are shown in Figure 5-7. Compared with discotic LCs, rod-shaped smectic LCs have lower viscosity, which means they are more liquidlike and are easier to align. From the viewpoint of synthesis, it is generally easier to achieve rod-shaped materials because the core moiety of the smectic molecule is smaller than that of discotic LC. The fluidity of the calamitic mesophases allows them to self-repair defects in the bulk easily. These important advantages make the smectic LCs attractive as self-organized organic semiconductors.

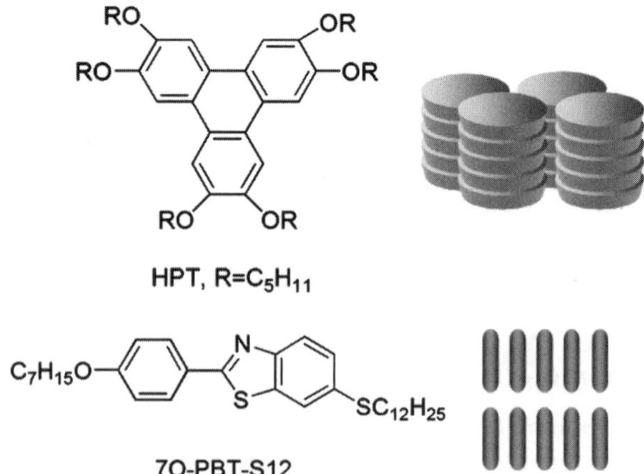

Figure 5-7   Schematic molecular shapes of the discotic LC HPT and the calamitic LC 7O-PBT-S12.

5.4.1. Materials and Charge-Carrier Mobilities

As suggested by the compound structures, many kinds of smectic LCs have been studied as semiconducting materials. Phenylbenzothiazole derivative **1** [2-(4′-heptyloxyphenyl)-6-dodecylthiobenzothiazole, 7O-PBT-S12] was reported to exhibit fast carrier conduction in the SmA phase (90–100°C) with a hole mobility of 5.0×10^{-3} cm^2/Vs by the TOF measurement method [9, 29]. The mobility of 7O-PBT-S12 showed independence of applied field and temperature in the range of smectic mesophase, whereas the carrier mobilities of amorphous organic semiconductor strongly depend on the electric field and temperature. This novel property of mobility in LC is interpreted by the hopping transport localized state and narrow distribution of density of state [30].

Phenylnaphthalene LCs **2a** [2-(4′-octylphenyl)-6-dodecyloxynaphthalene, 8-PNP-O12] and **2b** [2-(4′-octylphenyl)-6-butyloxynaphthalene, 8-PNP-O4] were subsequently synthesized and studied [31–33]. The hole and electron mobilities of **2a** were 2.5×10^{-4} cm^2/Vs in SmA phase (100–121°C) and increased to 1.6×10^{-3} cm^2/Vs in SmB phase (79–100°C). The hole and electron mobilities of **2b** were 4×10^{-4} cm^2/Vs in SmA phase (125–129°C) and further increased to 1×10^{-2} cm^2/Vs in SmE phase (55–125°C). These results indicate that charge-carrier mobilities increase with the increasing order of the smectic phases. Higher charge-carrier mobility can therefore be achieved in the higher-ordered smectic phases. This is because the charge-carrier transports are relevant to the molecular arrangements and determined by molecular distance in the smectic layer. When the charge-carrier transports take place within the smectic layer (perpendicular to the layer normal), the shorter molecular distance in the higher-ordered smectic materials is favorable to the charge-carrier transport. Therefore the mobility values are higher. The mobilities of **2a** and **2b** are also independent of applied electric field and temperature within a given phase.

Oxadiazole derivative **3** [2,5-hexyloxyphenyl-hexyloxybiphenyl-oxadiazole, HOBP-OXD] had an electron mobility of 10^{-4} cm^2/Vs in SmA phase (78–119°C) and a higher electron mobility of 8×10^{-4} cm^2/Vs in the higher-ordered SmX phase (not identified) at the temperature range 41–70°C. In **3**, the mobilities are almost temperature independent within the SmX phase [34]. Dialkylterthiophene derivatives **4a** (dioctylterthiophene, 8-TTP-8) and **4b** (hexyldodecylterthiophene, 6-TTP-12) having π-conjugated molecular systems exhibited smectic phases [35, 36]. The hole and electron mobilities of **4a** were 5×10^{-4} cm^2/Vs in SmC phase (87.8–91.3°C), 2×10^{-3} cm^2/Vs in SmF phase (72–87.8°C), and 1×10^{-2} cm^2/Vs in SmG phase (63.9–72°C). Similar to phenylnaphthalene **2**, terthiophene **4** has the trend of mobility increase with the order of smectic phases and the mobilities are independent of electric field and temperature within a certain smectic phase. Furthermore, the hole mobility of **4a** was further increased up to 0.3 cm^2/Vs in the polycrystalline phase formed by cooling the material slowly from the isotropic phase [37].

The mixture of terthiophene derivatives **5** (*trans*-pentylcyclohexylethylpropyl-terthiophene, 3-TTP-C2CH-5) and **6** [(2-methylbutyl)-dodecylterthiophene, 5*-TTP-12] in the ratio of 1:3 showed the SmB phase between -23 and $100°C$ [38]. The hole and electron mobilities of **5** and **6** were 1×10^{-3} cm^2/Vs in SmB phase (11–112°C) and 3×10^{-4} cm^2/Vs in SmC* phase (46–58°C), while the hole and electron mobilities of mixture were 2×10^{-3} cm^2/Vs in SmB phase (at 80°C). The carrier mobilities of both **5** and **6** showed the property of being independent of electric field and temperature. However, the mobilities of the mixture of **5** and **6** were only independent of electric field and temperature above 50°C. Below this temperature, the mobilities started to decrease and became dependent on temperature and electric field. This behavior is similar to the conduction of amorphous organic materials and may be induced by the small disorder of the smectic phase [38]. Figure 5-8 shows the chemical structures of materials **1–6**.

Asymmetrically substituted ethynyl-terthiophene and ethynyl-quaterthiophene derivatives **7** (5-hexyl-5′-hexynyl-2,2′:5′,2″-terthiophene, 6-TTP-yne-4) and **8** (5-propyl-5‴-hexynyl-2,2′:5′,2″:5″,2‴-quaterthiophene, 3-QTP-yne-4) were reported to have the smectic phase at room temperature [39]. **7** has SmB$_{cryst}$ between -20 and 98.6°C, while **8** has SmG or SmH phase between -20 and 191°C. The asymmetric design decreases the transition temperature and extends the temperature range of mesophase. The hole mobility at room temperature was 3×10^{-2} cm^2/Vs in **7** and further increased up to 1×10^{-1} cm^2/Vs in **8**, which is comparable to that in the polycrystals of aromatic compounds. The mobilities were independent of electric field and temperature above room temperature.

The hole mobility of symmetric quaterthiophene derivative **9** (dihexylquar-terthiophene, 6-QTP-6) in the SmX phase (not identified, 56–179°C) was 6×10^{-2} cm^2/Vs, slightly lower than that of **8** [40]. Compared with smectic phase temperature range of symmetric molecular structure **4a** (63.9–91.3°C) and **9** (56–179°C), asymmetric derivatives **7** (-20–96.8°C) and **8** (-20–191°C) showed a lower transition temperature and a wider temperature range.

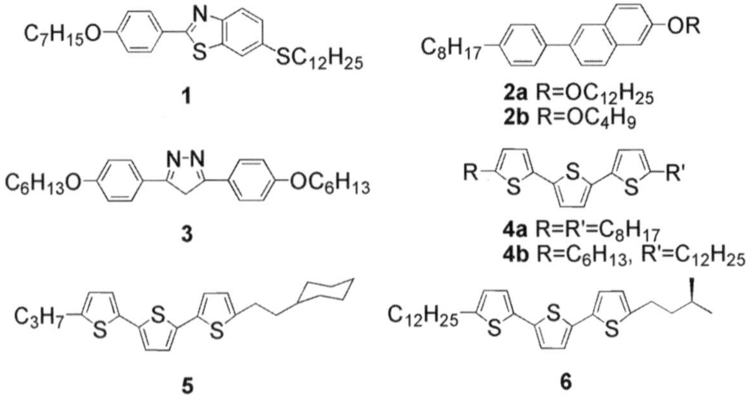

Figure 5-8 Chemical structures of materials **1–6**.

Symmetric bisthienylbenzene derivative **10** [1,4-di(5′-octyl-2′-thienyl)-benzene, 8-TPT-8] showed charge-carrier mobility on the order of 10^{-1}–10^{-2} cm^2/Vs in three smectic mesophases between 47 and 145°C [41]. The hole mobilities were independent of electric field but depended slightly on temperature. The values increased with decrease in temperature. The hole mobility of **10** at 60°C was up to 1×10^{-1} cm^2/Vs, close to the value of **8**. The results indicated that a thiophenephenylene-cored molecular structure, rather than terthiophene or quaterthiophene systems, also has high carrier mobilities, although only two sulfur atoms and two thiophene groups are in the structure. The two thiophene rings linked at the 1,4-position of the phenyl ring in **10** may produce closer packing of molecular arrangements; therefore the carrier mobilities are high. The chemical structures of **7–10** are shown in Figure 5-9.

Figure 5-10 illustrates the chemical structures of materials **11–15**. The phenylenethiophene derivative compounds have been studied extensively for application as field-effect transistors (FET). One example is phenylene-thiophene

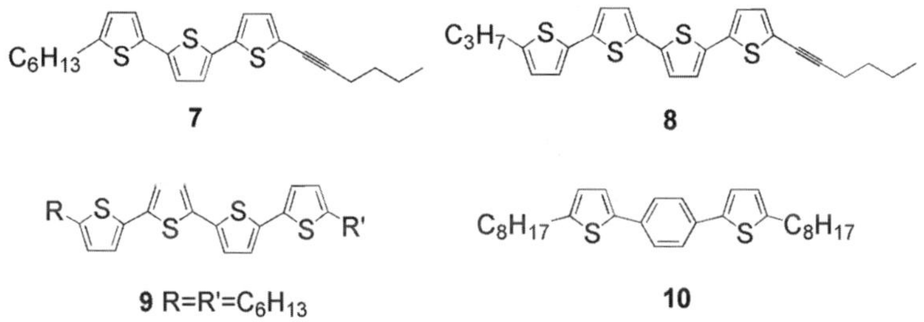

Figure 5-9 Chemical structures of materials **7–10**.

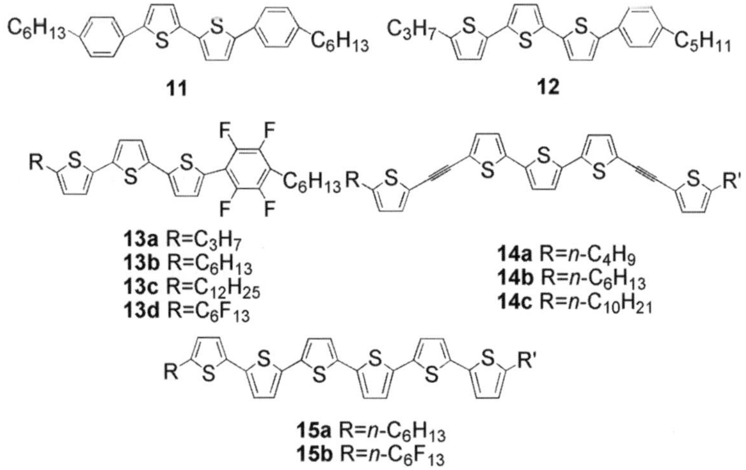

Figure 5-10 Chemical structures of materials **11–15**.

11 [5,5′-bis(4-n-hexylphenyl)-2,2′-bithiophene,   dH-PTTP]   with   field-effect hole mobility of 6×10^{-3} cm^2/Vs. This value was achieved from the device fabricated by the solution-cast method [42]. When the substitution of benzene ring in **11** was replaced with thiophene, the material structure was changed to phenylterthiophene derivative **12** [5-propyl-5″-(4-phentylphenyl)-2,2′:5′,2″-terthiophene, 3-TTPPh-5] with carrier mobilities of 7×10^{-2} cm^2/Vs (hole) and 2×10^{-1} cm^2/Vs (electron) at 30°C in SmX phase (not identified) [43, 44]. The hole mobility is almost independent of the electric field and temperature above the room temperature, and the value is close to that of **8**. The asymmetric structure of **12** makes it display the smectic phase at room temperature, and it did not crystallize even when the temperature was lowered to −50°C.

To increase the oxidation potential to enhance the LC stability, fluorine is introduced into the molecular structure as an electron-withdrawing functional atom. Fluorophenylterthiophene derivatives **13** with different terminal groups were studied [45]. **13a** [5-propyl-5″-(4-hexyltetrafluorophenyl)-2,2′:5′,2″-terthiophene, 6TFPTTP3) has a chemical structure similar to that of **12** except for the terafluorophenyl group. In SmA phase (75–106°C) and a highly ordered mesophase SmX (−50°C −75°C, not identified), **13a** showed a stable hole mobility of 7×10^{-2} cm^2/Vs between 50 and 100°C and 3×10^{-2} cm^2/Vs below 50°C. The decrease of the value was induced by the defects formed in the cooling process. The hole mobility was dependent of the electric field and temperature below 50°C. The electron mobility of **13a** was 2×10^{-2} cm^2/Vs at room temperature, which was only dependent of electric field. Compared with **12, 13a** shows higher-order smectic phase judging from the X-ray diffraction measurements, which may be induced by the fluorine atom. The terafluorophenyl group lowers the transition temperature of mesophase.

The compounds with five thiophene heterocycles were reported for the application as large-area LC monodomain field-effect transistor [46]. The hole mobility of **14b** [5,5″-bis(5-hexyl-2-thienylethynyl)-2,2′:5′,2″-terthiophene] in SmB phase (68–109°C) amounted to 2×10^{-2} cm^2/Vs measured by the TOF method, which was almost independent of applied field and temperature. The electron mobility of **14b** was 1.2×10^{-3} cm^2/Vs at 80°C, which increased with the temperature. The hole and electron mobilities of **14c** [5,5″-bis(5-decyl-2-thienylethynyl)-2,2′:5′,2″-terthiophene] were 1.9×10^{-2} cm^2/Vs and 2.4×10^{-3} cm^2/Vs in SmB phase at 100°C. The mixture of **14b** and **14c** with the weight ratio of 1:1 was examined. The disorder induced by the two different compounds did not change charge-carrier transport too much. The hole and electron mobilities of the mixture were 1.1×10^{-2} cm^2/Vs and 1.4×10^{-3} cm^2/Vs in SmB phase at 80°C. The LC monodomain area of **14b** was achieved up to 150 mm in diameter, showing the potential for application as a large-area field-effect transistor.

Compounds with six thiophene heterocycles **15a** (dihexylsexithiophene, DH-6T) and **15b** (diperfluorohexylsexithiophene, DFH-6T) were synthesized [47, 48]. **15b** was established as the first n-type thiophene-based conductor material for the thin-film transistor with optimized field-effect mobility of 2×10^{-2} cm^2/Vs. The perfluorinated group in the oligothiophene core of **15b** is electron withdrawing, which enhances the material stability in the air. A series of liquid crystalline oligothiophenes with six thiophene heterocycles and alkoxy chains were also synthesized [49]. The effects of the number and the length of the terminal alkoxy chains, which induced various LC phases including smectic, columnar, and micellar cubic phases, were studied. The maximum of the hole mobilities among these was 1×10^{-2} cm^2/Vs in SmX (not identified) at 160°C. The aromatic system with long-core molecular structure shows higher mobility and more stability in the smectic phase.

Although the charge-carrier mobilities are low, simple molecular structures and relatively small molecules such as biphenyl, terphenyl, and distyrylbenzene derivatives also exhibit charge-carrier mobilities in the smectic phases. The chemical structures of biphenyl derivatives with alkyl and alkoxy substitution are shown in Figure 5-11. At first, the low magnitudes of the photocurrents of biphenyls **16a** (5-BP-5) and **16b** (7-BP-11) in SmB phase (50 and 55°C, respectively) were observed by the TOF method, and the amounts of the photocurrents were changed to giant values upon repeated irradiation several orders of magnitude higher than the initial single light pulse. This was probably due to residual impurities, and the charge carrier did not transport through the sample [50]. For compounds with an oxygen atom in the alkyl chains, **16c** (8-BP-O6) and **16d** (8-BP-O8), the hole and electron mobilities in SmB phase (55 and 60°C, respectively) were both in the same range of 10^{-5} cm^2/Vs. The alkoxy group eliminated the trap-limited behaviors induced by the residual impurities [50]. These results in **15** and **16** indicate that the materials should be highly purified for further usage.

Typical terphenyl derivatives such as 4-4″-bis(dodecyloxy)-3-methyl-*p*-terphenyl **17** (12O-MeTP-O12) showed hole and electron mobilities up to 5×10^{-4} cm^2/Vs in SmC phase (133–158°C) and 0.9×10^{-2} cm^2/Vs in SmG phase (84–133°C). The methyl group on the terphenyl core makes the phase transition temperature lower [17]. A longer simple conjugated 4,4′-distyrylbenzene core derivative **18** was demonstrated to have higher charge-carrier mobilities in the smectic phase than **16** and **17**. The hole mobility

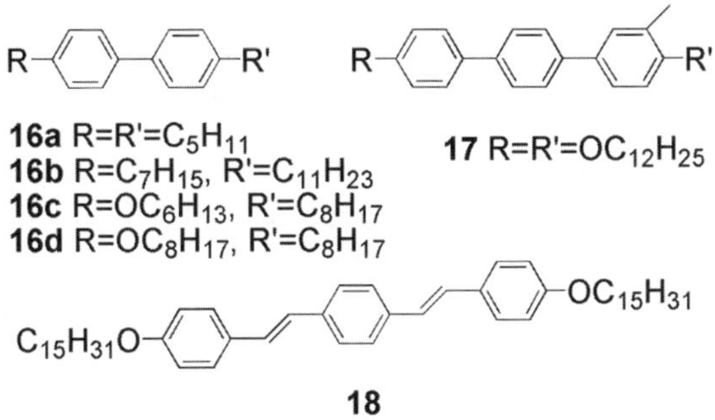

16a R=R'=C$_5$H$_{11}$
16b R=C$_7$H$_{15}$, R'=C$_{11}$H$_{23}$
16c R=OC$_6$H$_{13}$, R'=C$_8$H$_{17}$
16d R=OC$_8$H$_{17}$, R'=C$_8$H$_{17}$

17 R=R'=OC$_{12}$H$_{25}$

18

Figure 5-11   Chemical structures of **16–18**.

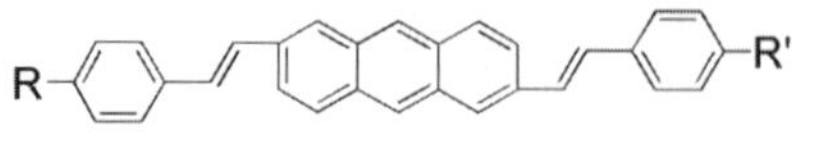

19 R=R'=C$_5$H$_{11}$

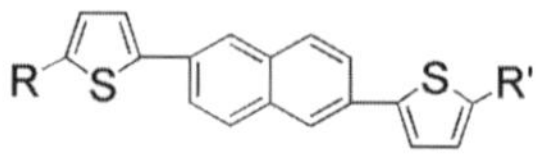

20 R=R'=C$_8$H$_{17}$

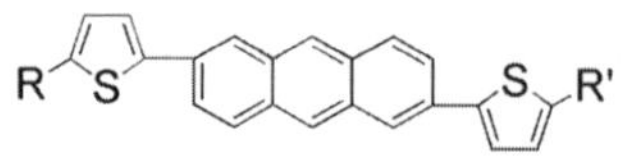

21 R=R'=C$_6$H$_{13}$

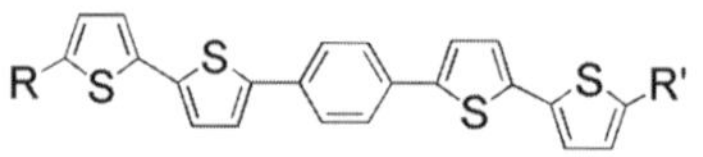

22 R=R'=C$_{10}$H$_{21}$

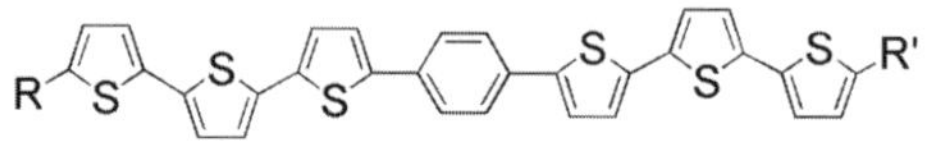

23 R=R'=C$_{10}$H$_{21}$

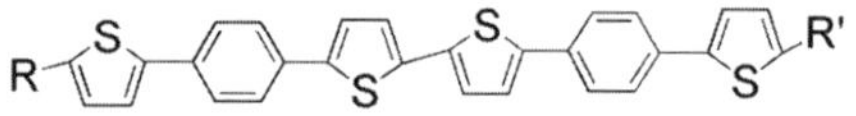

24 R=R'=C$_{10}$H$_{21}$

Figure 5-12   Chemical structures of **19–24**.

was 6.2×10^{-3} cm^2/Vs in SmF phase (170°C) and increased to 1.8×10^{-2} cm^2/Vs in SmG phase (150°C) [51].

Calamitic compounds with simple backbones such as anthracene, naphthalene, and thiophene-phenylene (**19–24**) were studied for application as FET. The chemical structures are shown in Figure 5-12. The crystalline solid FET of **19** {2,6-bis[2-(4-pentylphenyl)vinyl] anthracene, DPPVAnt} was obtained through the annealing of the material film from 100–140°C, giving rise to a new orientation [52]. The field-effect mobility of the device was in the range 0.1–1.28 cm^2/Vs, dependent on the temperature of the substrate. The temperature range of the maximum charge-carrier mobility was 60–80°C. **19** showed smectic phase beyond 200°C judging from differential scanning calorimetry (DSC) analysis. The order of the film could be optimized by annealing with proper temperature treatment.

High field-effect hole mobility of 1.4×10^{-1} cm^2/Vs was achieved using thiophene-naphthalene derivative **20** [2,6-di(5'-n-octyl-2'-thienyl)-naphthalene, 8-TNAT-8] in its highly ordered polycrystalline thin film at room temperature [27]. The charge-carrier mobility of the material was on the order of magnitude 10^{-1}–10^{-2} cm^2/Vs at highly ordered mesophase (not identified, 95–182°C) measured by the TOF method. Thiophene-anthracene derivative **21** [bis-(5'-hexylthiophen-2'-yl)-2,6-anthracene, DHTAnt] showed field-effect mobility of 5×10^{-1} cm^2/Vs [53]. If there was no alkyl group in the system, that is, bis-(thiophen-2'-yl)-2,6-anthracene (DTAnt), the mobility was one order of magnitude lower than that of **21**. The high mobility of DHTAnt may be induced by the self-organized property of the liquid crystalline mesophase, while there is no mesophase in the DTAnt because no alkyl groups are involved. Thiophene-phenylene derivatives **22–24** with different numbers and positions of

1,4-phenyl rings within a 2,5-oligothiophene conjugated chain for FET showed field-effect mobilities up to 4×10^{-1} cm^2/Vs [54]. Among these systems, **22** and **24** have liquid crystalline smectic phase behaviors. Moreover, the decyl-end-capped groups in the thiophene-phenylene oligomers improve the electric characteristics and oxidation stability of the thin film. **19–24** all showed high field-effect mobilities. This is important for practical application as FET.

Compounds of dithienothiophene (DTT) and benzothiadiazole (BTZ) derivatives **25–27** (Fig. 5-13) exhibited smectic phase and were used as FET [55]. This also shows the possibility of retaining molecular ordering arrangement states of the materials by cooling the material through the smectic phase to room temperature. Actually, in compound **25**, the current of the film obtained by the annealing process through SmB (137°C) was two orders of magnitude higher than that of the film casted. The annealing processed field-effect hole mobility of vacuum-evaporated film of **25** was 3.7×10^{-3} cm^2/Vs. It was found that a shorter oligomer exhibits a higher-order smectic mesophase easily whereas a longer oligomer exhibits an ordered mesophase with dificulty [55].

Non-rod-shaped molecules such as sanidic- and bent-shaped molecules also show smectic phases (Fig. 5-14). A sanidic LC with flat central aromatic core **28** ([1]benzothieno[3,2-*b*]benzothiophene-2,7-dicarboxylate, BTBT) showed a lamellar-columnar mesophase (SmA type) between 83.5 and 95.6°C. The hole and electron mobilities were on the order of 2×10^{-3} cm^2/Vs in the mesophase [56]. A bent-shaped compound 9-methyl-9*H*-carbazole derivative (**29**, CBZ7) showed SmG and SmX (not identified) phases between 129 and 139°C. The field-effect hole mobility was observed at 7×10^{-5} cm^2/Vs [55].

Some liquid crystalline polymers present smectic phases and charge-carrier transport properties. For example, the first photopolymerizable LC reactive mesogen, penta-1,4-dien-3-yl derivative **30**, exhibited smectic phase (SmC) at room temperature (Fig. 5-15). The hole and electron mobilities were

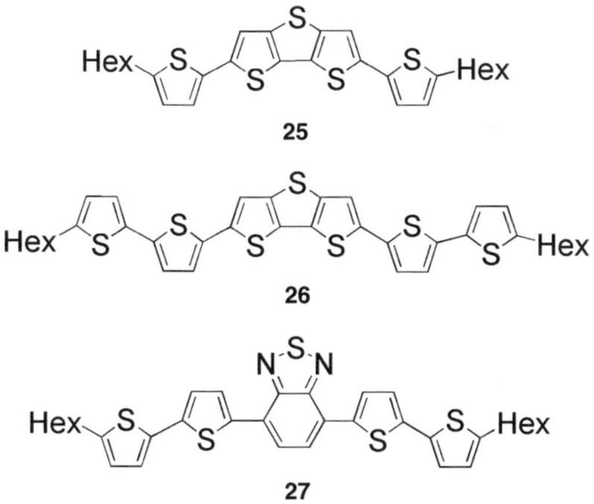

25

26

27

Figure 5-13 Chemical structures of **25–27**.

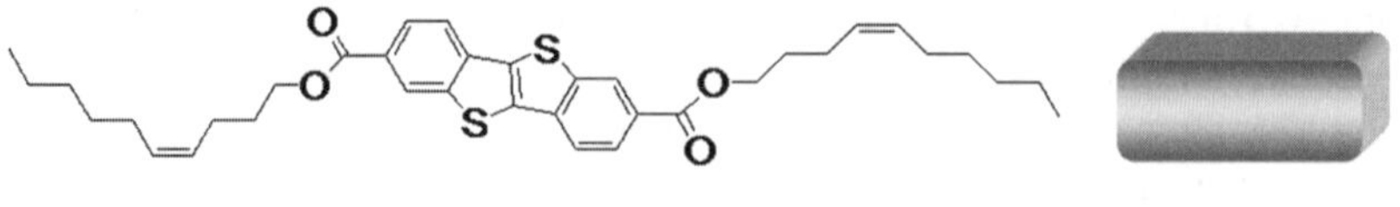

28

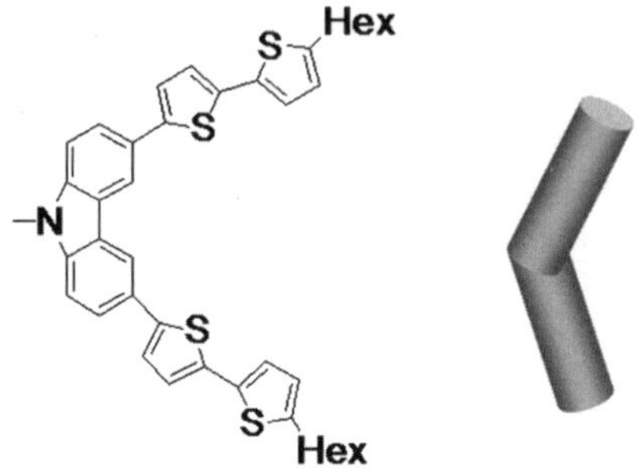

29

Figure 5-14   Chemical structures of **28** and **29**.

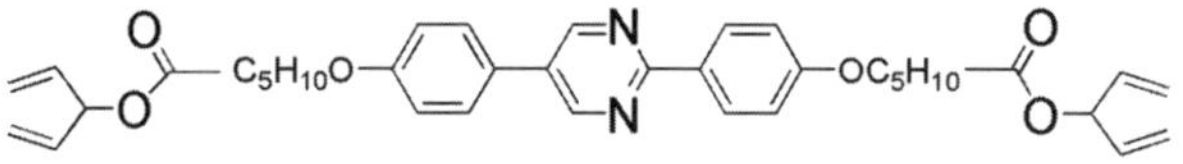

30

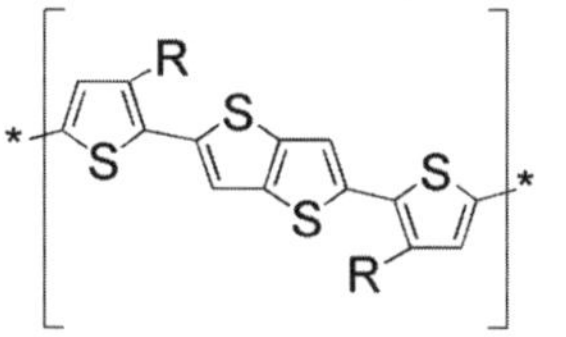

31a R=C$_{10}$H$_{21}$
31a R=C$_{12}$H$_{25}$
31a R=C$_{14}$H$_{29}$

Figure 5-15   Chemical structures of **30** and **31**.

1.8×10^{-5} cm^2/Vs and 1.5×10^{-5} cm^2/Vs in SmC phase measured by the TOF method [57]. The polymerization may retain the highly ordered molecular arrangements and some physical properties of the LC reactive monomer. Polymer networks are not soluble but stable, which are more suitable for the fabrications of multilayer devices. A very high field-effect mobility of 0.2–0.6 cm^2/Vs in the field-effect transistor devices was obtained with the liquid crystalline thieno[3,2-*b*]thiophene polymer **31** {poly(2,5-bis(3-alkylthiophen-2-yl)thieno[3,2-*b*]thiophenes), PBTTT} [58]. There is no smectic LC phase in **31**. However, highly organized morphology and large crystalline domain size were achieved by thermal processing through the LC mesophase. These properties

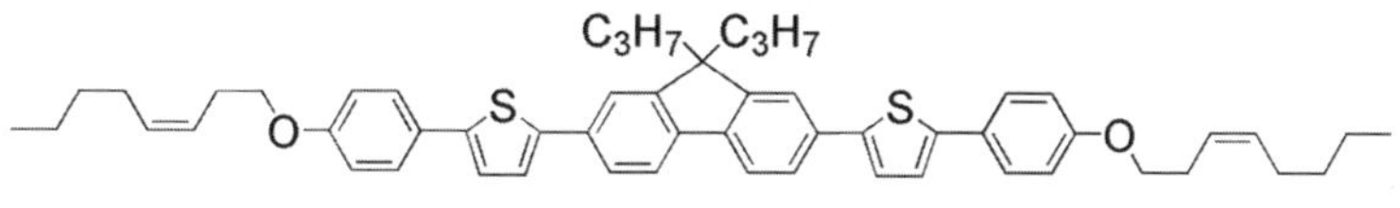

32

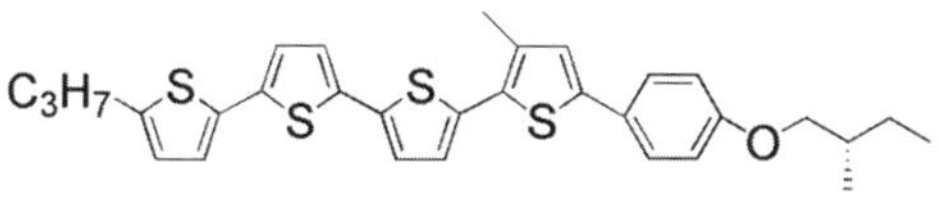

33

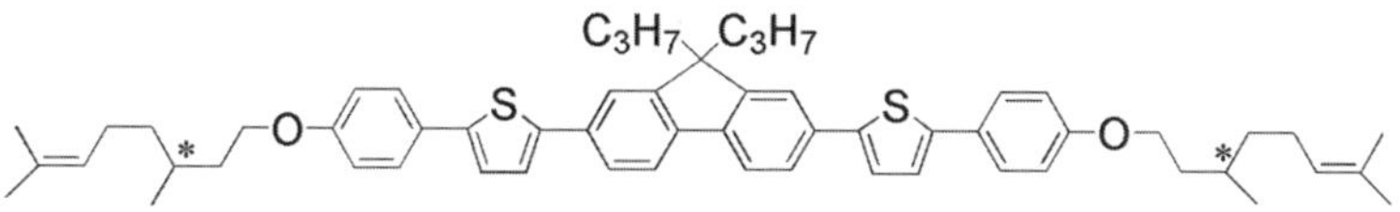

34

Figure 5-16 Chemical structures of **32–34**.

possibly come from the contribution of the thermal processing history, offering the fabrication of single-crystal polymer FET devices.

The charge transports in the rod-shaped nematic LCs and cholesteric LCs were also discovered. For example, nematic LC **32** had hole mobility up to 1.0×10^{-3} cm^2/Vs at room temperature [59]. Cholesteric LC **33** showed hole mobility of 2×10^{-4} cm^2/Vs and electron mobility of 1×10^{-4} cm^2/Vs in the cholesteric phase at 120°C [60]. Cholesteric LC **34** showed hole mobility of 4×10^{-4} cm^2/Vs in the glass phase at room temperature [61]. The chemical structures of **32–34** are shown in Figure 5-16. However, the mobilities of **32–34** are lower than the values of smectic LC materials. The reason is that mobility is related to the molecular arrangement and intermolecular distance. Smectic LCs have higher-ordered molecular arrangements and shorter intermolecular distances than nematic and cholesteric LCs, so the mobilities of smectic phases are higher.

5.4.2. Summary of Charge Transport Properties in Smectic Liquid Crystals

From the investigations of charge transports in smectic LCs, we find that smectic molecular systems have high carrier mobilities. The LCs have many appropriate properties popular for fast carrier transports and semiconducting devices, which are easy to process, align, and dynamically control, etc. The dynamic nature, that is, self-organization, of LCs allows them to self-repair defects such as disclinations and domain boundaries in the bulk to enhance mobility [62, 63].

There are two important properties of charge transports in smectic LCs. One property, in most cases, is that the charge transports are independent or less dependent on applied field and temperature in a certain LC mesophase or above a critical temperature. This behavior is much different from that of amorphous organic materials [64]. The mobilities of amorphous organic materials are dependent on applied field and temperature, which is not good for the stability of device performance. The other property is that the charge transports are related to the molecular structures and arrangements, namely, different smectic packing, order, and arrangement. The charge mobility is determined by the intermolecular distance. The higher the order which the smectic phase has, the closer the intermolecular distance, the better the orbital overlap, and the higher the charge-carrier mobility [65]. These diverse materials and smectic phases offer many opportunities to design and customize new devices.

As shown by the research on the relationships between semiconducting properties and molecular structures of materials, the aromatic π-conjugated system in the core is more suitable for the design of the LC material with high mobility. Some fast charge transport materials with high mobilities have been already observed in the smectic phases of alkynylquaterthiophene, phenylterthiophene, dithienylbenzene, bis(thienylethynyl)terthiophene derivatives, and so on. It is important for the LCs to exhibit mesophase at room temperature or a very wide temperature range. The molecules with linear or branched alkyl chains and with asymmetrical structures can lower the transition temperature and increase the temperature range of mesophase.

Furthermore, introduction of fluorine atom or perfluorinated group as an electron-withdrawing functional group can enhance the stability of the materials. The annealing process through mesophase or the polymerization of LC reactive monomer provides the feasibility of retaining molecular ordering arrangement and some physical properties of mesophase. These are favorable for the construction of liquid crystalline semiconductor-based practical devices. It appears that these new applications will be realized in the near future.

5.5. DEVICES AND APPLICATIONS

5.5.1. Fabrication Technologies

Because of the potential applications in organic semiconducting devices such as photoconduction devices, thin-film transistors (TFTs) or field-effect transistors (FETs), organic light-emitting diodes (OLEDs), and solar cells, the charge transports in the LCs create new challenges and opportunities in these areas. The commercialization of the low-cost solution-processed deposition technique has been used to fabricate some organic semiconducting devices. However, it is difficult to precisely control the crystal growth process and to obtain appropriate morphologies, good performances, and high mobilities. The grain boundary or crystallographic defects of the casted films influence the charge transport properties in the devices severely, whereas the smectic LCs with fluidlike low viscosity, self-organization, and highly ordered molecular structures are soluble in most common solvents, which make them more suitable for "wet-process" or "solution-process" production techniques. The self-organization of LCs with uniform alignments is often able to produce defect-free, highly ordered arrangements, high charge-carrier mobility, and large-area thin-film devices.

The ordered molecular arrangements of self-organized LCs in smectic mesophase are required to be maintained at room temperature. In addition, the stability of the LC thin film should be considered. One method to enhance the stability is to design and develop stable mesophases at room temperature. The other method is to use thermal- or photo-polymerization to stabilize the LCs [66] or to design reactive polymerizable end groups such as oxetane, diene, acrylate, and methacrylate in the molecular main chains via a flexible spacer to form reactive mesogens (RMs) [67, 68]. The schematic molecular structures of RM and polymerized liquid crystalline arrangements are shown in Figure 5-17. The "R" groups are the end groups with reactive radical and the "S" groups are optional side substituent groups that may adjust the solubility and stability of the materials [68]. The cross-linked liquid crystalline films after polymerization at room temperature are expected to freeze the orders of the mesophase with less crystallization and grain boundaries. The polymerized system has the advantage of a simple fabrication process, favorable film-building technology of multilayer, large-scale devices, ambient stability, and the capability to be patterned with etching technology through a photomask.

When smectic LCs are fabricated as semiconductor layers, uniform alignments of LC molecules may be essential. It is easy to align the LC by the well-established alignment method used in the LC display (LCD) industry. Figure 5-18 illustrates homeotropic and planar alignments of a calamitic LC, wherein the long axis of the molecule is perpendicular to or parallel with the substrate [69]. For the

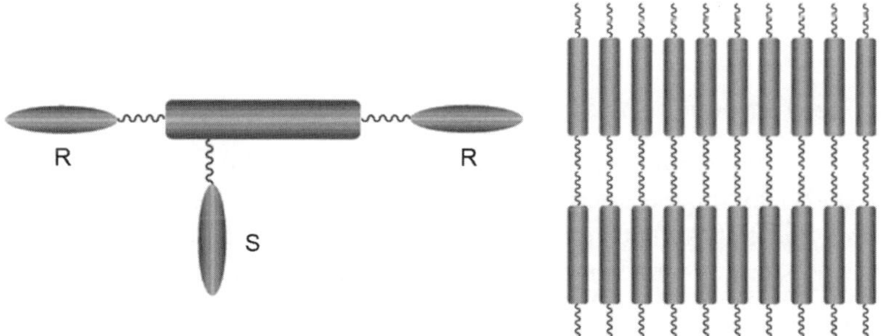

Figure 5-17 Schematic illustration of molecular structure of a reactive mesogen (RM) (left) and a polymerized liquid crystalline arrangement (right).

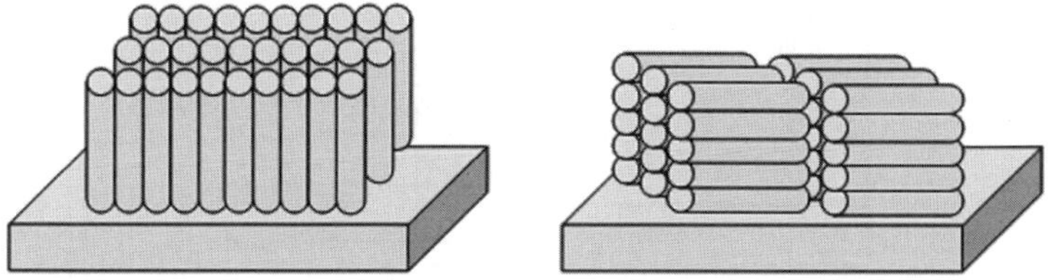

Figure 5-18 Schematic illustrations of a homeotropic alignment (left) and a planar alignment of a calamitic LC (right).

planar alignment mode, a homogeneous anti-parallel rubbed polyimide method is the most common commercial technique. However, it is easy to induce defects such as dust and scratching by the mechanical rubbing cloth. Noncontact photoalignment is a good option to avoid these drawbacks [70]. For the homeotropic alignment mode, no rubbing process is to be implemented. Because of the rod shape of the molecules, the fast carrier conductions in the calamitic LCs are anisotropic in nature. The photocurrent conducted through intermolecules (short axis of a LC molecule) is higher than that through the smectic layer direction (long axis of a LC molecule) detected by the TOF method [65]. Proper alignments of LCs can enhance the anisotropic property.

5.5.2. Photoconductors

Photoconductors that involve LC materials such as xerography photoreceptors for photocopiers and laser printers are charge-carrier-generation devices. For example, a 100-μm-thick layer of LC **2a** was covered with a polyester thin film (12 μm) as a xerographic photoreceptor [71], the layer structure of which is shown in Figure 5-19. When the photoreceptor is irradiated by a light, photoinduced discharge occurs. A contrast voltage is formed to obtain patterns of corona charging. LCs as photoconductors can be made of large area, high photosensitivity, and low cost. The carrier mobility of 10^{-3} cm^2/Vs is higher than the 10^{-6}–10^{-5} cm^2/Vs in amorphous organic materials. In addition, the property of charge mobilities independent of the applied field and temperature increases performance stability.

5.5.3. Field Effect Transistors

Smectic LCs can be used as field-effect transistors (FETs) or thin-film transistors (TFTs) [72]. In the previous discussions, we have introduced the molecular structures and field-effect mobilities of some materials as FETs or TFTs. FETs are the footstones of the modern semiconductor industry, while organic FETs

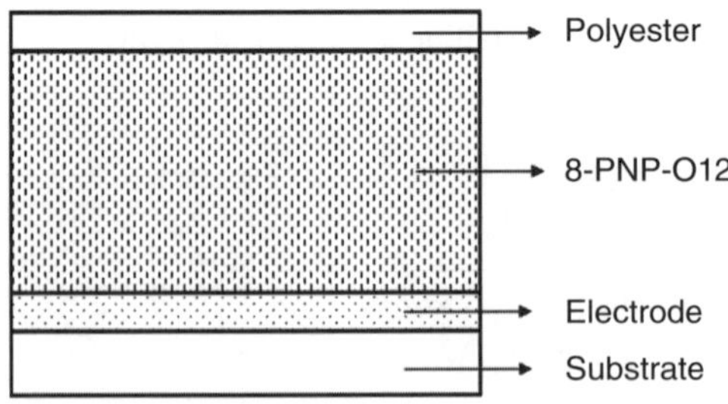

Figure 5-19   A example of smectic LC as a photoreceptor.

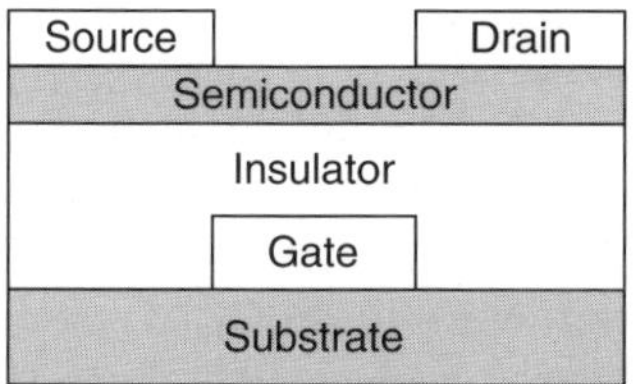

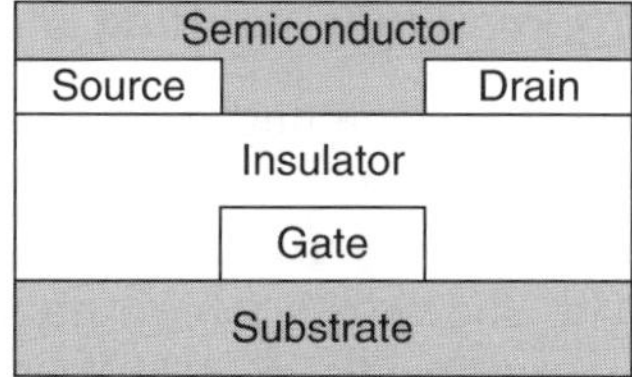

Figure 5-20   Schematic typical structures of organic field-effect transistor. Left: Top-contact geometry. Right: Bottom-contact geometry.

(OFETs) are the foundation of organic semiconductors. For LC solutions, the spin-coating method can be used to form organic semiconducting layers. Typical diagrams of a FET device with "top-contact" or "bottom-contact" geometry are shown in Figure 5-20. Usually, a layer of SiO_2 (~ 300 nm) as gate insulator is on a heavily n-doped silicon (gate electrode) wafer. Gold is evaporated and photolithographically defined to obtain source and drain electrodes. The distance between the source and the drain electrodes is the length of the channel, and the transverse dimension of the structure is the width of the channel. For the top-contact FET, the source and drain electrodes are deposited on the organic semiconductor layer, while for the bottom-contact FET, the semiconductor layer is formed on the source and drain electrodes [26, 73]. The current flows occur between the source and the drain electrodes, which are controlled by the applied voltage of the gate electrode. For example, for the "p-type" FET semiconductor, when the gate electrode is biased positive with respect to the source electrode, the carriers are depleted and charge flow is little. When the gate electrode is biased negative, the carriers are accumulated in the channel between source and drain electrode. A conducting flow is formed and transported through the channel [74, 75].

When liquid crystalline materials are used as semiconducting layers, physical properties such as self-organization, ambipolar charge transport, electrically inactive domain boundaries, and good solubility in the ordinary organic solvents make them become a new molecular system category in the semiconductor catalogs. The solution-processable FETs using LC materials are very economical, whereas inorganic semiconductors are fabricated by a costly vacuum deposition process. Some successful examples have already been realized using solution-processed smectic LC materials, such as the terthiophene derivatives **14b** with a field-effect hole mobility of 2×10^{-2} cm^2/Vs in the large-area thin-film transistor [46] and the first reactive mesogen demonstration of the phenylnaphthalene derivatives with a field-effect hole mobility of 10^{-3} cm^2/Vs in the solution-deposited mesophase and of 10^{-4} cm^2/Vs after photopolymerization in the bottom-contact FET [68]. Furthermore, the high field-effect mobilities reported reach values of 10^{-1} cm^2/Vs among compounds **20–24** [27, 54], 0.6 cm^2/Vs in compound **31** [58], even reaching 1.28 cm^2/Vs in compound **19** [52], revealing potential applications as FETs.

5.5.4. Organic Light-Emitting Diodes (OLEDs)

The organic light-emitting diode (OLED) is a light-emitting technology based on electroluminescence as a new generation display. Figure 5-21 shows a simple single-layer structure that consists of an anode, a cathode, and an emitter layer (EML). Also shown is a typical triple-layer structure of an OLED device. Here, a hole-transporting layer (HTL) and an electron-transporting layer (ETL) are sandwiched between cathode and anode. The EML is between HTL and ETL. The basic principle of light emitting in an OLED device is shown in Figure 5-22. When applying an external voltage across the two electrodes, electrons and holes are injected from the cathode and the anode into the lowest unoccupied molecular orbital (LUMO) and the highest occupied molecular orbital (HOMO) by a hopping transport mechanism [76–79]. The recombination of the electrons and the holes near the interfaces in the emitter layer generates electronically excited states or excitons followed by their deactivation, yielding an emitted light.

OLED technology has many advantages over other flat-panel displays such as LCDs. OLEDs are self-luminescent, having an extremely wide viewing angle and no requirement for the backlight and color filters. These properties make them thinner, lighter, and more efficient than other flat-panel displays [80]. When LC materials are used for OLED application, the anisotropic properties can generate polarized emitted light. In Section 5.1, the alignment mode of LCs is discussed. The homogeneous alignment mode can be used to generate a linearly polarized emission OLED device. Such a device has potential application as the backlight

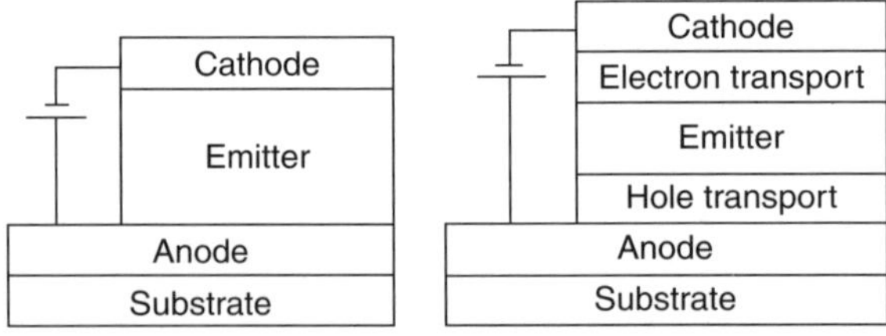

Figure 5-21   Schematic illustration of a typical structure of OLED. Left: a simple single-layer structure. Right: a typical triple-layer structure.

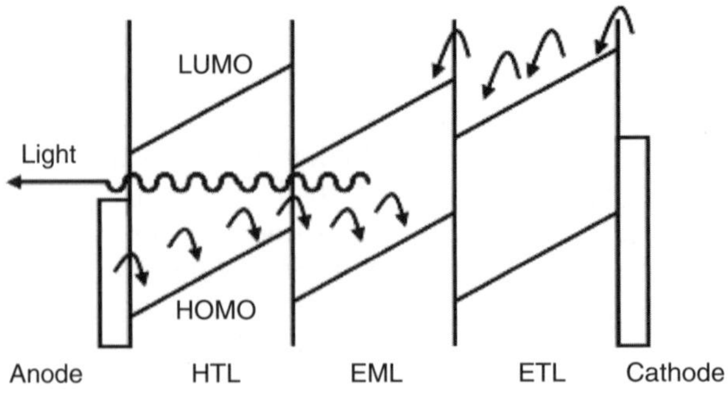

Figure 5-22   Principle of the light-emitting in an OLED device.

of the LCDs. Colored, polarized emitting light as the backlight would remove the polarizer and color filter.

The calamitic LC 2,5-hexyloxyphenyl-hexyloxybiphenyl-oxadiazole (**3**, HOBP-OXD) demonstrated polarized electroluminescence (EL) blue light from SmX phase (at 70°C) with a homogeneous alignment [81]. In this design, a copper phthalocyanine vacuum-deposited film was used as a hole injection layer and a tris-(8-hydroxyquinoline)aluminum vacuum-deposited film was used as an electron injection layer. The anisotropy of the EL emission can be determined by the order parameter $S_{EL} = (I_{EL\parallel} - I_{EL\perp})/(I_{EL\parallel} + 2I_{EL\perp})$, where $I_{EL\parallel}$ is the intensity of the EL parallel with the alignment direction and $I_{EL\perp}$ is the intensity of the EL perpendicular to the alignment direction. Here, $S_{EL} = 0.32$. The luminance was 0.8 cd/m^2. The HOBP-OXD exhibited capability of electron transport with emission. Compound **2a** also showed capability as a carrier transport layer in the SmB phase as the EL device [82]. A simple structure was used, which was composed of two glass substrates with a homogeneous alignment. **2a** and a certain amount of 3-(2-benzothiazolyl)-7-diethylamino-coumarin were sandwiched between the substrates, and the luminance of a green light was 0.7 cd/m^2. The coumarin dye is the recombination center for the injected carriers, and there would be no emission if the dye was removed from the LC host.

If the temperature range of the LC phase is well above room temperature, the device has to be carried out at high temperature, which is not suitable for practical applications. Therefore, the design and development of a molecular system with a thermally stable LC mesophase at room temperature should be considered. It is also essential to develop red, green and blue light-emitting systems to satisfy modern color displays. Such an application has not been fulfilled with smectic LCs as yet. However, a series of full-color electroluminescent devices have been achieved based on the calamitic nematic LC at room temperature [83]. The normalized EL spectra of red, green, and blue light and a prototype OLED device are shown in Figure 5-23. This demonstrates the possibility of a full-color, large-area, and high-information-content OLED device using LC materials.

5.5.5. Other Developments and Flexible Applications

Photoconduction in smectic LCs has other applications such as a one-dimensional position sensor and a fast-speed photosensor. The fluid property of LCs makes them easy to be drawn into a glass cell by capillary action at temperatures above their clearing point. A wedge cell filled with a smectic LC **2a** was used to demonstrate a one-dimensional position sensor (Fig. 5-24). As previously mentioned, the charge-carrier mobility is independent of the applied field, so the transit time is only decided by the transporting distance in this case. The relationship between the transit time and the cell gap is linear. The position, that is, the cell thickness, can be deduced by measurement of the transit time [84]. For a high-speed photosensor, a cell filled with **2a** and C$_{70}$ showed a fast photoresponse in the

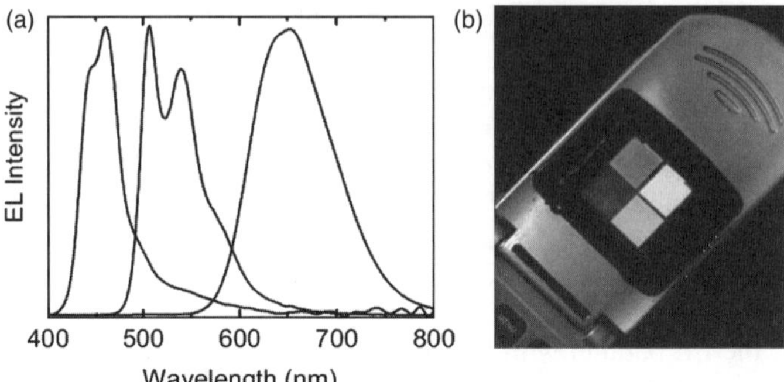

Figure 5-23 (a) Normalized EL spectra of red, green and blue light and (b) a prototype of an OLED device. From Ref. 83 (M. P. Aldred et al. *Adv. Mater*. **2005**, *17*, 1368–1371. Copyright Wiley-VCH Verlag GmbH & Co. KGaA. Reproduced with permission).

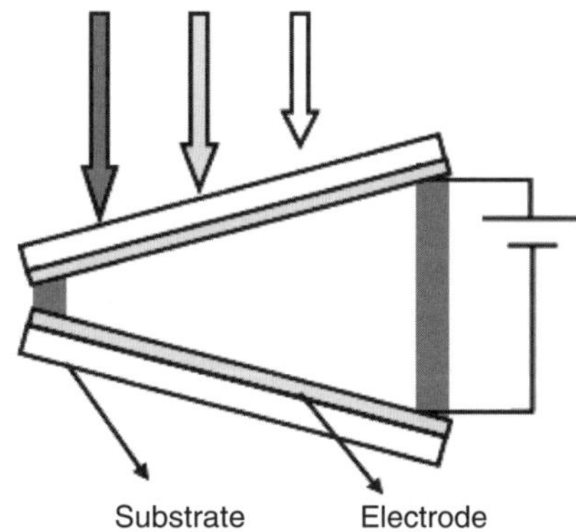

Figure 5-24 One-dimensional position sensor in a wedge cell.

microsecond range [85]. Here the function of the C_{70} was to create the spectra in the visible region (400–700 nm). Once irradiated by a light, the photocurrent response would be shown in the TOF setup. The current rise and decay rate were not influenced by the repeating times. This device suggested that the photosensor could be created by utilizing the photocurrent effect of the semiconducting LCs.

Another important application trend is to develop practical and flexible manufacturing processes using semiconducting LCs. The LC properties are well-suited for this technology since LCs are solution processable and fluid and it is easy to get uniform alignment even on large-scale substrate. The ultimate technology can be manufactured by a solution-deposition roll-to-roll flexible process, as illustrated in Figure 5-25. In this process, the LC can be wet processed and deposited on the flexible substrate to form various semiconducting layers. The wet process may be soaking, dropping or spin-coating. Some successful examples

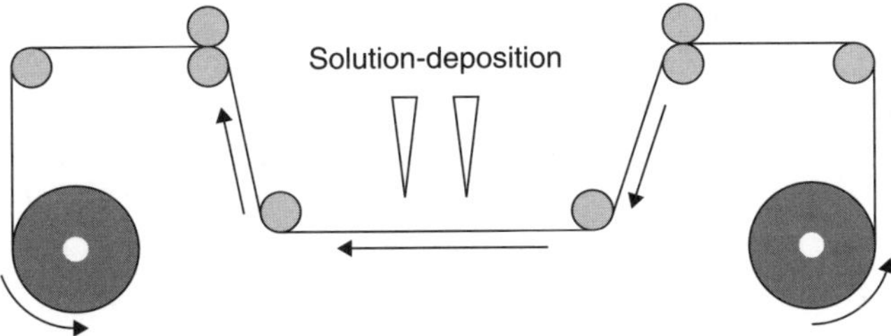

Figure 5-25   Schematic illustration of a solution-deposition roll-to-roll process to manufacture flexible semiconducting layers.

have been achieved, such as flexible FET devices on a polymer substrate by the spin-coating of the compound **12** [72]. This technology will be very economical and easy to manufacture in the future.

5.6. CONCLUSION AND OUTLOOK

In this chapter, the materials, characterization techniques, charge transport properties, device fabrications, and potential applications of self-organized semiconducting smectic LCs are summarized and discussed. Smectic LCs have significant advantages as semiconducting materials. (a) The charge-carrier mobility of 10^{-4}–10^{-3} and even $1\,cm^2$/Vs is higher than that of 10^{-6}–$10^{-5}\,cm^2$/Vs in the amorphous organic materials. (b) For most smectic materials, the property of mobility being independent of or less dependent on applied field and temperature increases the stability of the material performance, whereas for the organic amorphous materials, the mobility would be changed by applied field and temperature. (c) The dynamic self-organized properties of LCs allow them to self-repair defects such as disclinations and domain boundaries in the bulk. (d) They are easy to deposit onto a substrate and easy to align on that substrate. These characteristics allow them to be fabricated as large-area flexible devices with a wet process. (e) The various molecular arrangements of smectic LCs and various molecular systems provide more choices and opportunities to design and develop new molecular materials, devices, and applications.

More and more molecular systems and potential applications based on the LC mesophases will arise. However, this chapter cannot cover each aspect of smectic LC materials and their device applications. There is still much work to do in the future, such as design of new molecular structures with high charge-carrier mobilities, wide and appropriate mesophase temperature ranges, ambient, thermal, and optoelectric stabilities, and design and development of novel devices with practical applications. These technologies will involve much work in the area of chemical synthesis, physical characterization, electronic design, and manufacturing process. We believe that semiconducting devices and applications using LCs will be soon a reality.

ACKNOWLEDGMENTS

The preparation of this chapter benefited from the support to Quan Li by the Ohio Board of Regents under its Research Challenge program, the Air Force Office of Scientific Research (AFOSR FA9550-09-1-0193 and FA9550-09-1-0254), the Department of Energy (DOE DE-SC0001412), and the National Science Foundation (NSF IIP 0750379).

REFERENCES

1. P. G. de Gennes and J. Prost. *The Physics of Liquid Crystals*. New York, Oxford University Press, **1993**.

2. L. M. Blinov and V. G. Chigrinov. *Electrooptical Effects in Liquid Crystal materials*. New York, Springer, **1994**.

3. D. K. Yang and S. T. Wu. *Fundamentals of Liquid Crystal Devices*. New York, Wiley, **2006**.

4. I. Dierking. *Textures of Liquid Crystals*. Weinheim, Wiley-VCH, **2003**.

5. G. H. Heilmeier and P. M. Heyman. Note on transient current measurements in liquid crystals and related systems. *Phys. Rev. Lett*. **1967**, *18*, 583–585.

6. S. Kusabayashi and M. M. Labes. Conductivity in liquid crystals. *Mol. Cryst. Liq. Cryst*. **1969**, *7*, 395–405.

7. G. Derfel and A. Lipinski. Charge carrier mobility measurements in nematic liquid crystals. *Mol. Cryst. Liq. Cryst*. **1979**, *55*, 89–100.

8. D. Adam, F. Closs, T. Frey, D. Funhoff, D. Haarer, H. Ringsdorf, P. Schuhmacher and K. Siemensmeyer. Transient photoconductivity in a discotic liquid crystal. *Phys. Rev. Lett*. **1993**, *70*, 457–460.

9. M. Funahashi, and J. Hanna. Fast hole transport in a new calamitic liquid crystal of 2-(4-heptyloxyphenyl)-6-dodecylthiobenzothiazole. *Phys. Rev. Lett*. **1997**, *78*, 2184–2187.

10. S. Chandrasekhar. *Liquid Crystals*. New York, Cambridge University Press, **1992**.

11. D. Demus, J. W. Goodby, G. W. Grey, H. W. Spiess, and V. Vill, Ed. *Handbook of Liquid Crystals*. Weinheim, Wiley-VCH, **1998**.

12. L. Vicari, Ed. *Optical Applications of Liquid Crystals*. Bristol and Philadelphia, Institute of Physics Publishing, **2003**.

13. G. W. Gray, K. J. Harrison, and J. A. Nash. New family of nematic liquid crystals for displays. *Electron Lett*. **1973**, *9*, 130–131.

14. S. Singh and D. A. Dunmur. *Liquid Crystals Fundamentals*. Singapore, World Scientific, **2002**.

15. J. P. F. Lagerwall and F. Giesselmann. Current topics in smectic liquid crystal research. *ChemPhysChem* **2006**, *7*, 20–45.

16. G. W. Gray and J. W. Goodby. *Smectic Liquid Crystals: Textures and Structures*. Philadelphia, Heyden and Son Inc, **1984**.

17. Y. Takayashiki, H. Iino, T. Shimakawa, and J. Hanna. Ambipolar carrier transport in terphenyl derivative. *Mol. Cryst. Liq. Cryst*. **2008**, *480*, 295–301.

18. Q. Li and L. Li. Photoconducting discotic liquid crystals. In: *Thermotropic Liquid Crystals: Recent Advances*, A. Ramamoorthy, Ed., Dordrecht, Springer, **2007**.

19. P. G. Schouten, J. M. Warman, and M. P. de Haas. Effect of accumulated radiation dose on pulse radiolysis conductivity transients in a mesomorphic octa-n-alkoxy-substituted phthalocyanine. *J. Phys. Chem.* **1993**, *97,* 9863–9870.

20. A. M. van de Craats, J. M. Warman, M. P. de Haas, D. Adam, J. Simmerer, D. Haarer, and P. Schuhmacher. The mobility of charge carriers in all four phases of the columnar discotic material hexakis (hexylthio) triphenylene: combined TOF and PR-TRMC results. *Adv. Mater.* **1996**, *8,* 823–826.

21. A. M. van de Craats, J. M. Warman, A. Fechtenkoter, J. D. Brand, M. A. Harbison, and K. Mullen. Record charge carrier mobility in a room-temperature discotic liquid-crystalline derivative of hexabenzocoronene. *Adv. Mater.* **1999**, *11,* 1469–1472.

22. A. M. van de Craats, J. M. Warman, K. Mullen, Y. Geerts, and J. D. Brand. Rapid charge transport along self-assembling graphitic nanowires. *Adv. Mater.* **1998**, *10,* 36–38.

23. J. M. Warman, G. H. Gelinck, and M. P. Haas. The mobility and relaxation kinetics of charge carriers in molecular materials studied by means of pulse-radiolysis time-resolved microwave conductivity: dialkoxy-substituted phenylene-vinylene polymers. *J. Phys.: Condens Matter* **2002**, *14,* 9935–9954.

24. Z. An, J. Yu, S. C. Jones, S. Barlow, S. Yoo, B. Domercq, P. Prins, L. D. A. Siebbeles, B. Kippelen, and S. R. Marder. High electron mobility in room-temperature discotic liquid-crystalline perylene diimides. *Adv. Mater.* **2005**, *17,* 2580–2583.

25. M. A. Lampert and P. Mark. *Current Injection in Solids*. New York, Academic Press, **1970**.

26. Z. Bao and J. Locklin, Ed. *Organic Field-Effect Transistors*. Boca Raton, CRC Press, **2007**.

27. K. Oikawa, H. Monobe, K. Nakayama, T. Kimoto, K. Tsuchiya, B. Heinrich, D. Guillon, Y. Shimizu, and M. Yokoyama. High carrier mobility of organic field-effect transistors with a thiophenenaphthalene mesomorphic semiconductor. *Adv. Mater.* **2007**, *19,* 1864–1868.

28. S. Tiwari and N. C. Greenham. Charge mobility measurement techniques in organic semiconductors. *Opt. Quant. Electron.* **2009**, *41,* 69–89.

29. K. Tokunaga, H. Iino, and J. Hanna. Charge carrier transport properties in liquid crystalline 2-phenylbenzothiazole derivatives. *Mol. Cryst. Liq. Cryst.* **2009**, *510,* 241–249.

30. J. Hanna. Towards a new horizon of optoelectronic devices with liquid crystals. *Opto-Electron. Rev.* **2005**, 13, 259–267.

31. M. Funahashi and J. Hanna. Fast ambipolar carrier transport in smectic phases of phenylnaphthalene liquid crystal. *Appl. Phys. Lett.* **1997**, *71,* 602–604.

32. M. Funahashi and J. Hanna. Anomalous high carrier mobility in smectic E phase of a 2-phenylnaphthalene derivative. *Appl. Phys. Lett.* **1998**, *73,* 3733–3735.

33. H. Maeda, M. Funahashi, and J. Hanna. Electrical properties of domain boundary in photoconductive smectic mesophases and their crystal phases. *Mol. Cryst. Liq. Cryst.* **2001**, *366,* 369–376.

34. H. Tokuhisa, M. Era, and T. Tsutsui. Novel liquid crystalline oxadiazole with high electron mobility. *Adv. Mater.* **1998**, *10,* 404–406.

35. M. Funahashi and J. Hanna. High ambipolar carrier mobility in self-organizing terthiophene derivative. *Appl. Phys. Lett.* **2000**, *76*, 2574–2576.

36. M. Funahashi and J. Hanna. Probe of molecular ordering in photoconductive smectic mesophases by transient photocurrent measurement. *Mol. Cryst. Liq. Cryst.* **2001**, *368*, 303–310.

37. H. Iino and J. Hanna. Availability of liquid crystalline molecules for polycrystalline organic semiconductor thin films. *Jpn. J. Appl. Phys.* **2006**, *45*, L867–L870.

38. M. Funahashi and J. Hanna. Mesomorphic behaviors and charge carrier transport in terthiophene derivatives. *Mol. Cryst. Liq. Cryst.* **2004**, *410*, 529–540.

39. M. Funahashi and J. Hanna. High carrier mobility up to 0.1 $cm^2V^{-1}s^{-1}$ at ambient temperatures in thiophene-based smectic liquid crystals. *Adv. Mater.* **2005**, *17*, 594–598.

40. H. Iino, J. Hanna, and M. Yokohama. TOF and TFT mobilities in polycrystalline thin films of liquid crystalline material. *Proceedings of SPIE* **2008**, 70540Y.

41. K. Oikawa, H. Monobe, J. Takahashi, K. Tsuchiya, B. Heinrich, D. Guillon and Y. Shimizu. A novel calamitic mesophase semiconductor with the fastest mobility of charged carriers: 1,4-di(5-octyl-2-thienyl) benzene. *Chem. Commun.* **2005**, 5337–5339.

42. M. Mushrush, A. Facchetti, M. Lefenfeld, H. E. Katz, and T. J. Marks. Easily processable phenylene-thiophene-based organic field-effect transistors and solution-fabricated nonvolatile transistor memory elements. *J. Am. Chem. Soc.* **2003**, *125*, 9414–9423.

43. M. Funahashi, F. Zhang, and N. Tamaoki. High ambipolar mobility in a highly ordered smectic phase of a dialkylphenylterthiophene derivative that can be applied to solution-processed organic field-effect transistors. *Adv. Mater.* **2007**, *19*, 353–358.

44. F. Zhang, M. Funahashi, and N. Tamaoki. High-performance thin film transistors from semiconducting liquid crystalline phases by solution processes. *Appl. Phys. Lett.* **2007**, *91*, 063515.

45. F. Zhang, M. Funahashi, and N. Tamaoki. Thin-film transistors based on liquid-crystalline tetrafluorophenylter thiophene derivatives: thin-film structure and carrier transport. *Org. Electron.* **2009**, *10*, 73–84.

46. A. J. J. M. van Breemen, P. T. Herwig, C. H. T. Chlon, J. Sweelssen, H. F. M. Schoo, S. Setayesh, W. M. Hardeman, C. A. Martin, D. M. de Leeuw, J. J. P. Valeton, C. W. M. Bastiaansen, D. J. Broer, A. R. Popa-Merticaru, and S. C. J. Meskers. Large area liquid crystal monodomain field-effect transistors. *J. Am. Chem. Soc.* **2006**, *128*, 2336–2345.

47. A. Facchetti, Y. Deng, A. Wang, Y. Koide, H. Sirringhaus, T. J. Marks, and R. H. Friend. Tuning the semiconducting properties of sexithiophene by alpha, omega-substitution-alpha, omega-diperfluorohexylsexithiophene: the first n-type sexithiophene for thin-film transistors. *Angew. Chem. Int. Ed.* **2000**, *39*, 4547–4551.

48. A. Facchetti, M. Mushrush, H. E. Katz, and T. J. Marks. n-Type building blocks for organic electronics: a homologous family of fluorocarbon-substituted thiophene oligomers with high carrier mobility. *Adv. Mater.* **2003**, *15*, 33–38.

49. T. Yasuda, H. Ooi, J. Morita, Y. Akama, K. Minoura, M. Funahashi, T. Shimomura, and T. Kato. π-Conjugated oligothiophene-based polycatenar liquid crystals: self-organization and photoconductive, luminescent, and redox properties. *Adv. Funct. Mater.* **2009**, *19*, 411–419.

50. I. Shiyanovskaya, K. D. Singer, R. J. Twieg, L. Sukhomlinova, and V. Gettwert. Electronic transport in smectic liquid crystals. *Phys. Rev. E* **2002**, *65*, 41715.

51. Y. Haramoto, Y. Kawada, A. Mochizuki, M. Nanasawa, S. Ujiie, M. Funahashi, K. Hiroshima, and T. Kato. New possibility for an organic semiconductor: a smectic liquid crystalline semiconductor having a long conjugated core and two long alkyl chains. *Liq. Cryst.* **2005**, *32*, 909–912.

52. H. Meng, F. Sun, M. B. Goldfinger, F. Gao, D. J. Londono, W. J. Marshal, G. S. Blackman, K. D. Dobbs, and D. E. Keys. 2,6-Bis [2-(4-pentylphenyl) vinyl] anthracene: a stable and high charge mobility organic semiconductor with densely packed crystal structure. *J. Am. Chem. Soc.* **2006**, *128*, 9304–9305.

53. H. Meng, F. Sun, M. B. Goldfinger, G. D. Jaycox, Z. Li, W. J. Marshall and G. S. Blackman. High-performance, stable organic thin-film field-effect transistors based on bis-5-alkylthiophen-2-yl-2, 6-anthracene semiconductors. *J. Am. Chem. Soc.* **2005**, *127*, 2406–2407.

54. S. A. Ponomarenko, S. Kirchmeyer, A. Elschner, N. M. Alpatova, M. Halik, H. Klauk, U. Zschieschang, and G. Schmids. Decyl-end-capped thiophene-phenylene oligomers as organic semiconducting materials with improved oxidation stability. *Chem. Mater.* **2006**, *18*, 579–586.

55. M. Melucci, L. Favaretto, C. Bettini, M. Gazzano, N. Camaioni, P. Maccagnani, P. Ostoja, M. Monari, and G. Barbarella. Liquid-crystalline rigid-core semiconductor oligothiophenes: influence of molecular structure on phase behaviour and thin-film properties. *Chem. Eur. J*. **2007**, *13*, 10046–10054.

56. S. Mery, D. Haristoy, J. F. Nicoud, D. Guillon, S. Diele, H. Monobe, and Y. Shimizu. Bipolar carrier transport in a lamello-columnar mesophase of a sanidic liquid crystal. *J. Mater. Chem.* **2002**, *12*, 37–41.

57. P. Vlachos, S. M. Kelly, B. Mansoor, and M. O'Neill. Electron-transporting and photopolymerisable liquid crystals. *Chem. Commun.* **2002**, 874–875.

58. I. McCulloch, M. Heeney, C. Bailey, K. Genevicius, I. MacDonald, M. Shkunov, D. Sparrowe, S. Tierney, R. Wagner, W. Zhang, M. L. Chabinyc, R. J. Kline, M. D. McGehee, and M. F. Toney. Liquid-crystalline semiconducting polymers with high charge-carrier mobility. *Nature Mater.* **2006**, *5*, 328–333.

59. K. L. Woon, M. P. Aldred, P. Vlachos, G. H. Mehl, T. Stirner, S. M. Kelly, and M. O'Neill. Electronic charge transport in extended nematic liquid crystals. *Chem. Mater.* **2006**, *18*, 2311–2317.

60. M. Funahashi and N. Tamaoki. Electronic conduction in the chiral nematic phase of an oligothiophene derivative. *Chem Phys Chem* **2006**, *7*, 1193–1197.

61. S. R. Farrar, A. E. A. Contoret, M. O'Neill, J. E. Nicholls, G. J. Richards, and S. M. Kelly. Nondispersive hole transport of liquid crystalline glasses and a cross-linked network for organic electroluminescence. *Phys. Rev. B* **2002**, *66*, 125107.

62. R. J. Bushby and O. R. Lozman. Photoconducting liquid crystals. *Curr. Opin. Solid State Mater. Sci.* **2002**, *6*, 569–578.

63. H. Iino and J. Hanna. Ambipolar charge carrier transport in liquid crystals. *Opto-Electron. Rev.* **2005**, *13*, 295–302.

64. J. Hanna. Ten years after the discovery of electronic conduction in liquid crystals. *Proceedings of SPIE*, Warsaw, **2005**, 594703.

65. J. Hanna. Photoconductive properties of smectic liquid crystalline photoconductors. *Proceedings of SPIE*, Seattle, **2003**, 99–113.

66. N. Yoshimoto and J. Hanna. Preparation of a novel organic semiconductor composite consisting of a liquid crystalline semiconductor and crosslinked polymer and characterization of its charge carrier transport properties. *J. Mater. Chem.* **2003**, *13*, 1004–1010.

67. T. Kreouzis, R. J. Baldwin, M. Shkunov, I. McCulloch, M. Heeney, and W. Zhang. High mobility ambipolar charge transport in a cross-linked reactive mesogen at room temperature. *Appl. Phys. Lett.* **2005**, *87*, 172110.

68. I. McCulloch, W. Zhang, M. Heeney, C. Bailey, M. Giles, D. Graham, M. Shkunov, D. Sparrowe, and S. Tierney. Polymerisable liquid crystalline organic semiconductors and their fabrication in organic field effect transistors. *J. Mater. Chem.* **2003**, *13*, 2436–2444.

69. W. Pisula, M. Zorn, J. Y. Chang, K. Mullen, and R. Zentel. Liquid crystalline ordering and charge transport in semiconducting materials. *Macromol. Rapid Commun.* **2009**, *30*, 1179–1202.

70. A. E. A. Contoret, S. R. Farrar, P. O. Jackson, S. M. Khan, L. May, M. O'Neill, J. E. Nicholls, S. M. Kelly, and G. J. Richards. Polarized electroluminescence from an anisotropic nematic network on a non-contact photoalignment layer. *Adv. Mater.* **2000**, *12*, 971–974.

71. J. Hanna. Liquid crystals as a self-organizing molecular semiconductors. *Proceedings of SPIE*, Seattle, **2002**, 136–147.

72. M. Funahashi. Development of liquid-crystalline semiconductors with high carrier mobilities and their application to thin-film transistors. *Polymer J.* **2009**, *41*, 459–469.

73. Y. Shimizu, K. Oikawa, K. Nakayama, and D. Guillon. Mesophase semiconductors in field effect transistors. *J. Mater. Chem.* **2007**, *17*, 4223–4229.

74. I. Kymissis. *Organic Field Effect Transistors: Theory, Fabrication and Characterization*. New York, Springer, **2009**.

75. C. Kagan and P. Andry, Ed. *Thin-Film Transistors*. New York, Marcel Dekker, **2003**.

76. M. O'Neill and S. M. Kelly. Liquid crystals for charge transport, luminescence, and photonics. *Adv. Mater.* **2003**, *15*, 1135–1146.

77. K. Mullen and U. Scherf, Ed. *Organic Light Emitting Devices: Synthesis, Properties and Applications*. Weinheim, Weily-VCH, **2006**.

78. E. Schubert. *Light-Emitting Diodes*. New York, Cambridge University Press; **2006**.

79. S. M. Kelly. *Flat Panel Displays: Advanced Organic Materials*. In: RSC Materials Monographs, Series Ed., J. A. Connor, **2000**.

80. G. Held. *Introduction to Light Emitting Diode Technology and Applications*. Boca Raton, CRC Press, **2008**.

81. H. Tokuhisa, M. Era, and T. Tsutsui. Polarized electroluminescence from smectic mesophase. *Appl. Phys. Lett.* **1998**, *72*, 2639–2641.

82. K. Kogo, T. Goda, M. Funahashi, and J. Hanna. Polarized light emission from a calamitic liquid crystalline semiconductor doped with dyes. *Appl. Phys. Lett.* **1998**, *73*, 1595–1597.

83. M. P. Aldred, A. E. A. Contoret, S. R. Farrar, S. M. Kelly, D. Mathieson, M. O'Neill, W. C. Tsoi, and P. Vlachos. A full-color electroluminescent device and patterned photoalignment using light-emitting liquid crystals. *Adv. Mater.* **2005**, *17*, 1368–1371.

84. H. Maeda, M. Funahashi, and J. Hanna. New aspects of organic electric materials in calamitic liquid crystalline photoconductors. *Mater. Res. Soc. Symp. Proc*. **2000**, BB3.61.

85. M. Funahashi and J. Hanna. Microsecond photoresponse in liquid crystalline photoconductor doped with C_{70} under illumination of visible light. *Appl. Phys. Lett*. **1999**, *74*, 2584–2586.

CHAPTER 6

Self-Assembling of Carbon Nanotubes

LIMING DAI

Department of Chemical Engineering, Case School of Engineering, Case Western Reserve University, Cleveland, Ohio

6.1. INTRODUCTION

Carbon nanotubes (CNTs) can be classified into single-walled carbon nanotubes (SWNTs) and multiwalled carbon nanotubes (MWNTs) [1, 2]. While a SWNT may be conceptually viewed as a graphene sheet that is rolled into a nanoscale tube form, a MWNT consists of additional graphene coaxial tube(s) around the SWNT core. Depending on their diameter and the helicity of the orientation of graphene rings along the nanotube length, CNTs can exhibit semiconducting or metallic behavior [1, 2]. These interesting electronic properties, coupled with their unusual molecular symmetries, have made CNTs very attractive for many potential applications, including as new materials for electron field emitters in panel displays [3], single-molecular transistors [4], scanning probe microscope tips [5, 6], gas and electrochemical energy storage [7, 8], catalyst supports [9–12], molecular-filtration membranes [13, 14], artificial muscles [15], sensors [16, 17], energy storage and energy conversion devices [18, 19], and even dry adhesives mimicking gecko foot [20, 21]. The use of carbon nanotubes for most of these, and many other applications, often requires a precision loading of individual nanotubes into various functional structures on appropriate substrates. In this regard, the synthesis of aligned and/or micropatterned carbon nanotubes has played a critical role in facilitating the integration of carbon nanotubes into useful devices. However, a more general self-assembling approach would allow the construction of individual nanotubes into various functional structures with a molecular precision.

By definition, self-assembly is a thermodynamics process, in which small structural units are spontaneously associated into multidimensional architectures under a defined condition [22]. The resultant self-assemblies are characterized both by the spatial arrangement of their components and by the nature of the interactions that hold these components together. Various forces, including van der Waals, charge-transfer (e.g., π-π stacking), hydrogen bonding, electrostatic, and covalent forces, have been demonstrated to govern the interaction/bonding associated with the self-assembling process [22]. With recent developments in nanomaterials and nanotechnology [2], the self-assembling of CNTs to create functional structures and/or materials has attracted a great deal of interest. For this purpose, surface functionalization of CNTs in a controllable fashion is essential. In this article, I summarize our recent studies on the self-assembling of CNTs by controlled growth and solution functionalization, although reference is also made to other complementary work as appropriate. Topics discussed include the self-assembling of CNTs with CNTs or other nanomaterials (e.g., nanoparticles) via van der Waals forces, specific chemical interactions, charge-transfer interactions, DNA pairing, and asymmetric functionalization, along with the potential applications for the resultant self-assemblies. Since focus is given to our own work to demonstrate some rational concepts for self-assembling of CNTs into potentially useful structures, with no intention for a comprehensive literature survey of the subject, there will be no doubt that the examples to be presented in this article do not exhaust all significant work reported in the literature. Therefore, I apologize in advance to the authors of papers not cited here.

6.2. SELF-ASSEMBLING OF CNTs BY VAN DER WAALS FORCES

van der Waals (VdW) forces arising from correlations in the fluctuating polarizations (dipoles) of nearby atoms, molecules, particles, and surfaces create attractions between them, albeit much weaker than the covalent bonds and/or ionic forces [22]. The exact values of VdW forces vary considerably with the nature and size of the entities involved and their shapes. The large surface-to-volume ratio intrinsically associated with most nanomaterials ensures strong VdW forces for their interactions, especially when they are intimately contacted. A notorious example of the self-assembling of CNTs induced by VdW forces is the bundling of SWNTs, which makes the nanotube insoluble and unprocessable [2]. On the other hand, the same VdW forces can facilitate the growth of CNTs into vertically aligned arrays in either a patterned or a nonpatterned form [2]. Unlike the SWNT bundles, the vertically aligned carbon nanotube (VA-CNT) arrays self-assembled by VdW forces have been demonstrated to show many advantages for various applications, particularly in nanotube devices, as VA-CNTs can provide well-defined large surface/interface areas and be easily incorporated into device configurations [2, 23, 24].

CNTs synthesized by conventional techniques usually exist in a randomly entangled form [2, 23, 24]. However, several chemical vapor deposition (CVD) methods have been developed for large-scale production of self-assembled, vertically aligned single-walled, multiwalled, and super-long CNT arrays [8, 25–28].

The resultant aligned CNT arrays can be transferred onto various substrates of particular interest in either a patterned or a nonpatterned fashion. The well-aligned structure provides additional advantages for controlled surface modification, as we see below.

Among the CVD methods developed for the growth of VA-CNT self-assembled arrays [23, 24], the formation of VA-MWNT arrays from organic-metal complexes [29], containing both the metal catalyst and carbon source required for the nanotube growth, is of particular interest as it is a one-step process involving no pre-preparation of catalyst nanoparticles on the substrate used for the nanotube growth. In this context, we have prepared large-scale aligned CNTs perpendicular to the substrate surface by pyrolysis of iron (II) phthalocyanine, $FeC_{32}N_8H_{16}$ (designated as FePc hereafter), in an Ar/H_2 atmosphere in a dual furnace fitted with independent temperature controllers [26].

The *as-synthesized* nanotubes align almost normal to the substrate surface, and the constituent CNTs held together by VdW forces have a fairly uniform tubular length and diameter with a well-graphitized multiwall structure [30]. Subsequently, we have developed microfabrication methods for patterning the VA-MWNTs with a submicrometer resolution and for patterned/nonpatterned transferring such nanotube self-assemblies to various other substrates of particular interest (e.g., polymer films for organic optoelectronic devices or metal substrates for electrochemistry) [23, 24]. Along with other researchers in the field [31–36], we have also synthesized vertically aligned/micropatterned SWNT and super-long ($\sim$6 mm) VA CNT self-assemblies on flat substrates and even on carbon fibers (CFs) [25–27, 37–40], as exemplified in Figure 6-1.

6.3. SELF-ASSEMBLING OF CNTs BY SPECIFIC CHEMICAL INTERACTIONS

In our further investigation on the VA-CNT growth, we found that VA-SWNTs can be synthesized on an Al-activated iron substrate (e.g., an Fe-coated SiO_2/Si wafer) while VA-MWNTs grew on the Al-free iron substrate by plasma-enhanced CVD (PE-CVD) of C_2H_2 at $750°C$ [25]. On this basis, we proceeded to prepare multicomponent micropatterns with the Al-activated VA-SWNTs interposed within the patterned VA-MWNT arrays on the Fe-precoated SiO_2/Si substrate. Figure 6-2A(a) shows the steps for patterned deposition of an Al thin coating through a TEM grid (as the physical mask) onto a SiO_2/Si wafer precoated with a 3-nm Fe film to support the region-selective growth of VA-SWNTs over the Al covered areas and VA-MWNTs in the Al-free regions. Figure 6-2A(b) shows a top view of the resultant nanotube micropattern, in which the hexagonal windows of the TEM grid "mask" were replicated, while the side view in Figure 6-2A(c) clearly shows the shorter VA-SWNTs interposed within the hexagonal areas surrounded by longer VA-MWNTs. The constituent CNTs in both areas are self-assembled into vertically aligned arrays by VdW forces. The observed length difference between the region-selectively grown VA-SWNTs and VA-MWNTs resulted mainly from their different growth rates and catalyst lifetimes.

Figure 6-1   (A) Typical SEM micrographs of patterned films of aligned nanotubes prepared by the pyrolysis of FePc onto a photolithographically prepatterned quartz substrate (adapted from Ref. 37). (B) SEM images (a, b), Raman spectra (c), and TEM image (d) of VA-SWNTs synthesized on a SiO_2/Si wafer. The dashed rectangles in (c) define the approximate regions for metallic (M) and semiconducting (S) Raman features of SWNTs. Inset of (d) shows a higher-magnification TEM image of an individual nanotube (adapted from Ref. 25). (C) (a) Digital photograph, (b) SEM image, and (c) TEM image of the super-long double-walled CNTs (adapted from Ref. 8). (D) Cross-sectional and side-view SEM images of the CNT-CFs. Inset at the bottom right corner shows the amperometric responses to successive additions of 3 mM glucose for the glucose-oxidase-attached aligned CNT-CF sensor, compared to a corresponding CF electrode (adapted from Ref. 40). (A full color version of this figure appears in the color plate section.)

The single-walled and multiwalled nanotube characteristics of those shorter VA-SWNTs and longer VA-MWNTs within the three-dimensional (3-D) multi-component nanotube micropatterns shown in Figure 6-2A(b and c) were confirmed by area-selected micro-Raman spectra (spot size ≈ 2 μm). The multi-component 3-D VA-MWNT/VA-SWNT interposed micropatterns could serve as

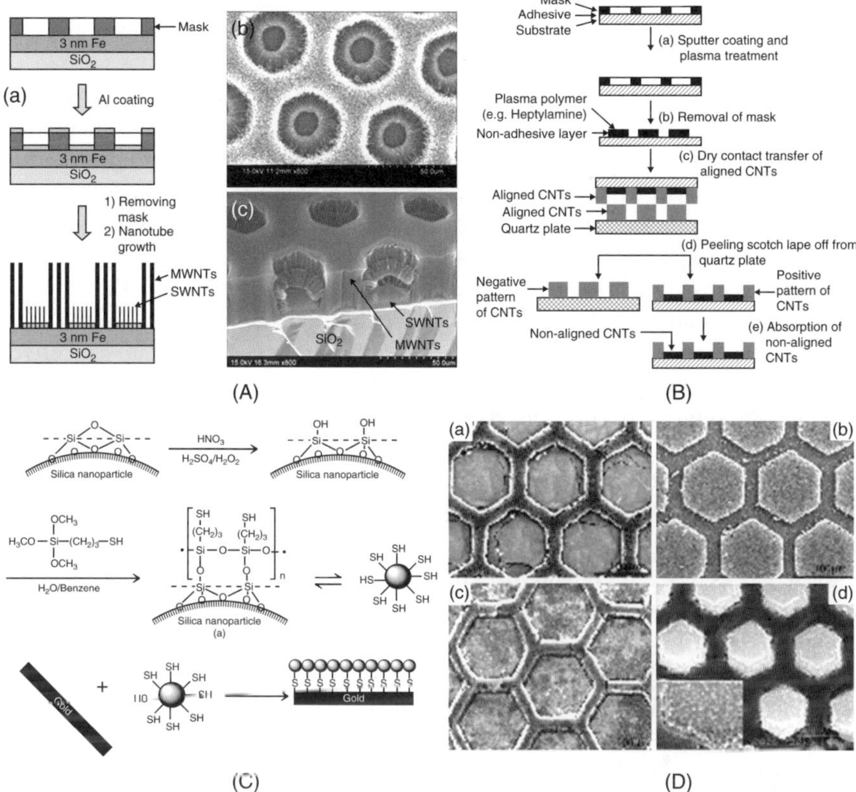

Figure 6-2   (A) (a) Schematic representation of Al patterning on a Fe-coated substrate for the patterned growth of three-dimensional (3-D), interposed VA-SWNTs and VA-MWNTs. (b, c) Top and side view of the 3-D VA-MWNT/VA-SWNT micropatterns (adapted from Ref. 25). (B) Schematic illustration of the procedures for fabricating multicomponent interposed carbon nanotube micropatterns by dry contact transfer, followed by region-specific self-assembling (adapted from Ref. 41). (C) Schematic representation for the surface modification reaction of silica nanoparticles with 3-mercaptopropyltrimethoxysilane and the subsequent self-assembling of the thiol-modified silica nanoparticles onto a gold surface (adapted from Ref. 42). (D) SEM images of (a) VA-CNT patterns after being transferred onto the Scotch tape (i.e., positive pattern), (b) VA-CNT patterns left on the quartz plate after the dry contact transfer (i.e., negative pattern), (c) the multicomponent interposed carbon nanotube micropatterns with the nonaligned carbon nanotubes region-specifically self-assembled between aligned carbon nanotube patterns within hexagonal regions (adapted from Ref. 41), (d) the multicomponent interposed carbon nanotube micropatterns with the thiol-modified silica nanoparticles region-specifically self-assembled between aligned carbon nanotube patterns within hexagonal regions. Inset shows a higher magnification (~6×) image of (d) (adapted from Ref. 42).

useful building blocks for fabricating multifunctional micro-/nanoelectronics with region-specific features, while the conducting substrate provides a direct electrical contact.

Multicomponent micropatterns of VA-MWNTs interposed with self-assembled nonaligned CNTs or nanoparticles have also been produced through contact transfer by simply pressing an adhesively sticky tape (e.g., Scotch tape) prepatterned with a nonadhesive layer onto a nonpatterned aligned CNT film, followed by peeling off the sticky tape from the quartz substrate in a dry state and subsequent region-specific self-assembling (Fig. 6-2(B–D)) [41, 42]. Figure 6-2B shows the steps used for micropatterning the Scotch tape with a thin layer of nonadhesive coating (e.g., sputter-coated gold), the subsequent plasma treatment of the nanoadhesive coating surface, the contact transfer to selectively transfer the nanotubes underneath the nonadhesive-free regions onto the Scotch tape, and the region-specific self-assembling of nonaligned acid-oxidized CNT or thiol-functionalized nanoparticles onto the nanotube-free gold-coated areas with or without the heptylamine-plasma treatment [41, 42]. Figure 6-2C shows the chemistry for modification of silica nanoparticles with surface thiol groups for self-assembling onto a gold substrate. As can be seen in Figure 6-2D, the integrity of those CNTs (e.g., alignment, packing density) transferred onto the Scotch tape (Figure 6-2D(a)) is almost the same as the *as-grown* nanotubes remaining on the quartz plate (Figure 6-2D(b)). While Figure 6-2D(c) shows a typical SEM image for the multicomponent micropattern of VA-MWNTs interposed with self-assembled nonaligned CNTs through region-specific interactions between-COOH groups of the acid-oxidized CNTs and the heptylamine-plasma-induced -NH$_2$ groups, the corresponding SEM image for the multicomponent micropattern of VA-MWNTs interposed with self-assembled thiol-modified silica nanoparticles via region-specific thiol-gold interactions is given in Figure 6-2D(d). SEM images in Figure 6-2D(c and d) show that both the contact-transferred aligned CNTs and the self-assembled nonaligned CNTs or silica nanoparticles are well registered in their respective areas. This, together with the soft nature of the Scotch tape used for the dry contact transfer, could facilitate the development of multicomponent, highly integrated, multifunctional nanotube optoelectronics even on flexible substrates.

Instead of the CVD growth, VA-CNTs have also been constructed from end-functionalized CNTs via solution self-assembling. For instance, Liu and coworkers [43, 44] have prepared VA-SWNTs by self-assembling of nanotubes end-functionalized with thiol groups on a gold substrate. As seen in Figure 6-3A, the carboxyl-terminated short SWNTs prepared by acid oxidation were used as the starting material for further functionalization with thiol-containing alkyl amines through the amide linkage. Then, the self-assembly of aligned SWNTs was obtained by dipping a gold (111) ball into the thiol-functionalized SWNT suspension in ethanol, followed by ultrasonication and drying in high-purity nitrogen. The resulting self-assembled aligned nanotube film is so stable that ultrasonication cannot remove it from the gold substrate. The packing density of the self-assembled aligned CNTs was found to strongly depend on the incubation

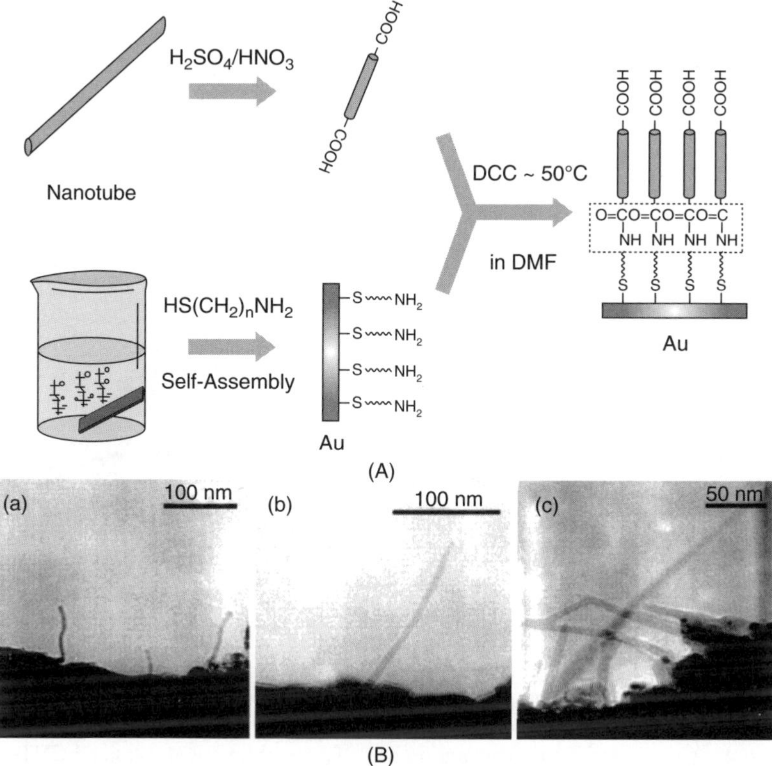

Figure 6-3 (A) Schematic representation for the end-functionalization of CNTs and subsequent self-assembling. (B) (a) A shorter nanotube assembly with a lower packing density was formed with a shorter adsorption time, while (b, c) longer adsorption time resulted in the formation of aligned nanotube array with a longer length and a higher packing density (adapted from Ref. 44).

time. As seen in the AFM images given in Figure 6-3B, a shorter nanotube assembly with a lower packing density was formed with a shorter adsorption time, whereas longer adsorption time resulted in the formation of VA-SWNT arrays with a longer length and a higher packing density.

Given that carboxylic acids could be deprotonated by various metal oxides (e.g., Ag, Al, Cu), Liu and coworkers [44] have further investigated the formation of VA-CNT arrays by self-assembling COOH-terminated CNTs onto certain metal oxide substrates (e.g., Ag). The specific chemical (deprotonation) interaction between the carboxylic groups on nanotubes and the metal surface effectively led to the anchoring of VA-CNTs on the metal substrate via their carboxylated anion head groups. Raman spectroscopic measurements confirmed the process shown in Figure 6-4 [44].

Later, Nan et al. [45] demonstrated that individual acid-oxidized SWNTs can also be deposited onto a gold substrate pretreated with $NH_2(CH_2)_{11}SH$ to form

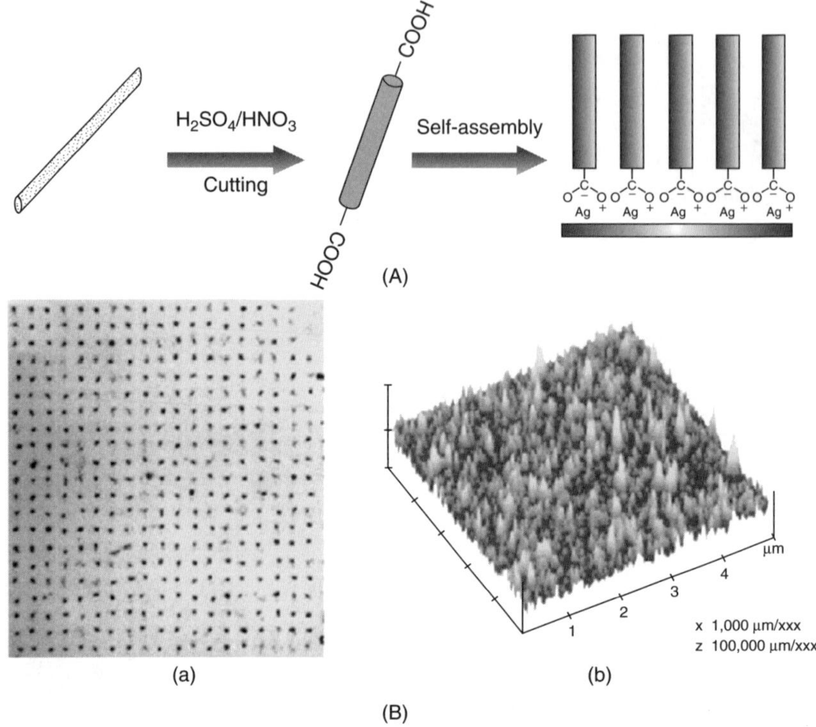

Figure 6-4   (A) Schematic representation of self-assembling acid-oxidized CNTs on Ag surface (adapted from Ref. 44). (B) Aligned carbon nanotubes formed by self-assembling of COOH-terminated nanotubes onto a gold surface prepatterned with thiol molecules. (a) Optical microscope image and (b) AFM image (adapted from Ref. 45).

a vertically aligned nanotube array on gold through the amide linkage [45]. In this case, the microcontact printing technique (μCP) technique was used to prepattern the thiol self-assembled monolayer on the gold surface for region-specific deposition of the VA-CNTs. As shown in the optical microscope image (Figure 6-4B(a)), the dark dot regions are covered by CNTs. The corresponding AFM image given in Figure 6-4B(b) clearly shows the aligned structure.

Just as the self-assembling induced by both VdW forces and the aforementioned specific chemical interactions have facilitated the formation of vertically aligned/micropatterned (either single or multiple component) CNT arrays by controlled nanotube growth and postgrowth functionalization, self-assembling induced by hydrogen bonding (H-bonding) has contributed new tools for the preparation of horizontally aligned carbon nanotubes (HA-CNTs) from solutions. For instance, Pan et al. [46] developed an interesting approach to HA-CNT films by self-assembling of functionalized CNTs with multiple hydroxyl groups (designated as carbon nanotubols). In particular, these authors prepared the nanotubols via a simple solid-phase mechanochemical reaction (ball milling) with potassium

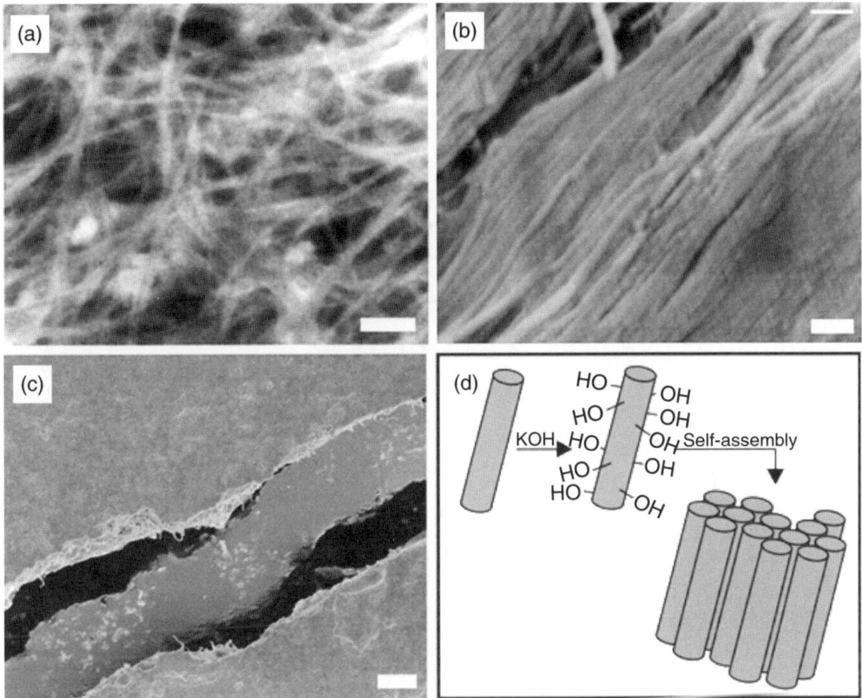

Figure 6-5 (a) SEM images of the starting materials (Tube@Rice), scale bar, 100 nm; (b) a higher-magnification view of the aligned tubes, scale bar, 100 nm; (c) top view of the aligned carbon nanotubols, scale bar, 1 μm, (d) a scheme of the self-assembly process (adapted from Ref. 46).

hydroxide (KOH) at room temperature. The multiwalled carbon nanotubols thus produced are highly soluble in water and can readily self-assemble into HA-CNTs upon drying.

As shown in Figure 6-5, the scanning tunneling electron microscopy (SEM) image for the starting HiPco SWNTs (Fig. 6-5a) shows a randomly entangled morphology. In contrast, the cross-section SEM image of a centrifuged nanotubol film, prepared by ball milling of a piece of buckypaper in the presence of KOH and subsequent self-assembling of the resultant nanotuols via H-bonding, reveals a well-aligned self-assembled structure (Fig. 6-5b). Figure 6-5c shows a large-area flat surface for the nanotubol film, indicating the occurrence of a large-scale self-assembling process. As schematically illustrated in Figure 6-5d, the strong hydrogen bonding interaction between the nanotubols is believed to be the driving force for the formation of the highly oriented, self-assembled nanotubol arrays.

On the other hand, Shimoda et al. [47] have demonstrated the preparation of ordered/micropatterned CNTs through self-assembling of preformed CNTs on glass by vertically immersing the substrate into an aqueous solution of shortened (acid oxidized) SWNTs [48]. Figure 6-6a shows the self-assembling process

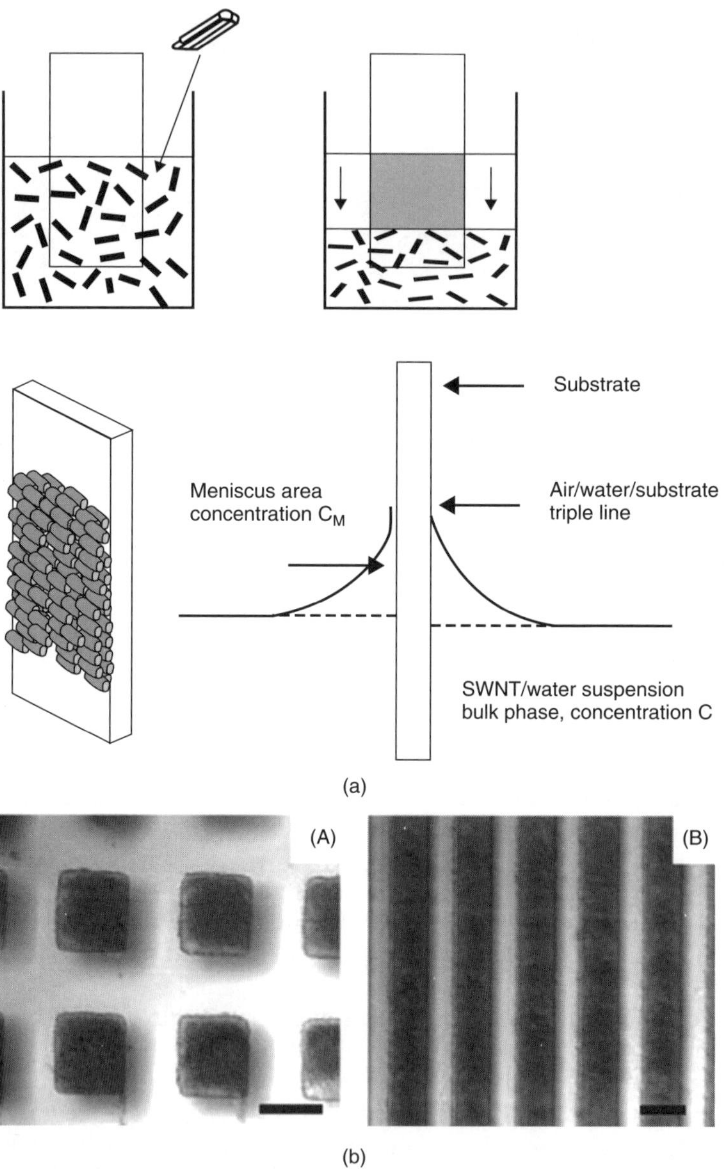

Figure 6-6 (a) A schematic representation of the self-assembling process. A hydrophilic glass slide was vertically immersed into an aqueous solution of acid-oxidized SWNTs, followed by the gradual evaporation of the water at room temperature. The SWNT bundles self-assembled on the glass substrate around the air/water/substrate interface. As the interface progressed downwards, a continuous SWNT film was formed on the glass substrate; (b) patterned SWNT structures formed by the self-assembling process on hydrophobic substrates prepatterned with periodic hydrophilic regions: (A) the squares are 100 μm × 100 μm, and (B) the strips are 100 μm in width. The shadows are due to reflections from the surface on which the samples were placed. The scale bars in both (A) and (B) are 100 μm (adapted from Ref. 47).

schematically. No CNT deposition was observed for a hydrophobic substrate (e.g., a polystyrene spin-coated glass slide) under the same conditions. This selectivity enabled the fabrication of CNT patterns by using glass substrates with prepatterned hydrophobic and hydrophilic regions. Figure 6-6b shows optical microscopic images for some of the SWNT micropatterns thus prepared.

More recently, LeMieux et al. [49] have reported an innovative approach to spin-assisted self-assembling of SWNTs onto surfaces with different chemical groups to produce a self-sorted SWNT network, in which nanotube chirality separation and simultaneous control of density and alignment occur in one step during the fabrication of field-effect transistors (FETs) (Fig. 6-7). In particular, these

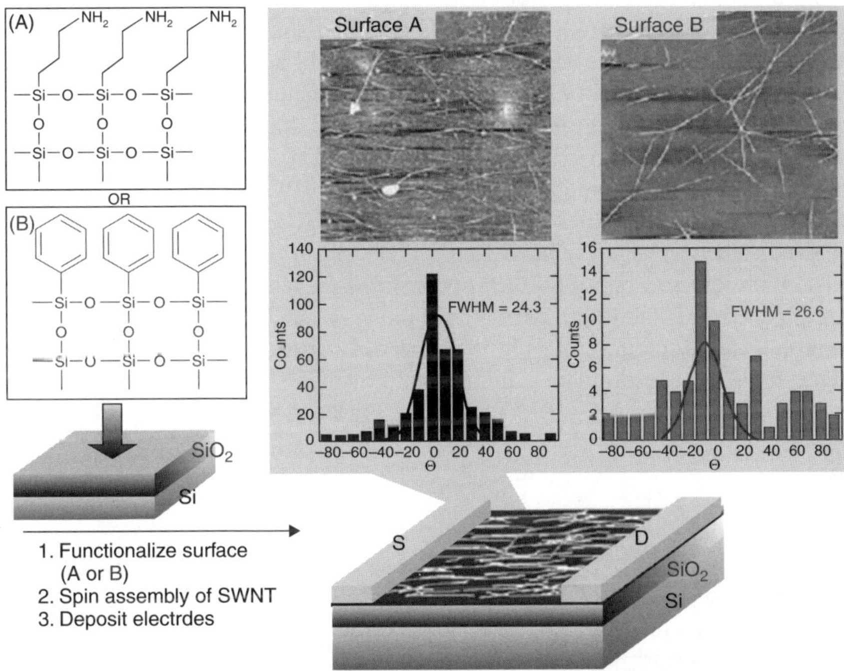

Figure 6-7   Schematic of SWNT FET fabrication and structure: The dielectric (300-nm SiO_2 on a heavily doped Si gate) is functionalized by either an amine-terminated (A) or a phenyl-terminated (B) silane. The SWNT solution is subsequently dispensed onto the spinning self-assembled monolayer-modified substrate and dried, followed by source (S) and drain (D) gold electrode deposition. Shown here is the top-contact device structure, although similar results are obtained in bottom-contact layouts where source and drain electrodes are deposited before spin assembly of the nanotubes. Upon spincoating, AFM tapping-mode topography images (10 μm by 10 μm, z scale = 10 nm) of the nanotubes applied under identical conditions on amine (top) and phenyl (bottom) surfaces reveals that density and alignment, represented by histogram (Θ is angle, in degrees, of variation from an arbitrary direction) below the corresponding AFM images, are a direct function of surface chemistry (adapted from Ref. 49). (A full color version of this figure appears in the color plate section.)

authors used aromatic molecules like the phenyl-terminated silane (Fig. 6-7B) to interact and bind selectively to metallic SWNTs and the aminosilane surface (Fig. 6-7A) to selectively self-assemble semiconducting SWNTs onto a FET via specific chemical interactions. FETs prepared by the one-step, spin-assisted self-assembling process show an on/off ratio as high as 900,000 without any postprocessing or device-level burn-off metallic nanotubes.

6.4. SELF-ASSEMBLING OF CNTs BY CHARGE TRANSFER INTERACTIONS

Having a conjugated structure of alternating single and double carbon bonds for delocalization of π-electrons [50], CNTs can also show strong π-π interactions with various conjugated molecules, including CNTs and conjugated polymers.

As shown schematically in Figure 6-8, conjugated polymers, such as poly (*m*-phenylenevinylene), PmPV, can wrap around the CNT sidewall via self-assembling induced by π-π interactions [2, 50, 51]. The wrapping of polymer chains around SWNTs could reduce the intertube VdW interaction for debundling, and hence enhance the solubility of SWNTs.

Indeed, Chen et al. [52] have found that certain aromatic molecules with a planar π moiety (e.g., pyrenylene) could strongly interact with the basal plane of graphite on the nanotube sidewall via the π-π stacking. The interaction is so strong that the aromatic molecule is irreversibly adsorbed onto the hydrophobic surface of CNTs, leading to a highly stable self-assembled structure in aqueous solutions. As shown schematically in Figure 6-9A, the long alkyl chains

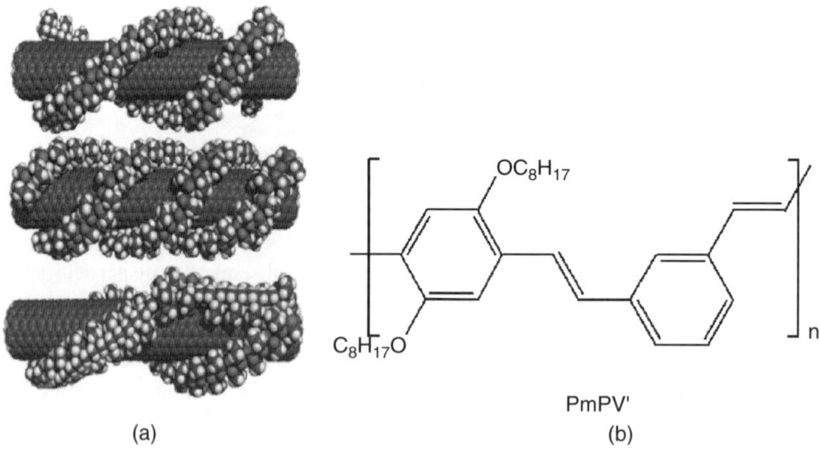

Figure 6-8   (a) Some possible wrapping arrangements of PmVP on a SWNT: a double helix (top) and a triple helix (middle). Backbone bond rotations can induce switch-backs, allowing multiple parallel wrapping strands to come from the same polymer chain (bottom). (b) Molecular structure of PmPV (adapted from Ref. 51).

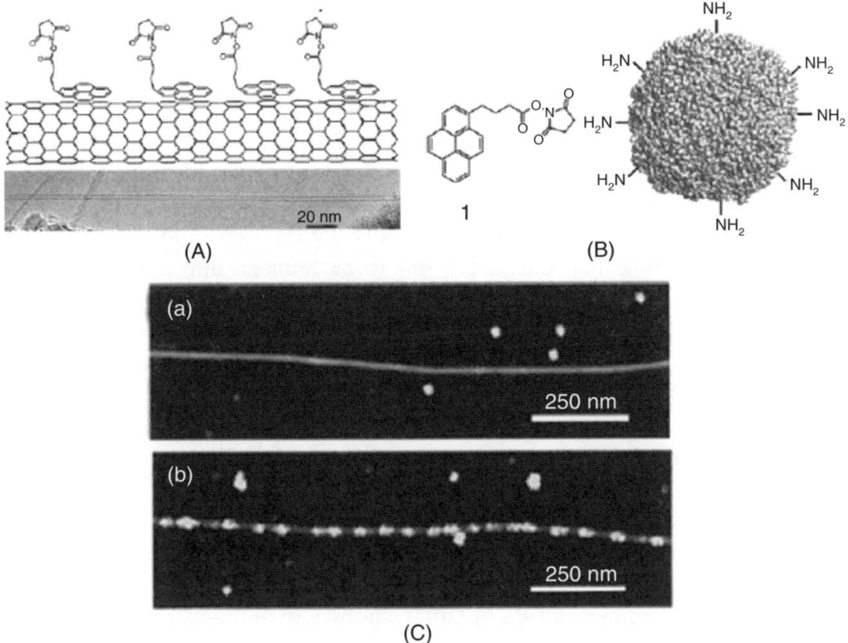

Figure 6-9 (A) Schematic representation of π-π stacking of pyrenylene derivatives on the nanotube sidewall. (B) H_2N-containing bioactive molecules (e.g., ferritin), which can be anchored onto the nanotube sidewall via a succinimidyl ester linkage from pyrenylene. (C) (a) An AFM image showing a SWNT bundle (diameter = 4.5 nm) free of absorbed ferritin after incubation in a ferritin solution. (b) An image showing ferritin molecules (apparent heights ∼10 nm) adsorbed on a SWNT bundle (diameter = 2.5 nm) functionalized by 1 and included in a ferritin solution (adapted from Ref. 52).

end-assembled onto the nanotube surface via the aromatic anchors could significantly improve the solubility of the nanotubes.

The above approach could be used to attach biochemically active molecules, such as DNA and protein chains, on the sidewall of SWNTs through appropriate anchoring molecules. For example, some H_2N-containing bioactive molecules, including ferritin, tavidin, and biotinyl-3,6-dioxactanediamine, have been attached onto the sidewall of SWNTs through a nucleophilic substitution of the amino functionality by N-hydroxysuccinimide group, which was preanchored onto the nanotube surface via a succinimidyl ester linkage to pyrenylene self-assembled onto the nanotube surface (Fig. 6-9) [52].

The large surface area, good chemical/mechanical stability, and excellent electronic property make CNTs, particularly VA-CNTs, very useful for supporting metal nanoparticles in many potential applications, ranging from advanced catalytic systems through very sensitive electrochemical sensors to highly efficient fuel cells [53, 54]. Therefore, functionalization of CNTs with metal nanoparticles has received ever-increasing interest in recent years, and a few interesting routes have now been devised for either covalently attaching or noncovalently self-assembling certain metal nanoparticles onto CNTs [55–61]. Among these, the electroless deposition is of particular interest because its simplicity could facilitate a large-scale production of the nanotube-nanoparticle hybrids through self-assembling. General applications of the electroless deposition, however, are limited by the fact that only metal ions of a redox potential higher than that of a reducing agent or CNT can be reduced into nanoparticles to self-assemble on the nanotube support. By simply supporting CNTs with a metal substrate of a redox potential lower than that of the metal ions to be reduced into nanoparticles, we have recently developed a facile yet versatile and effective substrate-enhanced electroless deposition (SEED) method for decorating CNTs with various metal nanoparticles, including those otherwise impossible by more conventional electroless deposition methods, in the absence of any additional reducing agent [9, 10]. In particular, we have used the SEED method to deposit Cu, Ag, Au, Pt, and Pd nanoparticles onto both SWNTs and MWNTs, as exemplified by Figure 6-10.

Asymmetric sidewall modification by self-assembling metal nanoparticles of different shapes onto the inner wall and outer wall of CNTs was also achieved by the SEED method [10]. As shown in Figure 6-11A(a), VA-CNTs were first deposited into a commercially available alumina membrane directly. Then Pt nanospheres were deposited onto the inner wall of the *template-synthesized* VA-CNTs supported by a Cu foil (Fig. 6-11A(b–e)), in which the template acts as a protective layer for the nanotube outer wall. Upon completion of the inner wall modification, the inner wall-modified nanotubes were released by dissolving the alumina template in aqueous HF (Fig. 6-11A(f)). As expected, Figure 6-11B (a and b) shows a very smooth outer wall for the inner wall-modified CNTs. The presence of nanospheres within the nanotube inner wall is clearly seen in Figure 6-11B(b). To deposit Pt nanocubes onto the outer wall, the innerwall-modified CNTs were supported in a sidewall-on configuration (Fig. 6-11A(g-j)) on a Cu foil and exposed to an aqueous K_2PtCl_4 solution containing $CuCl_2$ for 1 min. While Figure 6-11B(c) clearly shows the presence of nanocubes on the nanotube outer wall, Figure 6-11B(d) reveals the asymmetrically modified CNTs with nanocubes and nanospheres self-assembled onto the nanotube outer wall and inner wall, respectively.

Using electrophoresis for titanium dioxide (TiO_2) coating and photoexicited electron transfer for metal nanoparticle self-assembling, we have also developed a facile yet versatile and effective method for the development of vertically aligned coaxial nanowires of VA-CNTs sheathed with TiO_2 and well-defined TiO_2 nanomembranes decorated with and without metal nanoparticles [62]. The self-assemblies based on the TiO_2-coated VA-CNT coaxial nanowires, TiO_2 nanotubes, and TiO_2 nanomembranes showed very fast photocurrent responses to a pulsed light beam ($\lambda = 254$ nm, 4 W) with a good repeatability. The photoexcited electrons from TiO_2 have been further used to self-assemble various metal nanoparticles (e.g., Au, Ag, Pd, Pt) onto the TiO_2 nanostructures via the aforementioned SEED method without involving any reducing agent [9, 10]. The

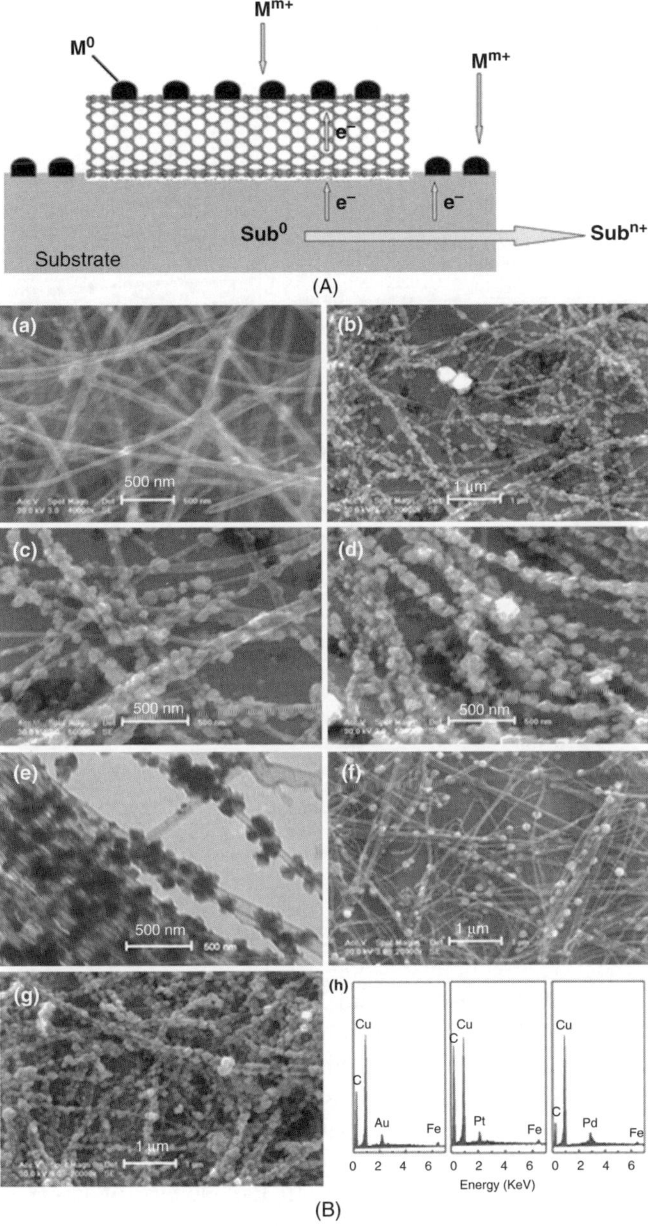

Figure 6-10 (A) Schematic illustration of metal nanoparticle deposition on carbon nanotubes via the SEED process. (B) SEM images of MWNTs supported by a copper foil after being immersed in an aqueous solution of HAuCl$_4$ (3.8 mM) for different periods of time: (a) 0 s, (b)10 s, (c) same as for (b) under a higher magnification, (d) 30 s, (e) TEM image of Au nanoparticle-coated MWNTs, (f) the Cu-supported MWNTs after being immersed in an aqueous solutions of K$_2$PtCl$_4$ (4.8 mM) for 10 s, (g) the Cu-supported MWNTs after being immersed in an aqueous solution of (NH$_4$)$_2$PdCl$_4$ (7.0 mM) for 10 s, (h) EDX spectra for the Au, Pt, and Pd nanoparticle-coated MWNTs on Cu foils (adapted from Ref. 9).

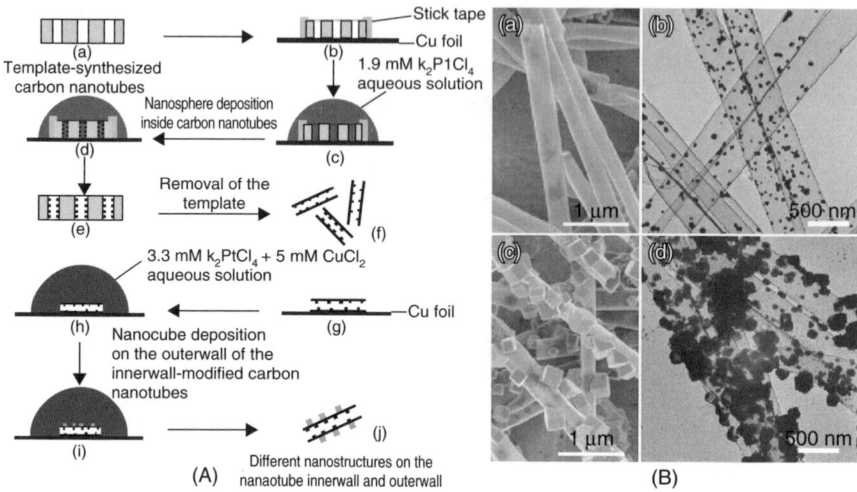

Figure 6-11   (A) Procedures for the CNT innerwall modification and the asymmetric modification of the nanotube innerwall with Pt nanospheres and the outerwall with nanocubes. (B) (a) SEM and (b) TEM images of CNTs having their innerwall modified with Pt nanospheres by immersing the synthesized template. (c) SEM and (d) TEM images of the innerwall-modified CNTs followed by outerwall modification in an aqueous solution of K_2PtCl_4 containing $CuCl_2$ with the innerwall-modified CNTs being supported by a Cu foil in a sidewall-on configuration (adapted from Ref. 10). (A full color version of this figure appears in the color plate section.)

resultant hybrid nanomaterials are expected to be useful in solar cells and other optoelectronics.

6.5. SELF-ASSEMBLING OF CNTs BY DNA PAIRING

In collaboration with Moghaddam and McCall [55], we have also developed a photochemical method to functionalize CNTs with photoreactive reagents (e.g., aziridothymidine, AZT), followed by the coupling of single-strand DNA (ssDNA) chains onto the CNTs through the photo-adduct and coating the nanotube sidewalls with cDNA-modified gold nanoparticles via self-assembling through DNA hybridization. As can be seen in Figure 6-12a, the UV irradiation caused the formation of very active nitrene groups in the vicinity of CNTs (Fig. 6-12a, step (i)). These nitrene groups were then coupled to the nanotube structure via a cycloaddition reaction to form aziridine photo-adducts (Fig. 6-12a, step (i)). After the nanotube photo-adducts were thoroughly rinsed with pure water, the nanotube sample was placed in an empty reaction chamber of an automated DNA synthesizer and a DNA oligonucleotide of defined base sequence was built in situ by the sequential addition of protected nucleotides with standard phosphoramidite chemistry (Fig. 6-12a, steps (ii)-(iv)). Here, the free hydroxyl group

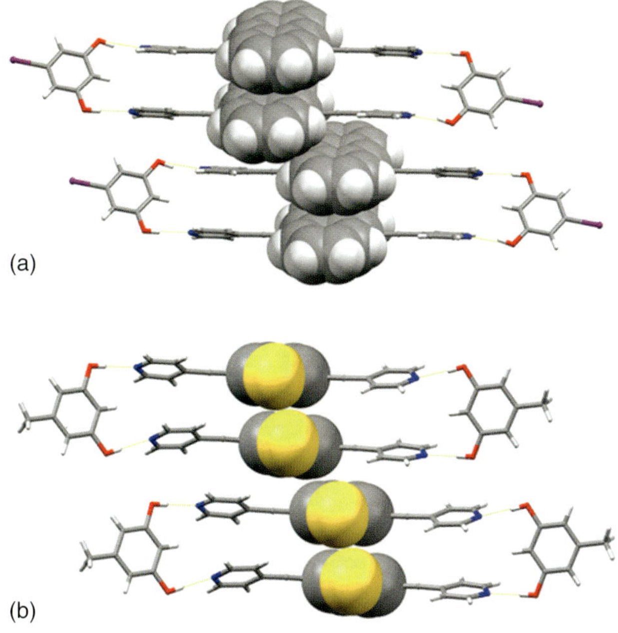

Figure 1-15 (See text for details.)

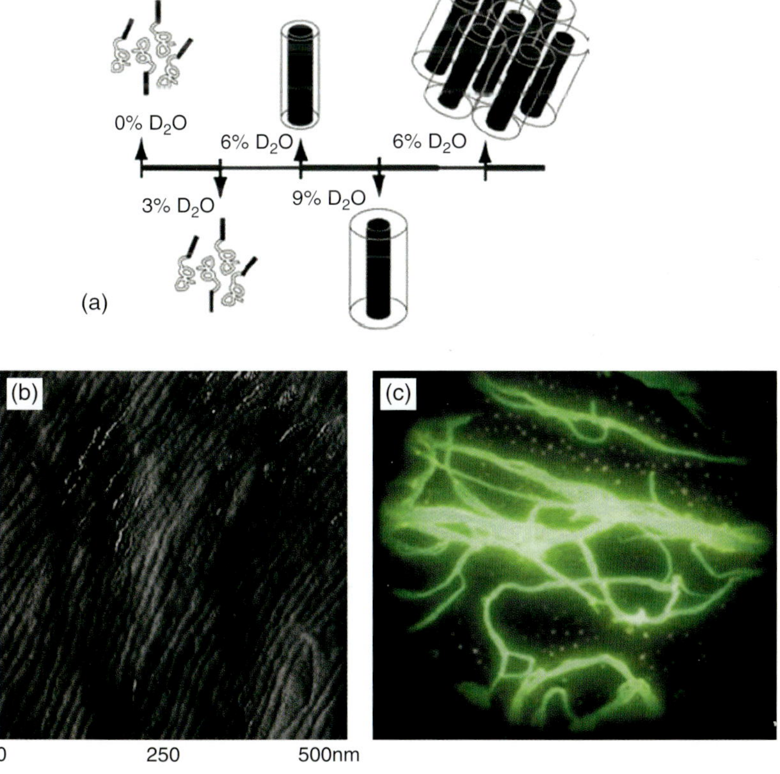

Figure 2-10 (See text for details.)

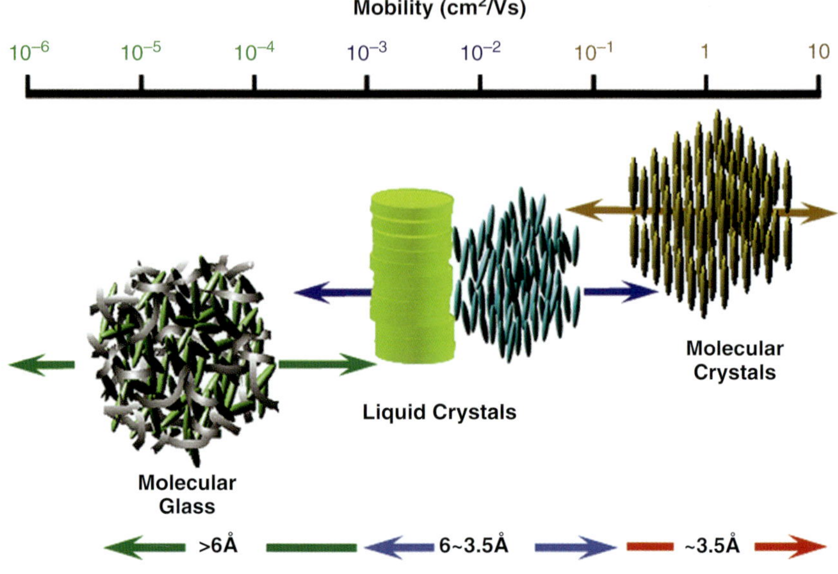

Figure 3-1   (See text for details.)

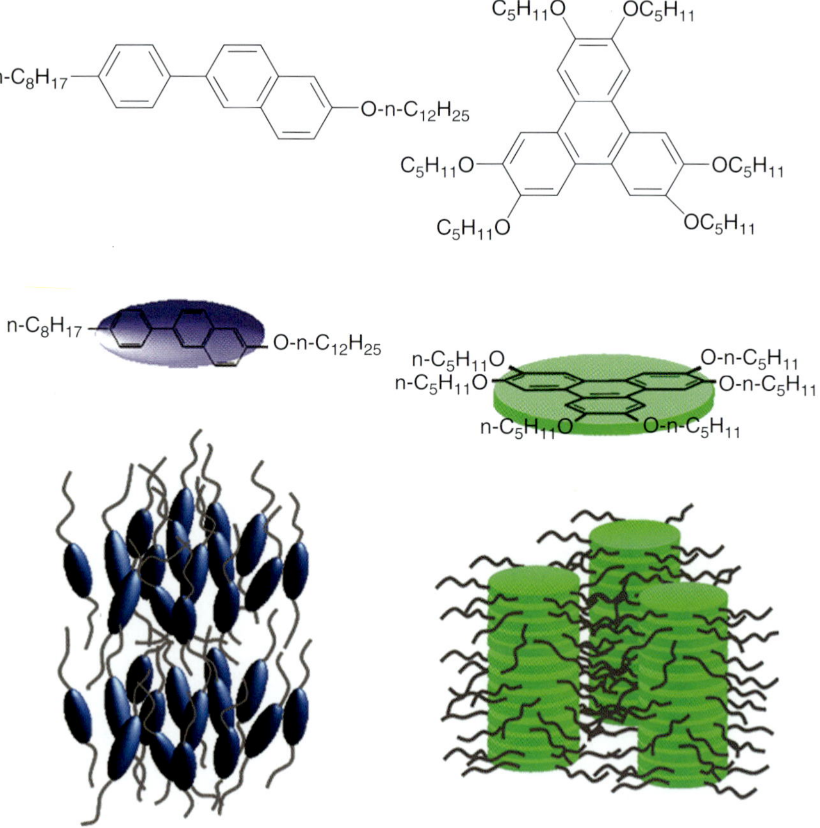

Figure 3-3   (See text for details.)

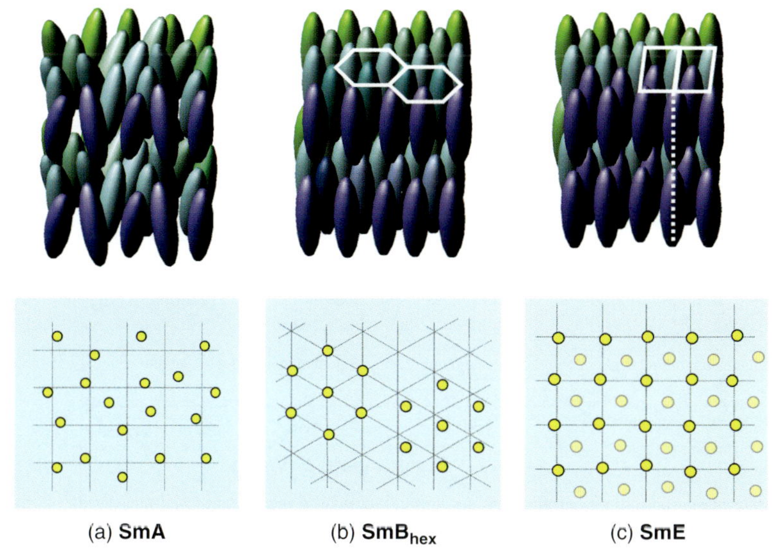

(a) **SmA**   (b) **SmB**$_{hex}$ (c) **SmE**

Figure 3-10   (See text for details.)

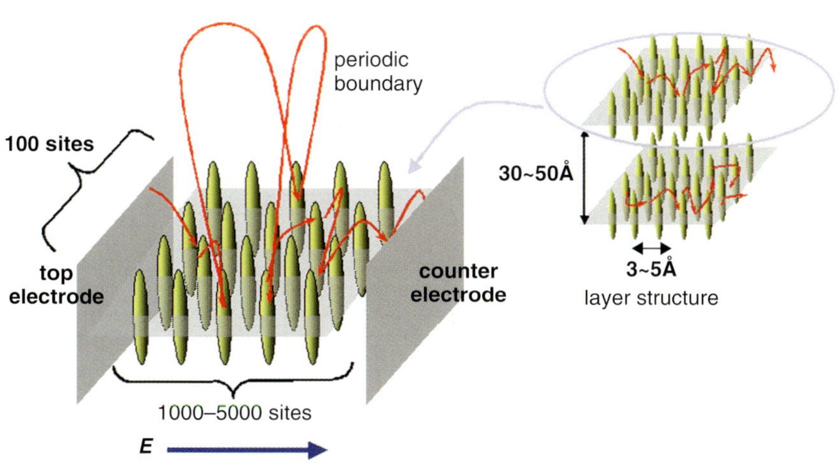

Figure 3-11   (See text for details.)

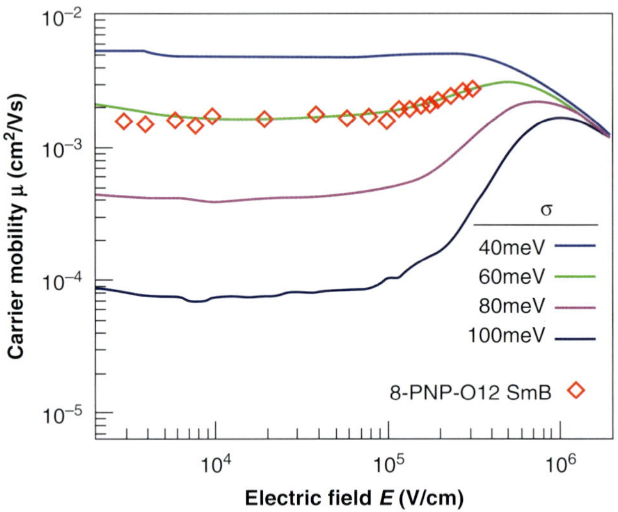

Figure 3-12   (See text for details.)

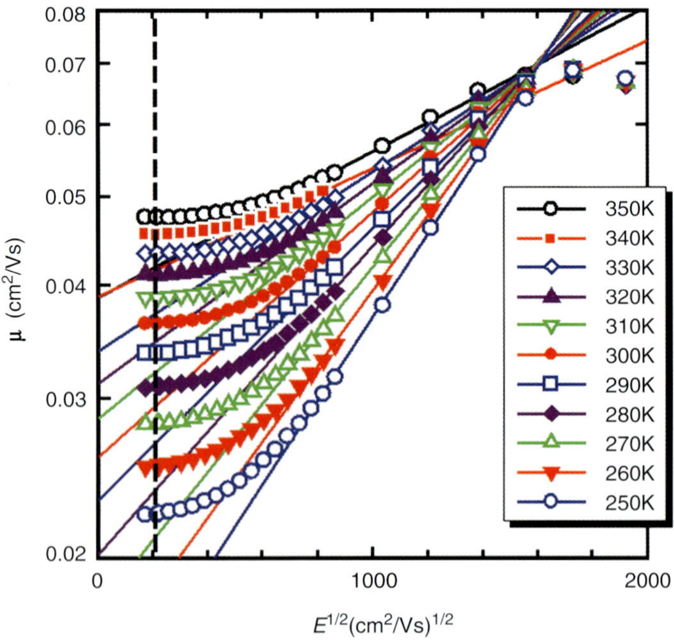

Figure 3-16   (See text for details.)

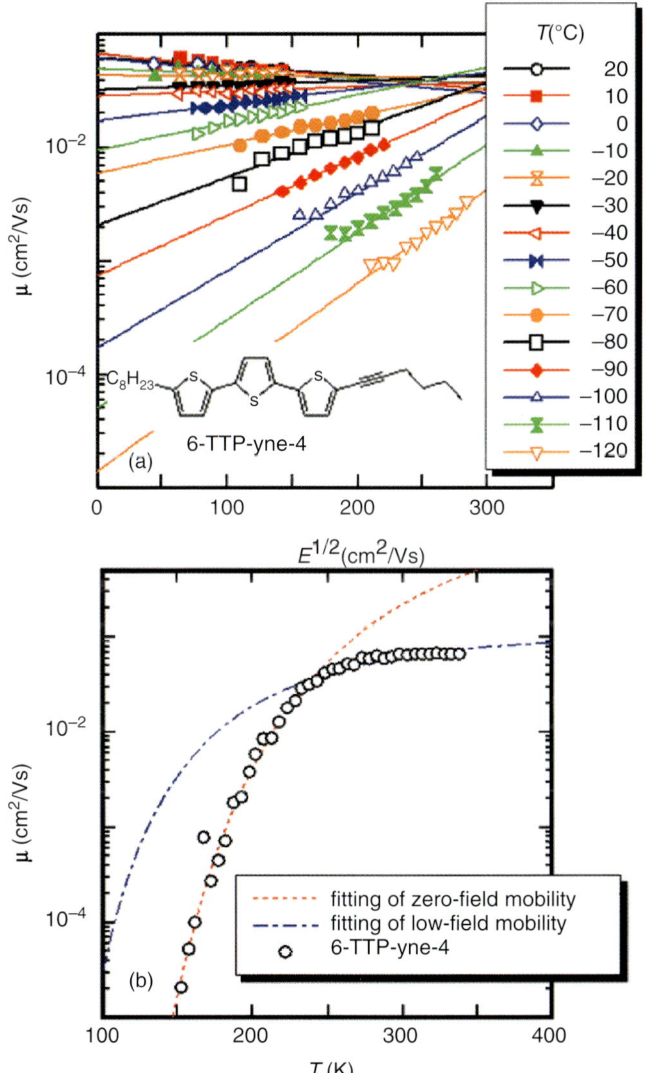

Figure 3-18   (See text for details.)

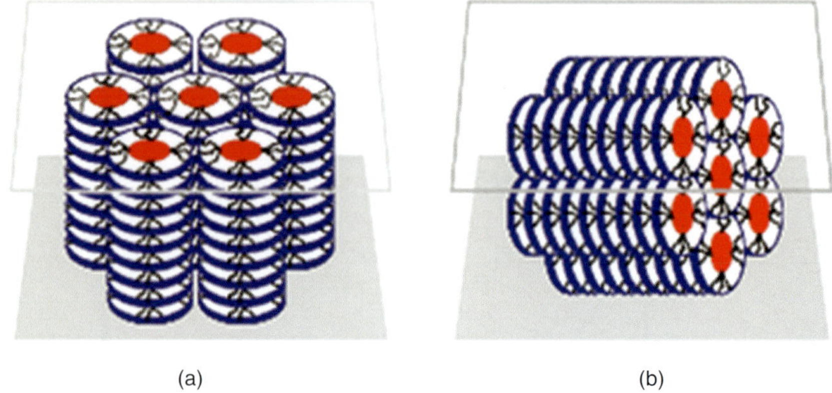

Figure 4-15   (See text for details.)

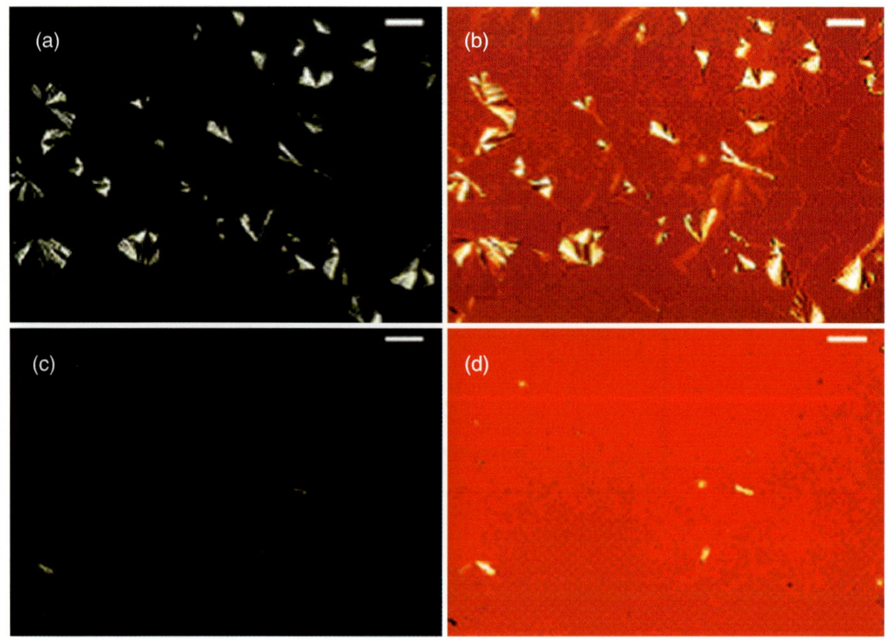

Figure 4-17   (See text for details.)

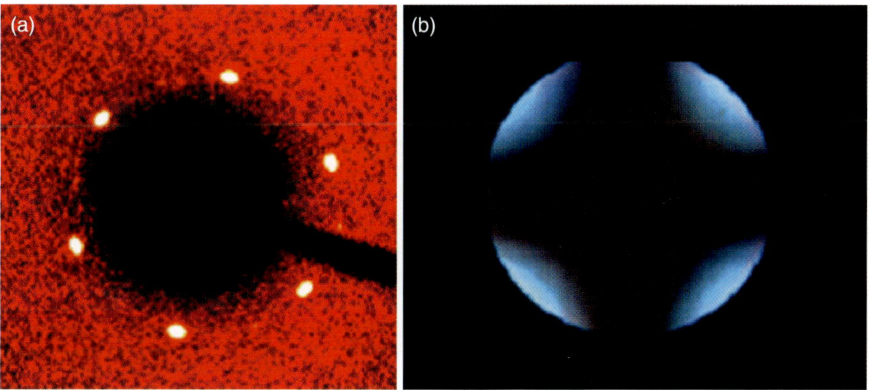

Figure 4-18 (See text for details.)

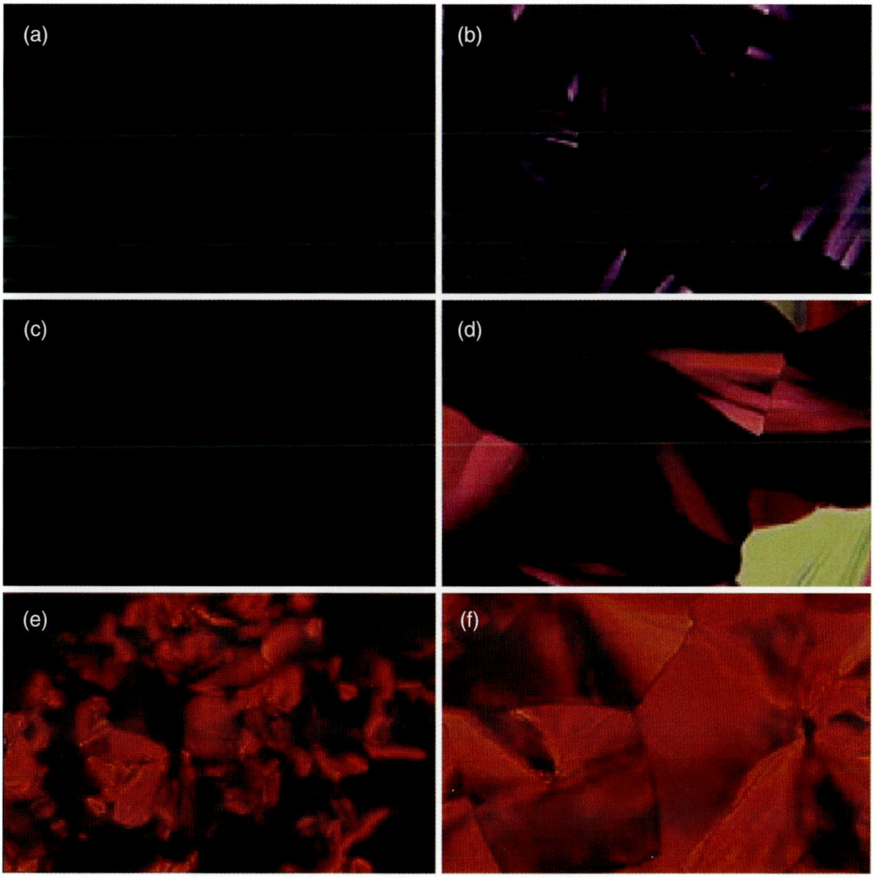

Figure 4-19 (See text for details.)

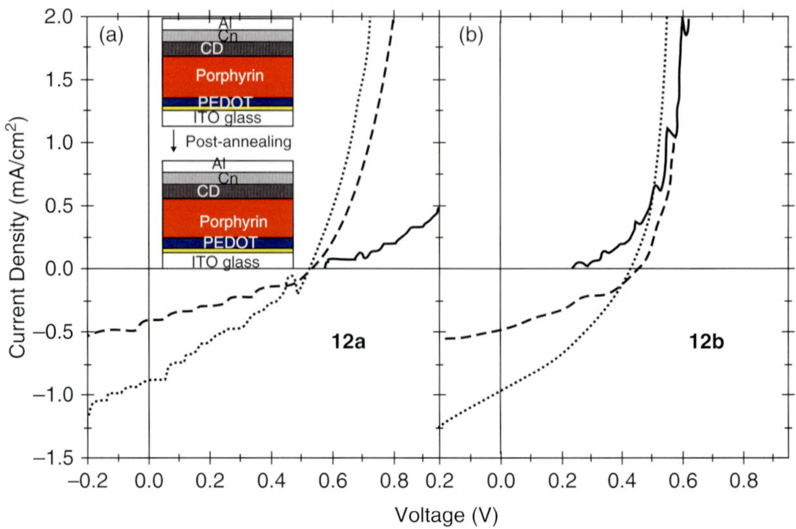

Figure 4-23 (See text for details.)

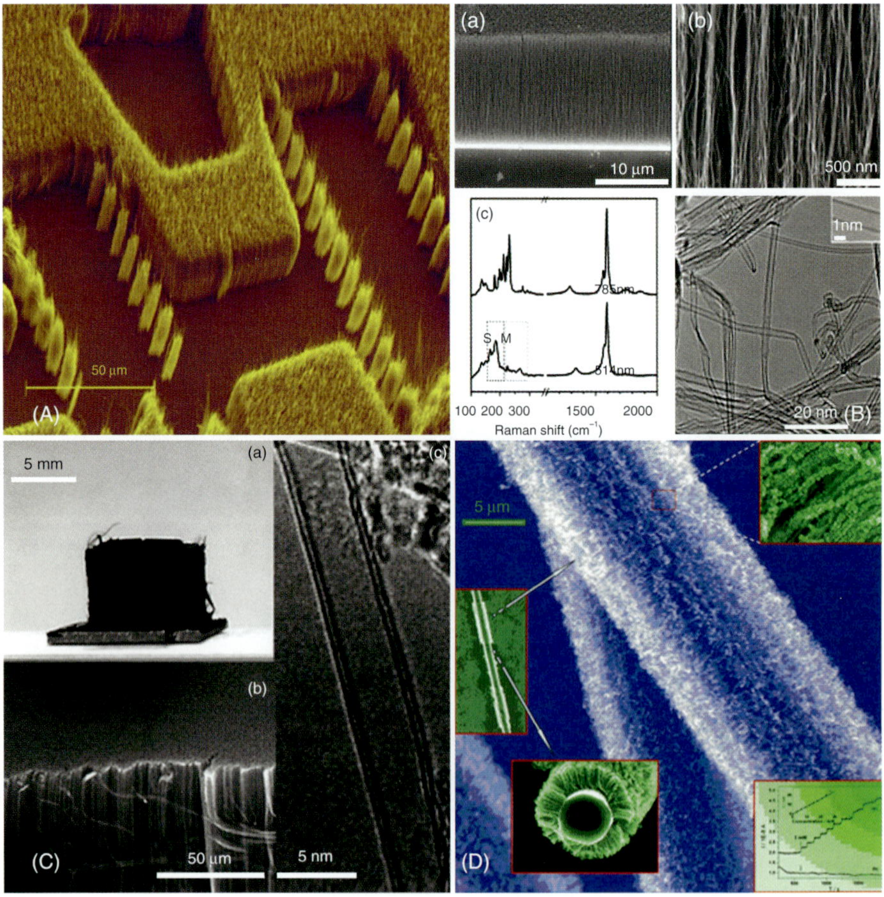

Figure 6-1 (See text for details.)

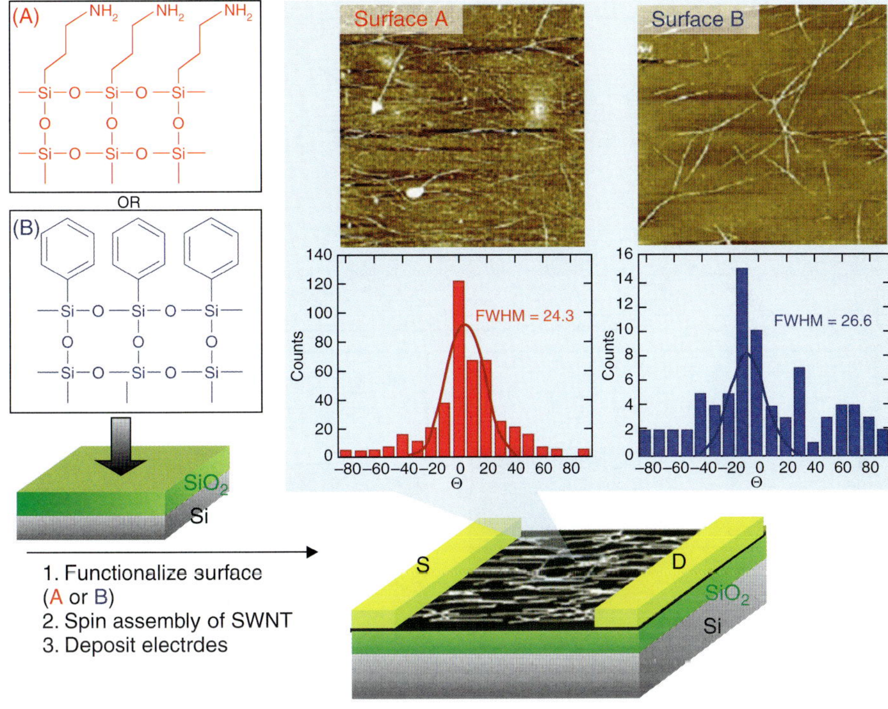

Figure 6-7 (See text for details.)

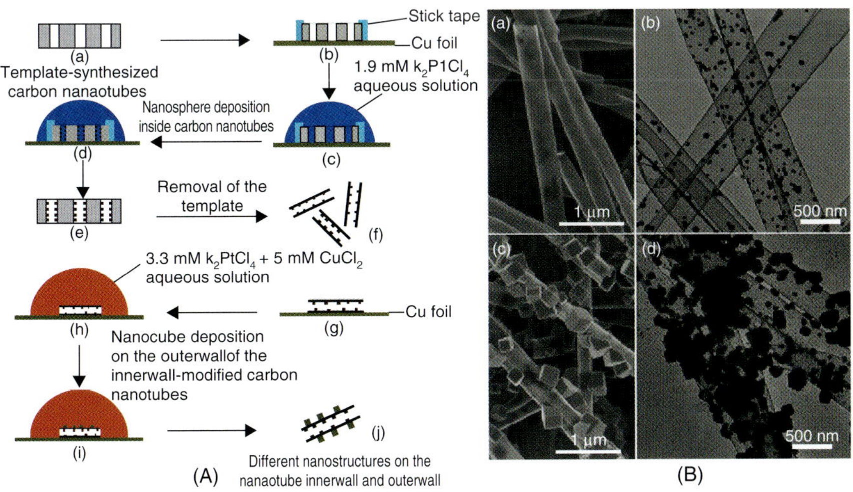

Figure 6-11 (See text for details.)

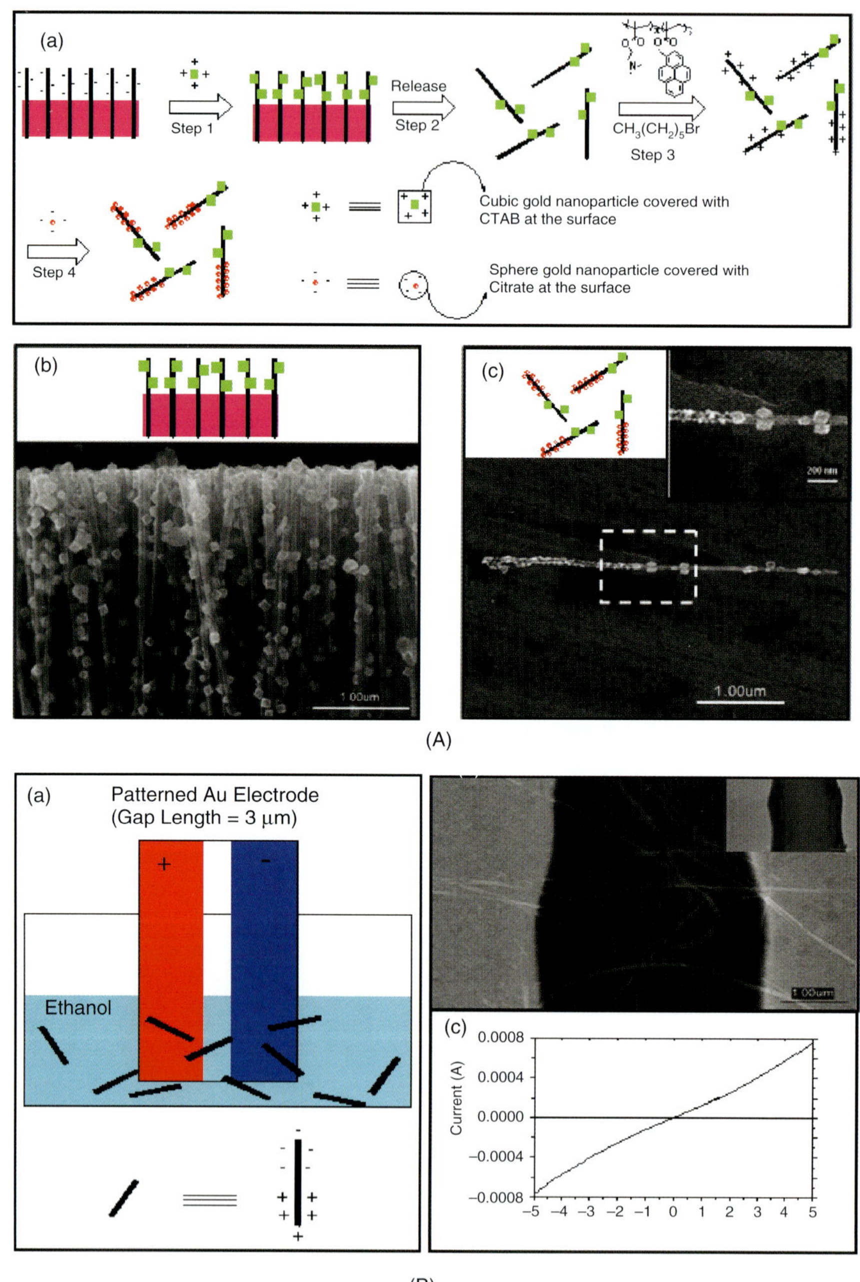

Figure 6-16 (See text for details.)

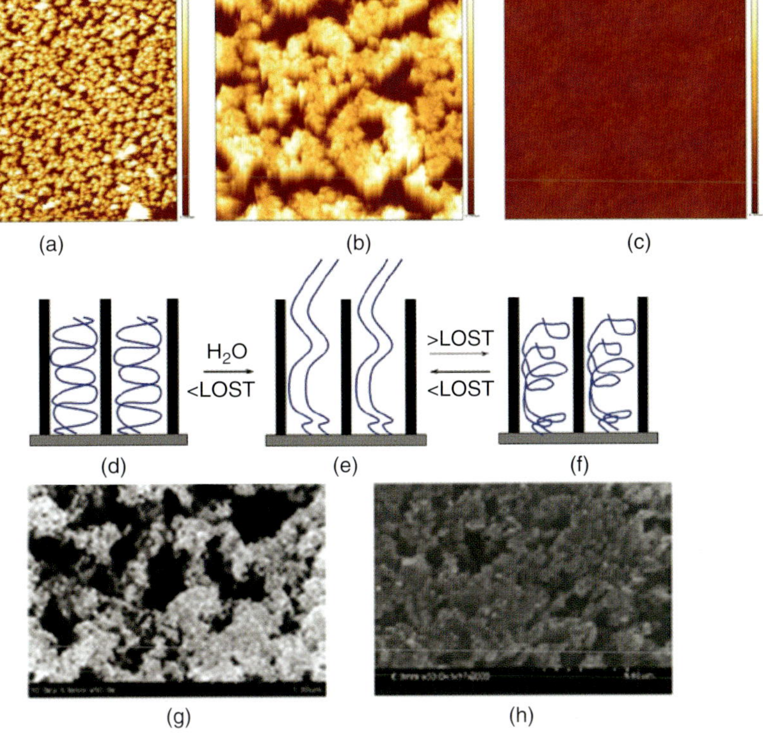

Figure 6-17 (See text for details.)

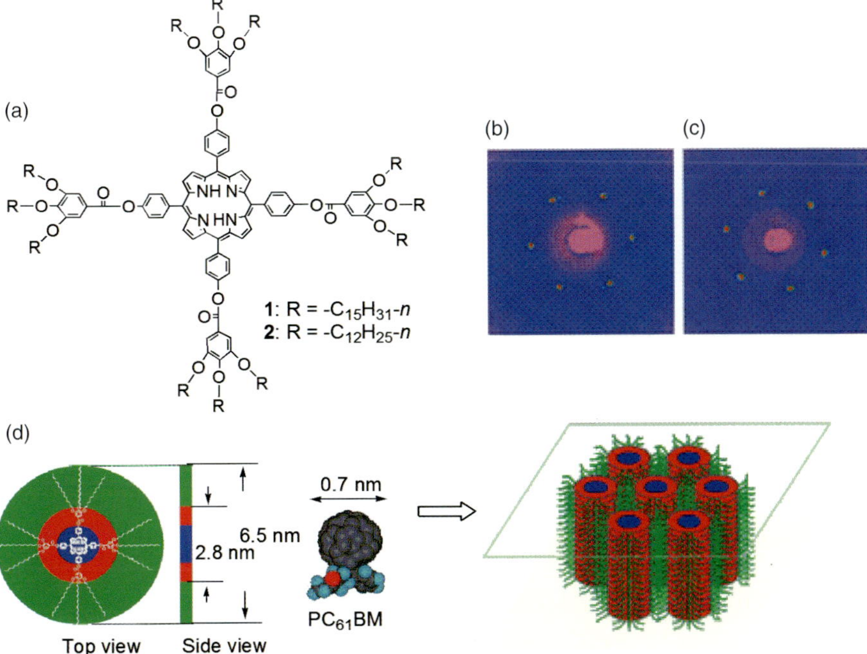

Figure 7-2 (See text for details.)

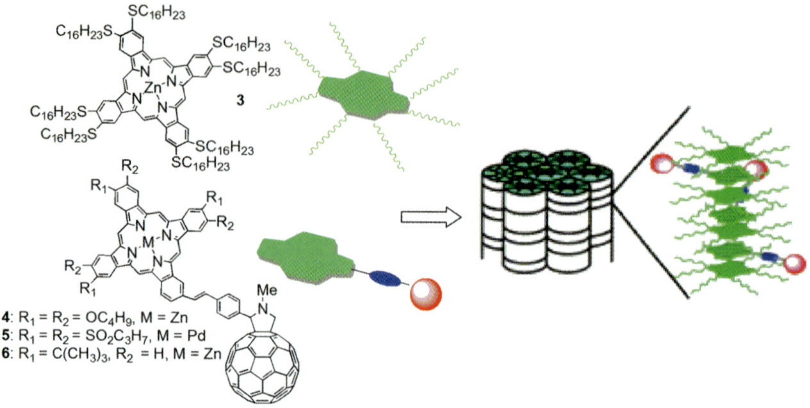

Figure 7-3   (See text for details.)

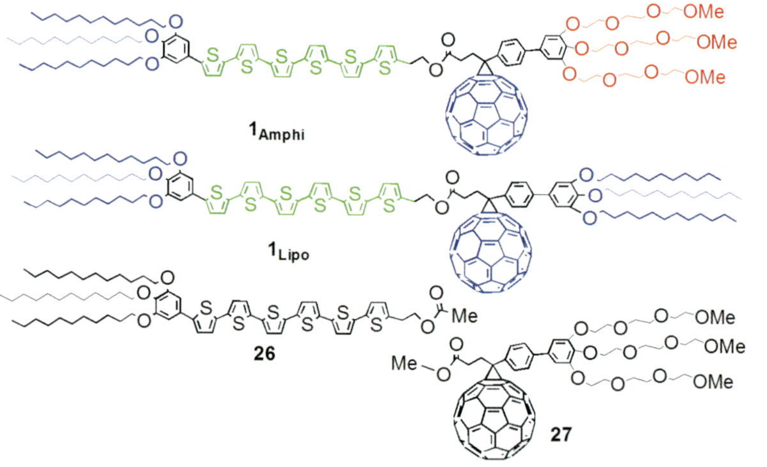

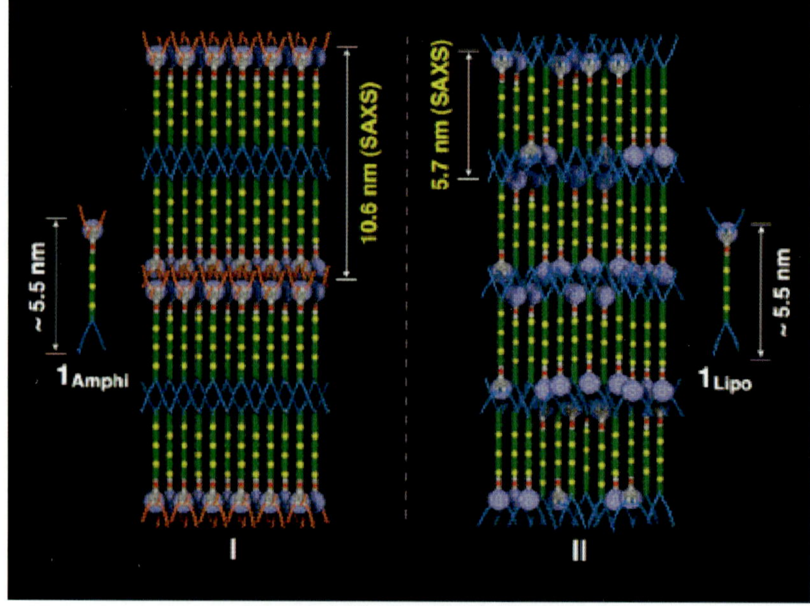

Figure 7-12   (See text for details.)

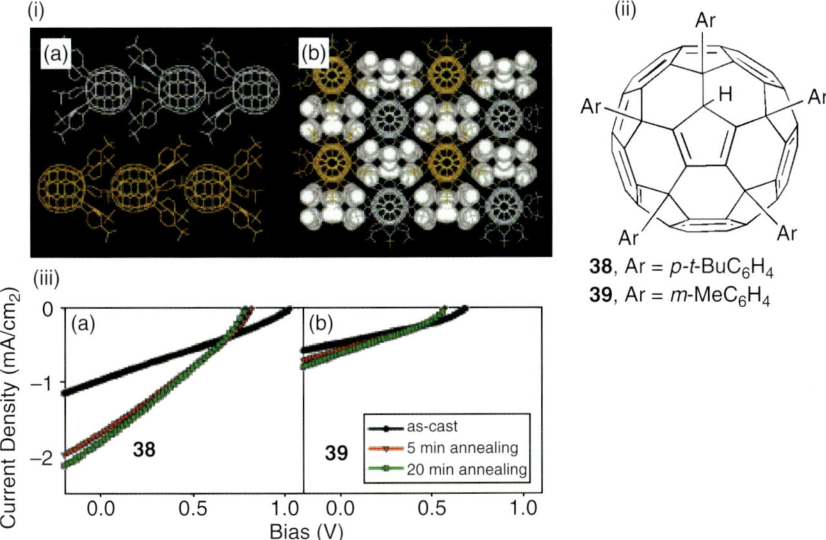

Figure 7-18   (See text for details.)

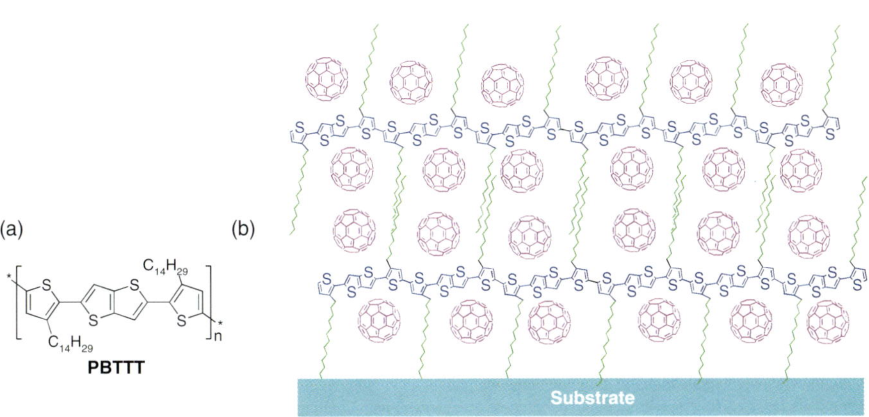

Figure 7-19   (See text for details.)

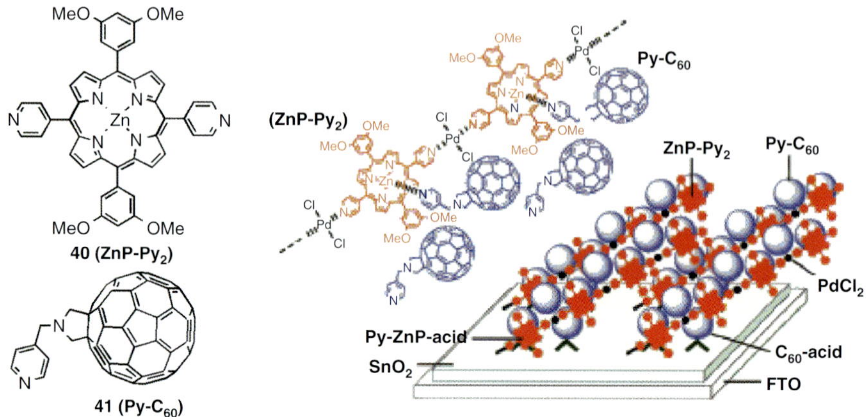

Figure 7-20   (See text for details.)

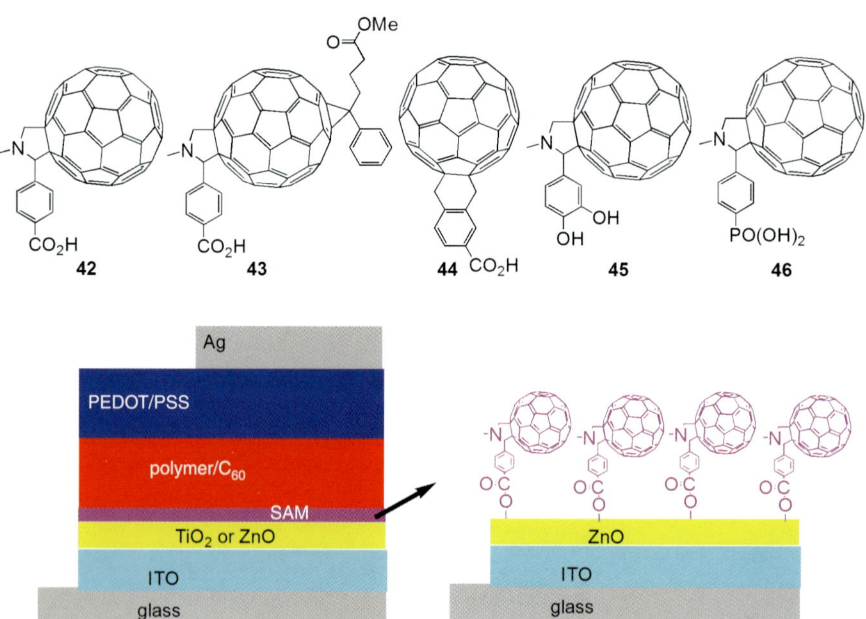

Figure 7-22   (See text for details.)

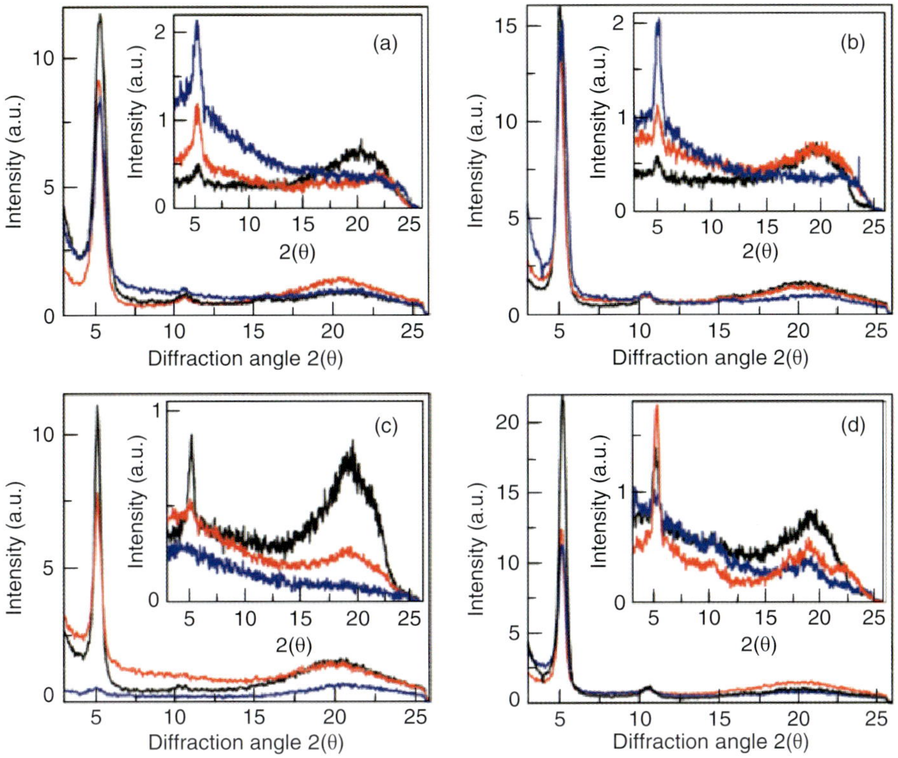

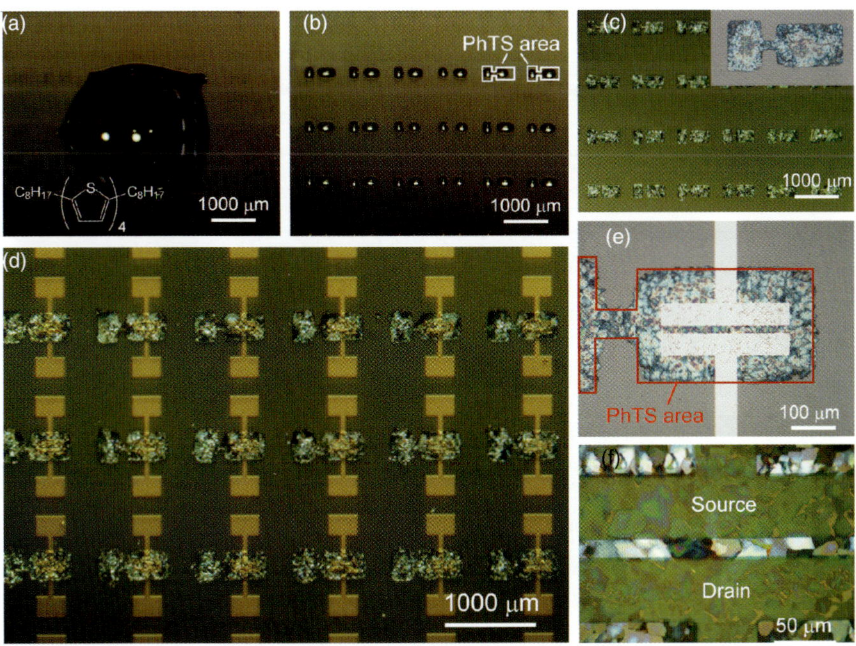

Figure 8-22 (See text for details.)

Figure 9-3 (See text for details.)

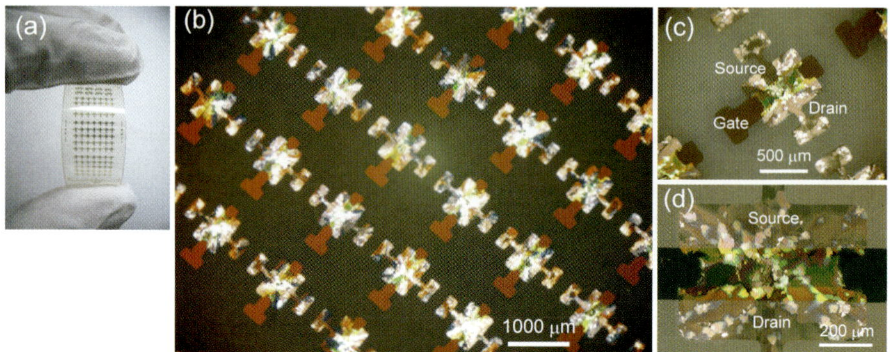

Figure 9-8   (See text for details.)

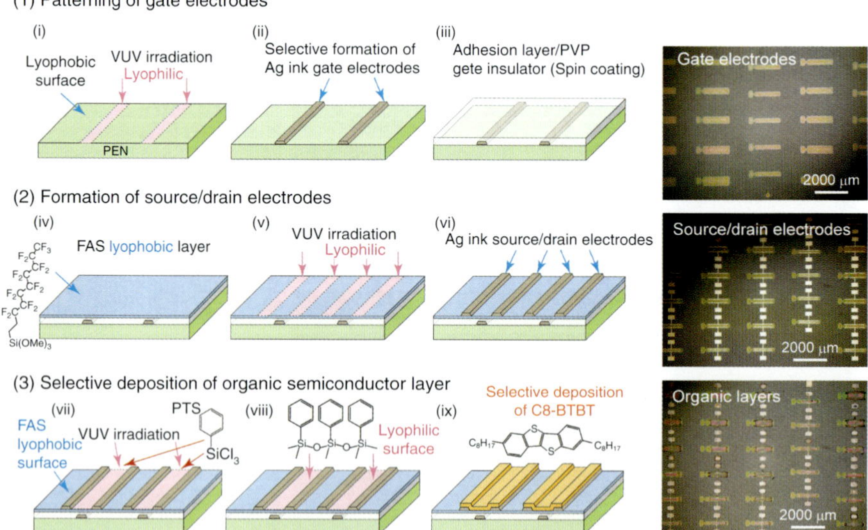

Figure 9-12   (See text for details.)

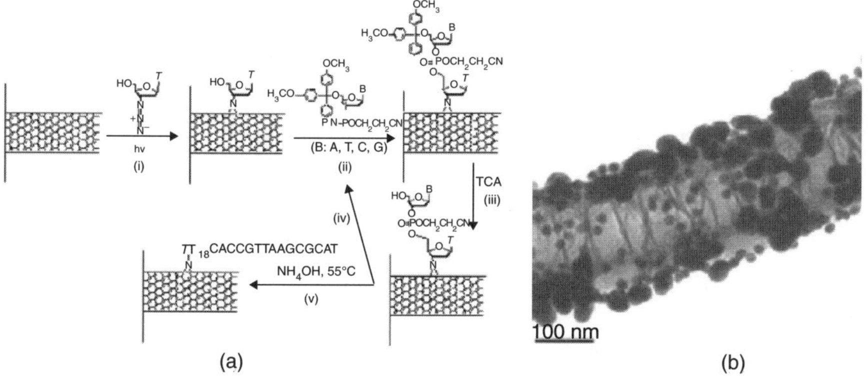

Figure 6-12   (a) In situ DNA synthesis from sidewalls of CNTs photo-etched with azidothymidine. Aligned MWNTs on a solid support are coated with a solution of azidothymidine (AZT) and UV-irradiated to produce photo-adducts each with a hydroxyl group (i). The hydroxyl group reacts with a phosphoramidite mononucleotide to initiate synthesis of the DNA molecule (ii). Trichloroacetic acid deprotects the hydroxyl group (iii) for reaction with the next nucleotide (iv), and the cycle is repeated until the molecule with desired base-sequence is made. Finally, the supported nanotubes are heated in ammonia solution to remove blocking groups from the nucleotides (v), to produce DNA-coated nanotubes. (b) DNA-directed modification of the surface of CNTs with gold nanoparticles of different diameters. The sample represented by this TEM image was prepared by binding the remaining unbound cDNA (on 16-nm gold nanoparticles precoated onto the nanotube) to complementary DNA attached to gold nanoparticles 38 nm in diameter (Scale bar is 100 nm; adapted from Ref. 55).

at the 5′-position of the deoxyribose moiety in each aziridothymidine group was used as the site of attachment from which the DNA molecule was built. After the DNA synthesis, the blocking groups on the DNA strand were removed by heating in an ammonia solution (Fig. 6-12a, step (v)). The resultant CNT sample attached with ssDNA chains was washed with water until the supernatant was neutral, and then stored in water at 4°C for further use. The locations of the DNA molecules thus attached onto the CNTs can be determined by a visual assay using the self-assembled cDNA-modified gold nanoparticles and TEM. Figure 6-12b shows gold nanoparticles positioned in close proximity to the surfaces of the nanotubes; the sample was prepared by self-assembling cDNA-attached gold nanoparticles with a 16-nm diameter to the ssDNA chains grafted on the nanotubes, followed by a further hybridizing of the remaining unbound cDNA on the 16-nm gold nanoparticles to their complementary DNA chains attached to gold nanoparticles 38 nm in diameter.

Recently, we have also demonstrated that a wide range of multicomponent structures of CNTs can be constructed by DNA-direct self-assembling of carbon nanotubes and gold nanoparticles (Fig. 6-13) [63]. Figure 6-13a shows the reaction steps for the DNA-directed self-assembling of multiple CNTs using the gold nanoparticles as a linkage, which involves the acid (HNO_3) oxidation of carbon

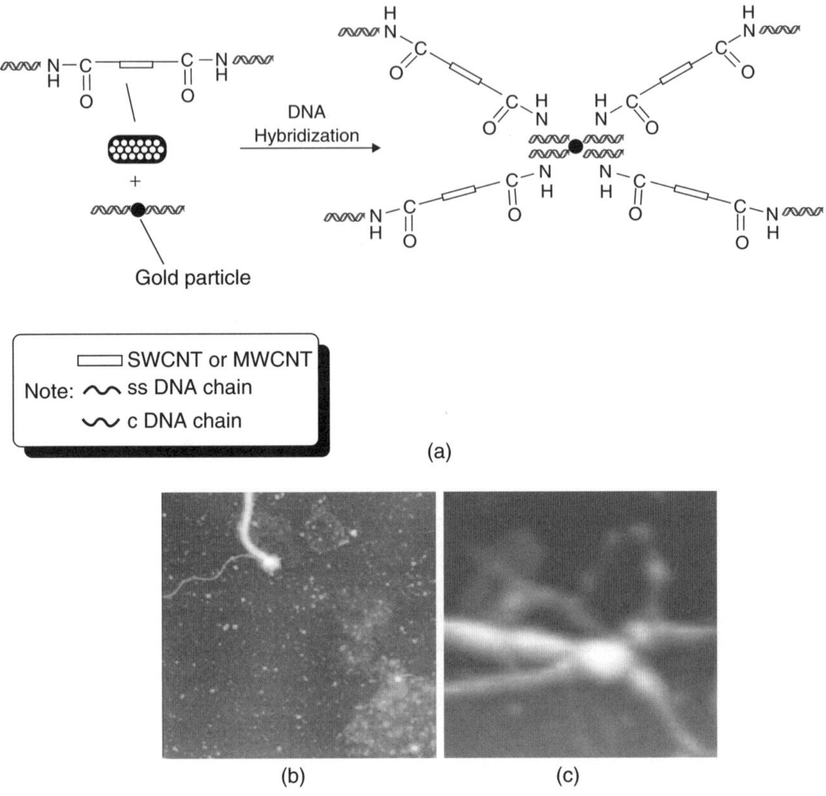

Figure 6-13 (a) Schematic representation of procedures for DNA-directed self-assembling of multiple carbon nanotubes and nanoparticles. Typical AFM images of (b) the self-assembly of a ssDNA-SWNT and ssDNA-MWNT through a cDNA-Au nanoparticle (scanning area: 5.60 μm × 5.60 μm), (c) multiple ssDNA-SWNTs and ssDNA-MWNTs connected with a cDNA-Au nanoparticle core (scanning area: 1.10 μm × 1.10 μm; adapted from Ref. 63).

nanotubes to introduce carboxylic end groups [48] for ssDNA grafting. The ssDNA-attached CNTs were then subjected to hybridization with cDNA chains grafted on gold nanoparticles through the highly specific thiol-gold interaction [64]. Figure 6-13(b and c) clearly shows the formation of various self-assembled multiple CNT structures, including one SWNT connected with one MWNT through a nanoparticle (Fig. 6-13b) and multinanotubes connected to a single nanoparticle core (Fig. 6-13c), depending on the reaction and/or self-assembling conditions. In view of the availability of various CNTs and DNA chains of different base sequences, these results represent an important advance in constructing many multiple CNT self-assembled structures for various applications (e.g., sensor chips, electronic circuits).

More recently, we have developed a simple but effective photochemical approach to directly grafting different chemical reagents onto the opposite tube-ends of individual CNTs by sequentially floating the substrate-free VA-CNT film on two different photoreactive solutions with only one side of the VA-CNT film being contacted with a photoreactive solution and exposed to UV light each time, as schematically shown in Figure 6-14A [65]. This methodology can be used to asymmetrically end-functionalize CNTs with ssDNA and cDNA chains onto their opposite tube-ends, which should be very useful for site-selective self-assembling of CNTs into many novel functional structures. In a similar but independent study, asymmetric end-functionalization has also been achieved through sidewall protection by impregnating a VA-MWNT array with polystyrene [66], as schematically shown in Figure 6-14B.

6.6. SELF-ASSEMBLING OF CNTs BY ASYMMETRIC FUNCTIONALIZATION

As mentioned above, characteristics of the self-assemblies depend on both the spatial arrangement of their components and the nature of the interactions that hold these components together. To control the architectures of CNT self-assemblies, it is paramount important to functionalize CNTs with various tailor-made surface characteristics with submolecular precision. In this regard, we have recently developed an effective and versatile method for tube-length-specific functionalization of CNTs through a controllable embedment of VA-CNTs into polymer matrices [67]. As schematically shown in Figure 6-15A(a), an appropriate polymer thin film [e.g., PS; $M_w = 350,000$; T_g (glass transition temperature) $\approx 105°C$; T_m (melting point) $\approx 180°C$; T_c (decomposition temperature) $\approx 350°C$; thickness: $\sim 50\ \mu m$] was first placed on the top surface of a VA-CNT array. Upon heating the SiO_2/Si substrate by an underlying hot plate to a temperature above T_m and below T_c, the melted PS film gradually filtrated into the nanotube forest (Fig. 6-15A(a)) through a combined effect of the gravity and capillary forces. As expected, the infiltration depth (i.e., the embedment length) of PS into the nanotube forest depends strongly on the temperature and heating time. After a predetermined heating time, the polymer-infiltrated nanotube array was peeled off from the SiO_2/Si substrate in an aqueous solution of HF (10% wt) to generate a free-standing film of VA-CNTs embedded into the PS matrix (Fig. 6-15A(a)). While Figure 6-15A(b)) shows a typical SEM image of the pristine VA-CNTs, Figure 6-15(c–g) reproduce SEM images of the VA-CNTs embedded into PS films after being heated at different temperatures for 1 min. For the given infiltration time of 1 min, the embedment length increased with the heating temperature from $\sim 10\%$ coverage of the nanotube length at $180°C$ up to the full-length coverage at $245°C$ (Fig. 6-15A(c–g)). A more quantitative relationship between the embedment length and the heating temperature is given in Figure 6-15A(h), which shows a pseudo-linear relationship over a wide range of the temperature. For a given heating temperature (e.g., $195°C$), the embedment length was found to be directly proportional to the PS filtration time (i.e., heating time),

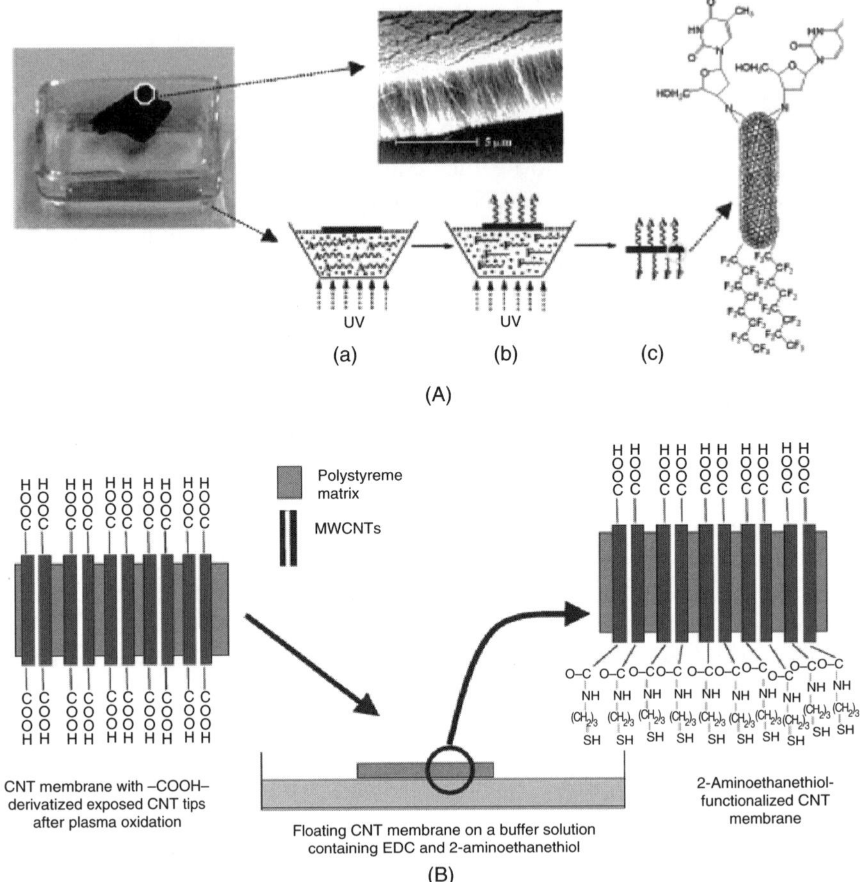

Figure 6-14   (A) A freestanding film of VA-CNTs floating on the top surface of (a) a $3'$-azido-$3'$-deoxythymidine solution in ethanol for UV irradiation at one side of the nanotube film for 1 h, and (b) a perfluorooctyl iodide solution in 1,1,2,2-tetrachloroethane for UV irradiation at the opposite side of the nanotube film (adapted from Ref. 65). (B) Plasma-oxidized VA-MWNT membrane (cross-sectional view) with carboxylic acid-derivatized MWNTs when floated on a buffer solution containing EDC and 2-aminoethanethiol, resulting in bifunctional CNTs in the membrane structure (adapted from Ref. 66).

as exemplified by Figure 6-15A(i). These results clearly show the feasibility for the tube-length-specific functionalization of CNTs through the controllable embedment of VA-CNTs into the polymer matrix. This is because the portion of the nanotube sidewall embedded into the polymer matrix is physically masked, whereas the polymer-free sidewall surface along the nanotube length can be selectively functionalized via various physicochemical processes.

In view of the ease with which shape/size-controlled metal nanoparticles can be deposited onto CNT structures by the aforementioned SEED method [9, 10]

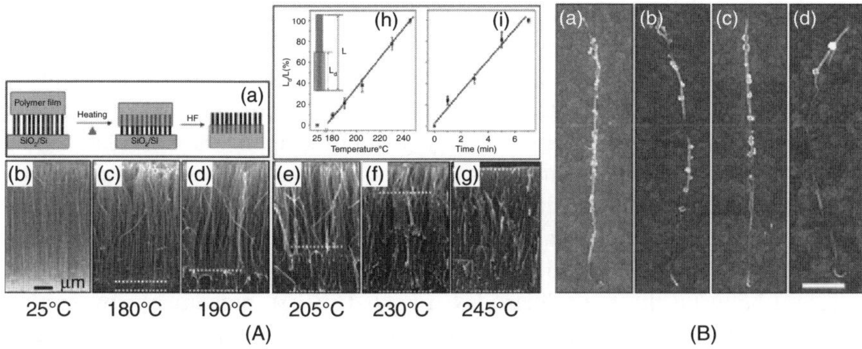

Figure 6-15 (A) (a) Schematic representation of VA-CNTs embeded into a polymer matrix by thermal infiltrating the melted polymer into the nanotube forest; (b–g) SEM images of (b) the pristine VA-CNT array and (c–g) the VA-CNT array after being embedded into PS films by heating at different temperatures for 1 min. The dashed-line gaps crossing the polymer coated regions show the approximate embedment length for each of the PS-embedded VA-CNTs; (h and i) Temperature and time dependence of the embedment length for VA-CNTs embedded into the PS matrix. (L: the nanotube length ($\sim$6 μm), L_d: the embedment length, which was estimated from the distance between the two dashed lines in each of the images shown in c–g). (B) SEM images of individual CNTs released out from the PS-embedded VA-CNTs by THF washing after the Au nanoparticle deposition by SEED. (a–d) correspond to samples (c–f) in (A), respectively (Scale bar: 1 μm; adapted from Ref. 67).

and the effective action of the deposited nanoparticles as labeling centers for characterization, we have demonstrated the tube-length-specific functionalization of CNTs by SEED deposition of Au nanoparticles onto each of the VA-CNTs partially embedded into the PS matrices shown in Figure 6-15A(c–f). In so doing, we first used the SEED method to deposit Au nanoparticles onto the constituent nanotubes in the PS-embedded VA-CNTs at the polymer-free region along their sidewalls. Thereafter, the polymer support was washed off with THF to yield individual nanotubes region-selectively attached with the Au nanoparticles, as shown in Figure 6-15B. The portions of the nanotube length covered by the Au nanoparticles seen in Figure 6-15B(a–d) consist well with the corresponding nanotube lengths extending out from the polymer matrices shown in Figure 6-15A(c–f). This indicates that this polymer-masking technique can be effectively used to precisely control the portion(s) of the CNT length to be functionalized.

Using a similar polymer-masking method, we have asymmetrically modified the sidewall of CNTs with oppositely charged moieties by plasma treatment and π-π stacking self-assembling, as shown schematically in Figure 6-16A [68]. The *as-prepared* asymmetrically sidewall-functionalized CNTs can be used as a platform for bottom-up self-assembly of complex structures or can be charge-selectively self-assembled onto and/or between electrodes with specific biases under an appropriate applied voltage for device applications [68]. As shown in Figure 6-16B(a), the asymmetrically sidewall-charged CNTs, with their half tube-length functionalized by pyrene-DMAEMA copolymer chains containing R_3HN^+ moieties and the other half tube-length by the plasma polymer containing COO^- moieties, were dispersed in ethanol within an electrochemical cell containing a silica wafer photolithographically patterned with two parallel electrodes less than 3 μm apart (Fig. 6-16B(a)). After the application of a DC voltage (10 V) between the two electrodes for 10 min, we took the silica wafer out of the electrochemical cell and thoroughly washed it with pure ethanol, followed by air drying at room temperature. SEM examination of the resultant electrodes revealed that some of the asymmetrically sidewall-charged CNTs had been well registered between the two electrodes (Fig. 6-16B(b)), presumably by simultaneously self-assembling their positively charged side onto the negative electrode and their negatively charged side onto the positive electrode.

The charge-induced self-assembling of the asymmetrically charged nanotubes across the two electrodes is, at least partially, confirmed by the absence of CNT adsorption in control experiments, in which either noncharged CNTs were used or no external voltage was applied onto the two electrodes immersed in the asymmetrically charged CNT solution (cf. insert of Fig. 6-16B(b)) under the same conditions. The current (I)-voltage (V) characteristics for the CNT self-assembled across the two electrodes show a pseudo-linear I-V curve (Fig. 6-16B(c)) consistent with the metallic nature of the multiwalled CNTs [2].

Apart from the above-mentioned *asymmetric* functionalization of nanotube sidewalls by sequentially masking VA-CNTs twice with only half of the nanotube sidewall being modified each time, the polymer-masking approach has also been used to impart magnetic properties to CNTs simply by region-selectively sputter-coating a thin layer of magnetic material (e.g., iron) onto the top-tips of VA-CNTs embedded into a polymer matrix (e.g., PS film) [67]. Therefore, the polymer-masking methodology, in conjunction with various functionalization chemistries, should allow for functionalization of CNTs in a region-specific fashion with a large variety of self-assembling reagents (e.g., ssDNA chains of specific sequences) for developing a wide range of novel nanotube-based multifunctional materials and devices.

By self-assembling temperature-responsive polymers [e.g., Poly(N-isopropylacrylamide), PNIPAAm] into VA-CNT arrays, we have created synergetic effects to allow us to develop polymer-CNT smart composites for various multifunctional applications [69]. As can be seen in Figure 6-17a, a uniform nanotube length and surface density for the pristine CNT array were observed. The dry surface of the PNIPAAm/aligned CNT film clearly shows features associated with the water-induced CNT bundles extruding out from the PNIPAAm matrix up to ~300 nm (Fig. 6-17b). For the wet sample, a much smoother featureless surface, with a reduced roughness of ~10 nm, is shown in Figure 6-17c. This is an indication that the PNIPAAm chains self-assembled around and between the VA-CNTs had expanded out from the gaps in the CNT forest to cover the nanotube top surface, as shown schematically by Figure 6-17(d–f). This provided the basis for the development of smart nanocomposite films with temperature-induced self-cleaning and/or controlled release capabilities [69]. We chose gold nanoparticles

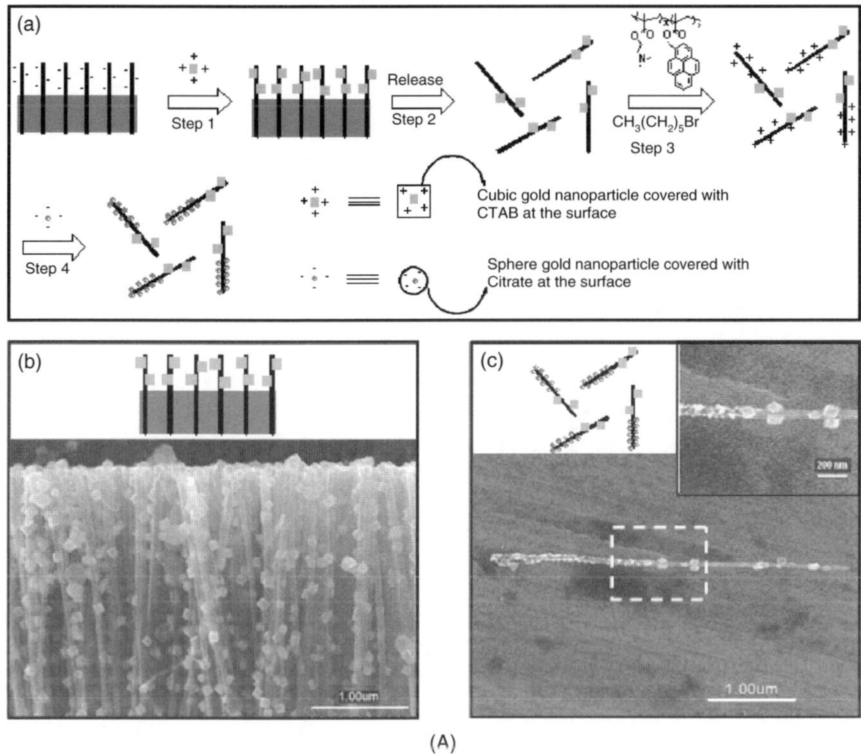

Figure 6-16 (A) Asymmetric functionalization of CNTs with opposite charges. (a) A schematic representation of procedures for asymmetric functionalization of CNTs with opposite charges, followed by tube-length-specific deposition of gold nanoparticles via electrostatical interactions. (b) Schematic representation and SEM image of the CNT array partially functionalized with cubic gold nanoparticles. (c) Schematic representation and SEM image of the resultant asymmetrically sidewall-functionalized CNTs with half of the nanotube length covered by gold nanocubes and the other half by spherical gold nanoparticles by electrostatical assembly (inset shows a higher-magnification SEM image for the squared area). Scale bars (b, c): 1 μm; Scale bar (right inset of c): 200 nm. (B) Self assembly of asymmetric functionalization of CNTs. (a) schematic illustration of the experimental setup for self-assembling CNTs, with their half tube-length functionalized by pyrene-DMAEMA copolymer chains containing R_3HN^+ moieties and the other half tube-length by the plasma polymer containing COO^- moieties (see text), onto two parallel electrodes less than 3 μm apart under a DC voltage. (b) SEM image of the patterned Au electrodes after having been immersed in the CNT solution shown in (a) for 10 min under 10 V. Inset at the top right corner shows the corresponding SEM image obtained from a control experiment with noncharged CNTs. (c) Typical current-voltage curve for the CNTs self-assembled between the two Au electrodes shown in (b) (Scale bar (b): 1 μm; adapted from Ref. 68). (A full color version of this figure appears in the color plate section.)

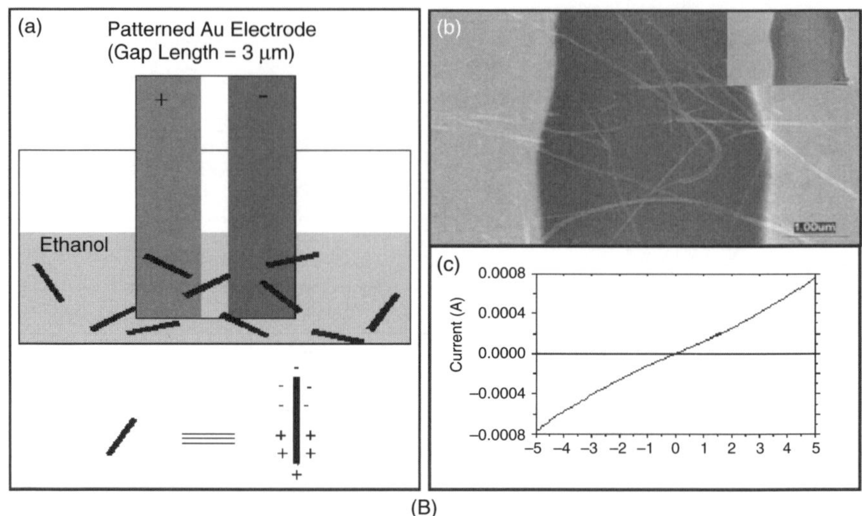

Figure 6-16   (*Continued*)

as model "contaminants" to demonstrate the self-cleaning capability of these self-assembled PNIPAAm/VA-MWNT nanocomposite films. To start with, we immersed the PNIPAAm/VA-MWNT self-assembled film into an aqueous solution of gold nanoparticles ($\sim$20 − 50 nm in diameter) at 50°C to allow for adsorption of the nanoparticles onto the nanotube top surface (Fig. 6-17g). By simply reducing the solution temperature down to 25°C (<LCST $\sim$32°C) (Fig. 6-17h), we found that almost all of the preadsorbed gold nanoparticles were dissociated from the top surface of the PNIPAAm/VA-MWNT self-assembled film.

6.7. CONCLUDING REMARKS

I have summarized our recent studies on the self-assembling of CNTs by controlled growth and solution functionalization, along with discussion of some complementary work in the literature and the potential applications for the resultant self-assemblies. Although the examples presented are highly selectively and mainly focused on our own work, they show that a large variety of functional structures and smart materials could be developed by self-assembling CNTs with controlled structure integrities and surface functionalities. With so many functionalization methods and controlled nanotube growth techniques already reported and more to be developed, there will be vast opportunities for developing numerous multifunctional structures and materials by self-assembling region-specifically functionalized CNTs. Continued research and development efforts in this exciting field would surely be of great value.

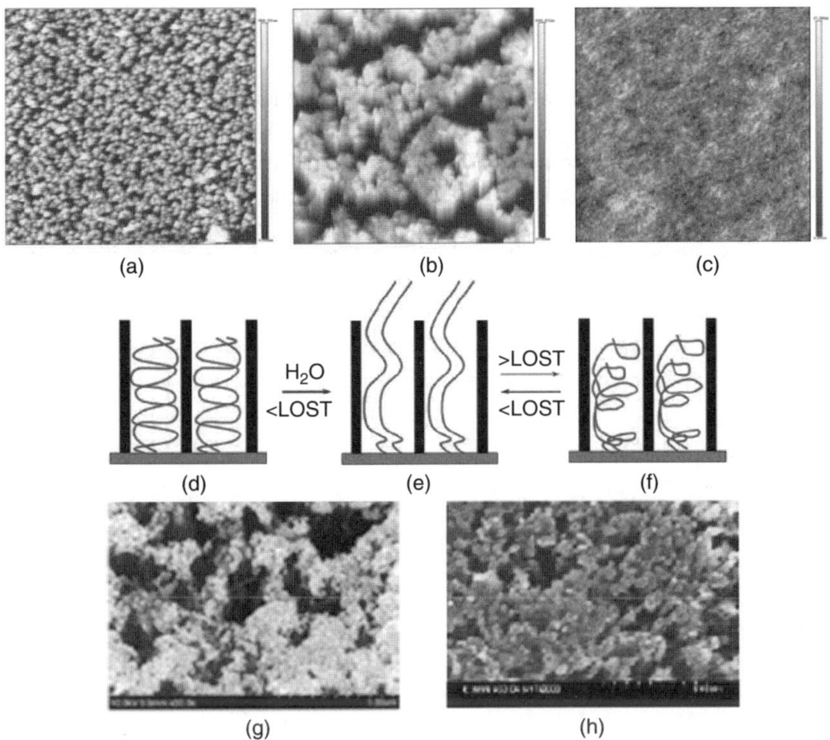

Figure 6-17 AFM images of (a) a pristine VA-CNT array and the PNIPAAm/VA-MWNT nanocomposite film: (b) in the dry state and (c) in the wet state. xy-Scale: (a–c) 5 μm × 5 μm, z-scale: (a) 445 nm; (b) 667 nm; (c) 17 nm. Schematic representation of the PNIPAAm/VA-MWNT film: (d, e) in the dry and wet state, respectively, at room temperature (20°C), (f) in the wet state at a temperature above the LCST (~32°C), and SEM images of the PNIPAAm/VA-MWNT nanocomposite film (g) after adsorbing gold nanoparticles on its top surface from an aqueous solution at 50°C and (h) after removal of the adsorbed gold nanoparticles by reducing the solution temperature down to 25°C (adapted from Ref. 69). (A full color version of this figure appears in the color plate section.)

ACKNOWLEDGMENTS

I thank my colleagues, including Wei Chen, Zhixin Guo, Pinggang He, Kung Min Lee, Sinan Li, Minoo Moghaddam, Maxine McCall, Shaoming Huang, Albert Mau, Qiang Peng, Liangti Qu, Ajit Roy, Tia Benson-Tolle, Richard Vaia, Junbing Yang, and Yongyuang Yang, for their contributions to the work cited. I am also grateful for financial support from the NSF (CMS-0609077, CTS0438389, CMMI-1000768, CCF-0403130) and AFOSR (FA9550-06-1-0384, FA9550-09-1-0331, FA8650-07-D-5800, FA2386-10-1-4071, FA9550-10-1-0546).

REFERENCES

1. P. J. F. Harris. *Carbon Nanotubes and Related Structures: New Materials for the Twenty-First Century*. Cambridge University Press, New York, **1999**.

2. L. Dai (Ed.). *Carbon Nanotechnology: Recent Developments in Chemistry, Physics, Materials Science and Device Applications*, Elsevier, Amsterdam, **2006**.

3. W. A. de Heer, J.-M. Bonard, K. Fauth, A. Châtelain, L. Forró, and D. Ugarte. Electron field emitters based on carbon nanotube films. *Adv. Mater*. **1997**, *9*, 87. and references cited therein.

4. R. H. Baughman, A. A. Zakhidov, and W. A. de Heer. Carbon nanotubes—the route toward applications. *Science* **2002**, *297*, 787.

5. H. Dai, J. H. Hafner, A. G. Rinzler, D. T. Colbert, and R. E. Smalley. Nanotubes as nanoprobes in scanning probe microscopy. *Nature* **1996**, *384*, 147.

6. S. S. Wong, E. Joselevich, A. T. Woolley, C. L. Cheung, and C. M. Lieber. Covalently functionalized nanotubes as nanometre-sized probes in chemistry and biology. *Nature* **1998**, *394*, 52.

7. G. Che, B. B. Lakshmi, E. R. Fisher, and C. R. Martin. Carbon nanotubule membranes for electrochemical energy storage and production. *Nature* **1998**, *393*, 346.

8. K. Gong, S. Chakrabarti, and L. Dai. Electrochemistry at carbon nanotube electrodes: Is the nanotube tip more active than the sidewall? *Angew. Chem. Int. Ed*. **2008**, *47*, 5446.

9. L. T. Qu and L. M. Dai. Substrate-enhanced electroless deposition of metal nanoparticles on carbon nanotubes. *J. Am. Chem. Soc*. **2005**, *127*, 10806.

10. L. T. Qu, L. M. Dai, and E. Osawa. Shape/size-controlled synthesis of metal nanoparticles for site-selective modification of carbon nanotubes. *J. Am. Chem. Soc*. **2006**, *128*, 5523.

11. X. H. Peng, J. Y. Chen, J. A. Misewich, and S. S. Wong. Carbon nanotube-nanocrystal heterostructures. *Chem. Soc. Rev*. **2009**, *38*, 1076.

12. G. G. Wildgoose, C. E. Banks, and R. G. Compton. Metal nanoparticles and related materials supported on carbon nanotubes: Methods and applications. *Small* **2006**, *2*, 182.

13. B. J. Hinds, N. Chopra, T. Rantell, R. Andrews, V. Gavalas, and L. G. Bachas. Aligned multiwalled carbon nanotube membranes. *Science* **2004**, *303*, 62.

14. M. Majumder, N. Chopra, R. Andrews, and B. J. Hinds. Nanoscale hydrodynamics: Enhanced flow in carbon nanotubes. *Nature* **2005**, *438*, 44.

15. R. H. Baughman, C. Cui, A. A. Zakhidov, Z. Iqbal, J. N. Barisci, G. M. Spinks, G. G. Wallace, A. Mazzoldi, D. De Rossi, A. G. Rinzler, O. Jaschinski, S. Roth, and M. Kertesz. Carbon nanotube actuators. *Science* **1999**, *284*, 1340.

16. L. Dai. Electrochemical sensors based on architectural diversity of the π-conjugated structure: Recent advancements from conducting polymers and carbon nanotubes. *Aust. J. Chem*. **2007**, *60*, 472.

17. C. Wei, L. Dai, A. Roy, and T. B. Tolle. Multifunctional chemical vapor sensors of aligned carbon nanotube and polymer composites. *J. Am. Chem. Soc*. **2006**, *128*, 1412.

18. W. Lu, L. Qu, K. Henry, and L. Dai. High performance electrochemical capacitors from aligned carbon nanotube electrodes and ionic liquid electrolytes. *J. Power Source* **2009**, *189*, 1270.

19. K. P. Gong, F. Du, Z. H. Xia, M. Durstock, and L. M. Dai. Nitrogen-doped carbon nanotube arrays with high electrocatalytic activity for oxygen reduction. *Science* **2009**, *323*, 760.

20. L. Qu, L. Dai, M. Stone, Z. Xia, and Z. L. Wang. Carbon nanotube arrays with strong shear binding-on and easy normal lifting-off. *Science* **2008**, *322*, 238.

21. L. Qu and L. Dai. Gecko-foot-mimetic aligned single-walled carbon nanotube dry adhesives with unique electrical and thermal properties. *Adv. Mater.* **2007**, *19*, 3844.

22. Y. S. Lee (Ed.). *Self-Assembly and Nanotechnology: A Force Balance Approach*, Wiley, Hoboken, NJ, **2008**.

23. L. M. Dai, A. Patil, X. Y. Gong, Z. X. Guo, L. Q. Liu, Y. Liu, and D. B. Zhu. Aligned nanotubes. *Chemphyschem* **2003**, *4*, 1150, and references cited therein.

24. Y. Yan, M. B. Chan-Park, and Q. Zhang. Advances in carbon-nanotube assembly. *Small* **2007**, *3*, 24, and references cited therein.

25. L. Qu and L. Dai. Direct growth of three-dimentional multicomponent micropatterns of vertically aligned singled-walled carbon nanotubes interposed with their multi-walled counterparts on Al-activated iron substrates. *J. Mater. Chem.* **2007**, *17*, 3401 and references cited therein.

26. S. M. Huang, L. Dai, and A. W. H. Mau. Patterned growth and contact transfer of well-aligned carbon nanotube films. *J. Phys. Chem. B* **1999**, *103*, 4223.

27. S. Chakrabarti, K. Gong, and L. Dai. Structural evaluation along the nanotube length for super-long vertically aligned double-walled carbon nanotube arrays. *J. Phys. Chem. C* **2008**, *112*, 8136.

28. L. Qu, Q. Peng, L. Dai, G. Spinks, G. Wallace, and R. H. Baughman. Carbon nanotube electroactive polymer materials: Opportunities and challenges. *MRS Bull.* **2008**, *33*, 215, and references cited therein.

29. C. N. R. Rao, R. Sen, B. C. Satishkumar, and A. Govindaraj. Large aligned-nanotube bundles from ferrocene pyrolysis. *Chem. Commun.* **1998**, 1525.

30. D. C. Li, L. Dai, S. M. Huang, A. W. H. Mau, and Z. L. Wang. Structure and growth of aligned carbon nanotube films by pyrolysis. *Chem. Phys. Lett.* **2000**, *316*, 349.

31. K. Hata, D. N. Futaba, K. Mizuno, T. Namai, M. Yumura, and S. Iijima. Water-assisted highly efficient synthesis of impurity-free single-walled carbon nanotubes. *Science* **2004**, *306*, 1362.

32. Y. Murakami, S. Chiashi, Y. Miyauchi, M. H. Hu, M. Ogura, T. Okubo, and S. Maruyama. Growth of vertically aligned single-walled carbon nanotube films on quartz substrates and their optical anisotropy. *Chem. Phys. Lett.* **2004**, *385*, 298.

33. E. Gyula, A. A. Kinkhabwala, H. Cui, D. B. Geohegan, A. A. Puretzky, and D. H. Lowndes. Molecular beam-controlled nucleation and growth of vertically aligned single-wall carbon nanotube arrays. *J. Phys. Chem. B* **2005**, *109*, 16684.

34. T. Iwasaki, G. F. Zhong, T. Aikawa, T. Yoshida, and H. Kawarada. Direct evidence for root growth of vertically aligned single-walled carbon nanotubes by microwave plasma chemical vapor deposition. *J. Phys. Chem. B* **2005**, *109*, 19556.

35. Y. Q. Xu, E. Flor, M. J. Kim, B. Hamadani, H. Schmidt, R. E. Smalley, and R. H. Hauge. Vetical array growth of small diameter single-walled carbon nanotubes. *J. Am. Chem. Soc.* **2006**, *128*, 6560; G. Y. Zhang, D. Mann, L. Zhang, A. Javey, Y. M. Li, E. Yenilmez, Q. Wang, J. P. McVittie, Y. Nishi, J. Gibbons, and H. J. Dai. Ultra-high-yield growth of vertical single-walled carbon nanotubes: Hidden roles of hydrogen and oxygen. *Proc. Natl. Acad. Sci. USA* **2005**, *102*, 16141.

36. P. B. Amama, C. L. Pint, L. McJilton, S. M. Kim, E. A. Stach, P. Murray, R. H. Hauge, and B. Maruyama. Role of water in super growth of single-walled carbon nanotube carpets. *Nano Lett.* **2009**, *9*, 44.

37. Y. Yang, S. Huang, W. He, A. W. H. Mau, and L. Dai. Patterned growth of well-aligned carbon nanotubes: A photolithographic approach. *J. Am. Chem. Soc.* **1999**, *121*, 10832.

38. S. Huang, A. W. H. Mau, T. W. Turney, P. A. White, and L. Dai. Patterned growth of well-aligned carbon nanotubes: A soft-lithographic approach. *J. Phys. Chem. B* **2000**, *104*, 2193.

39. Q. Chen and L. Dai. Plasma patterning of carbon nanotubes. *Appl. Phys. Lett.* **2000**, *76*, 2719.

40. L. Qu, Y. Zhao, and L. Dai. Carbon microfibers sheathed with aligned carbon nanotubes: Towards multidimensional, multicomponent, and multifunctional nanomaterials. *Small* **2006**, *2*, 1052.

41. J. Yang, L. Dai, and R. A. Vaia. Multicomponent interposed carbon nanotube micropatterns by region-specfic contact transfer and self-assembling. *J. Phys. Chem. B* **2003**, *107*, 12387.

42. J. Yang, L. Qu, Y. Zhao, Q. Zhang, L. Dai, J. W. Baur, B. Maruyama, R. A. Vaia, E. Shin, P. T. Murray, H. Luo, and Z. Guo. Multicomponent and multidimensional carbon nanotube micropatterns by dry contact transfer. *J. Nanosci. Nanotechnol.* **2007**, *7*, 1573.

43. Z. Liu, Z. Shen, T. Zhu, S. Hou, L. Ying, Z. Shi, and Z. Gu. Organizing single-walled carbon nanotubes on gold using a wet chemical self-assembling technique. *Langmuir* **2000**, *16*, 3569.

44. B. Wu, J. Zhang, Z. Wei, S. Cai, and Z. Liu. Chemical alignment of oxidatively shortened single-walled carbon nanotubes on silver surface. *J. Phys. Chem. B* **2001**, *105*, 5075.

45. X. Nan, Z. Gu, and Z. Liu. Immobilizing shortened single-walled carbon nanotubes (SWNTs) on gold using a surface condensation method. *J. Col. Inter. Sci.* **2002**, *245*, 311.

46. H. Pan, L. Liu, Z.-X. Guo, L. Dai, F. Zhang, D. Zhu, R. Czerw, and D. L. Carroll. Carbon nanotubols from mechanochemical reaction. *Nano. Lett.* **2003**, *3*, 29.

47. H. Shimoda, S. J. Oh, H. Z. Geng, R. J. Walker, X. B. Zhang, L. E. McNeil, and O. Zhou. Self-assembly of carbon nanotubes. *Adv. Mater.* **2002**, *14*, 899.

48. S. C. Tsang, Y. K. Chen, P. J. F. Harris, and M. L. H. Green. A simple chemical method of opening and filling carbon nanotubes. *Nature* **1994**, *372*, 159.

49. C. LeMieux, M. Roberts, S. Barman, Y. W. Jin, J. M. Kim, and Z. Bao. Self-sorted, aligned nanotube networks for thin-film transistors. *Science* **2008**, *321*, 101, and references cited therein.

50. L. Dai. *Intelligent Macromolecules for Smart Devices: From Materials Synthesis to Device Applications*, Springer-Verlag, New York, **2004**.

51. M. J. O'Connell, P. Boul, L. M. Ericson, C. Huffman, Y. Wang, E. Haroz, C. Kuper, J. Tour, K. D. Ausman, and R. E. Smalley. Reversible water-solubilization of single-walled carbon nanotubes by polymer wrapping. *Chem. Phys. Lett.* **2001**, *342*, 265.

52. R. J. Chen, Y. Zhang, D. Wang, and H. Dai. Noncovalent sidewall functionalization of single-walled carbon nanotubes for protein immobilization. *J. Am. Chem. Soc.* **2001**, *123*, 3838.

53. X. H. Peng, J. Y. Chen, J. A. Misewich, and S. S. Wong. Carbon nanotube-nanocrystal heterostructures. *Chem. Soc. Rev.* **2009**, *38*, 1076.

54. G. G. Wildgoose, C. E. Banks, and R. G. Compton. Metal nanoparticles and related materials supported on carbon nanotubes: Methods and application. *Small* **2006**, *2*, 182.

55. M. J. Moghaddam, S. Taylor, M. Gao, S. Huang, L. Dai, and M. J. McCall. Highly efficient binding of DNA on the sidewalls and tips of carbon nanotubes using photochemistry. *Nano Lett.* **2004**, *4*, 89.

56. B. Quinn, C. Dekker, and S. Lemay. Electrodeposition of noble metal nanoparticles on carbon nanotubes. *J. Am. Chem. Soc.* **2005**, *127*, 6146.

57. J. Li, M. Moskovits and T. L. Haslett. Nanoscale electroless metal deposition in aligned carbon nanotubes. *Chem. Mater.* **1998**, *10*, 1963.

58. H. C. Choi, M. Shim, S. Bangsaruntip, and H. J. Dai. Spontaneous reduction of metal ions on the sidewalls of carbon nanotubes. *J. Am. Chem. Soc.* **2002**, *124*, 9058.

59. Y. C. Xing. Synthesis and electrochemical characterization of uniformly-dispersed high loading Pt nanoparticles on sonochemically-trated carbon nanotubes. *J. Phys. Chem. B* **2004**, *108*, 19255.

60. R. Azamian, K. Coleman, J. Davis, N. Hanson, and M. Green. Directly observed covalent coupling of quantum dots to single-wall carbon nanotubes. *Chem. Commun.* **2002**, 366.

61. B. Kim and W. M. Sigmund. Functionalized multiwall carbon nanotube/gold nanoparticle composites. *Langmuir* **2004**, *20*, 8239.

62. Y. D. Yang, L. T. Qu, L. M. Dai, T. S. Kang, and M. Durstock. Electrophoresis coating of titanium dioxide on aligned carbon nanotubes for controlled syntheses of photoelectronic nanomaterials. *Adv. Mater.* **2007**, *19*, 1239.

63. S. Li, P. He, J. Dong, Z. Gao, and L. Dai. DNA-directed self-assembling of carbon nanotubes. *J. Am. Chem. Soc.* **2005**, *127*, 14.

64. A. Ulman. Formation and structure of self-assembled monolayers. *Chem. Rev.* **1996**, *96*, 1533 and references therein.

65. K. M. Lee, L. C. Li, and L. M. Dai. Asymmetric end-functionalization of multi-walled carbon nanotubes. *J. Am. Chem. Soc.* **2005**, *127*, 4122.

66. N. Chopra, M. Majumder, and B. J. Hinds. Bifunctional carbon nanotubes by sidewall protection. *Adv. Funct. Mater.* **2005**, *15*, 858.

67. L. T. Qu and L. Dai. Polymer-masking for controlled functionalization of carbon nanotubes. *Chem. Commun.* **2007**, 3859.

68. Q. Peng, L. T. Qu, L. Dai, K. Park, and R. A. Vaia. Asymmetrically charged carbon nanotubes by controlled functionalization. *ACS Nano* **2008**, *2*, 1833.

69. W. Chen, L. Qu, D. Chang, L. Dai, S. Ganguli, and A. Roy. Vertically-aligned carbon nanotubes infiltrated with temperature-responsive polymers: Smart nanocomposite films for self-cleaning and controlled release. *Chem. Commun.* **2008**, 163.

CHAPTER 7

Self-Organized Fullerene-Based Organic Semiconductors

LI-MEI JIN AND QUAN LI

Liquid Crystal Institute, Kent State University, Kent, Ohio

7.1. INTRODUCTION

Because of the promise of low cost and the possibility of roll-to-roll processing at ambient temperature and pressure, organic semiconductors hold great promise in applications such as organic photovoltaics (OPVs), organic light-emitting diodes (OLEDs), and organic field-effect transistors (OFETs). Among the known organic semiconductors, fullerene and its analogs are certainly a most important family that benefit from their elegant electron-accepting capability [1, 2]. The fullerene C_{60} was discovered in 1985 by Kroto, Curl, and Smalley. The significance of this discovery is marked by the awarding of the 1996 Nobel Prize in Chemistry. Since then, numerous electron donor-acceptor complexes with fullerene as the electron acceptor core have been synthesized to demonstrate the electron transfer process. The electron donor-acceptor complex or charge-transfer complex is a chemical association of two or more molecules, or of different parts of one large molecule. The attraction between the molecules or parts is created by an electron transition into an excited electronic state such that a fraction of electronic charge is transferred between the molecules or fragments.

It is well known that fullerene as an electron acceptor can accelerate photoinduced charge separation and decelerate charge recombination in the donor-acceptor system. Therefore, one of the most actively pursued research topics related to fullerene is development of OPV cells leading to low-cost photovoltaics. The soluble $PC_{61}BM$ [(1-(3-methoxycarbonyl)propyl-1-phenyl [6, 6]C_{61})] is a most widely used fullerene derivative in OPVs [3]. To date, numerous efforts have been devoted to the development of high-performance

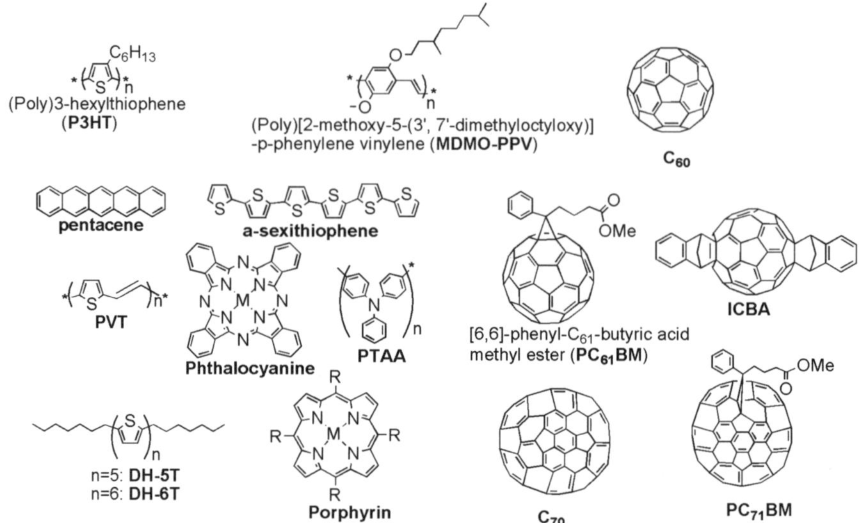

Figure 7-1 Examples of organic donors and fullerene acceptors used in OPV cells.

OPVs to propel this technology toward commercialization. Examples of organic donors and fullerene acceptors used in OPVs are shown in Figure 7-1. It is well established that donor and acceptor components in OPVs must be mixed or close together on a length scale that is shorter than the exciton diffusion length, typically ~10 nm, to ensure an efficient disassociation of excitons [4]. The formation of a bicontinuous interpenetrating network from donor and acceptor components as a result of self-organization on the nanometer scale is another key issue for efficient carrier transport of the separated charges. Since fullerene and its derivatives are the most widely used electron acceptor in OPVs, discovering new techniques and designing new molecules to control nanoscale morphology of a fullerene-based active layer are particularly important to improve the efficiency of OPVs. To improve the device performance, film formation techniques such as electrostatic layer-by-layer deposition, Langmuir–Blodget film, and self-assembled monolayers are considered. The use of fullerene liquid crystals or noncovalent supramolecular interactions to form a self-organized active thin layer in OPVs can be another important approach to improve device performance [5–15]. In this chapter, we give a brief overview of self-organization of fullerene [60]-based materials, with a focus on their OPV application.

7.2. FULLERENE-BASED LIQUID CRYSTALLINE DONOR-ACCEPTOR BLENDS

Liquid crystals (LCs) are an intermediate state of matter between the crystalline and the isotropic liquid phases seen in some highly anisotropic organic molecules.

Self-organization of the LC molecules allows us to control the alignment of the molecules and the minimization of defects, which are inevitable drawbacks for charge transportation in crystalline materials. As mentioned in Chapters 3 and 4, columnar (Col) LCs, which are formed by the self-assembly of discotic molecules, have unique one-dimensional charge-carrier mobility through their long columnar axis [16, 17]. The discotic LCs capable of the formation of defect-free thin film driven by π-π interactions provide tremendous opportunities in OPVs [18–29]. It is established that the efficiency of discotic LCs as an active layer in photovoltaic (PV) cells is critically dependent on the supramolecular arrangement of the blend made from an electron donor component and an electron acceptor component. Compared with a donor and acceptor bilayer PV cell, a heterojunction blend can offer a much larger interface between donor and acceptor and therefore offer more efficient dissociation of excitons in the device [30, 31]. Of all the donor-acceptor blends, the porphyrin-fullerene blend is supposed to be an excellent marriage since fullerene is an excellent electron acceptor, and porphyrin, which is the basic structure of the best photoreceptor in nature, chlorophyll, is a superior electron donor.

Recently, we have found that the blend of porphyrin **1** and $PC_{61}BM$ can self-organize into a highly ordered thin film just by thermal annealing, as confirmed by polarizing optical microscopy (POM) and X-ray diffraction (XRD) measurement (Fig. 7-2) [32]. Based on the fact that the diameter of porphyrin is approximately 6.5 nm whereas the distance between the two neighboring columns from the XRD experiment is about 4.4 nm, we can deduce that interdigitation of the long flexible chains exists since the lipid chain intercalating length is about 1.6 nm. Considering the size of $PC_{61}BM$ (about 0.7 nm) and its lipophilicity, one possible molecular arrangement is the hexagonal columnar phase of porphyrin with $PC_{61}BM$ at interstices of the columns (Fig. 7-2). Another possible molecular arrangement is that $PC_{61}BM$ is sandwiched between two porphyrin cores because of the strong π-donor and π-acceptor interaction. Regardless of the packing arrangement, the porphyrin and $PC_{61}BM$ complex self-organizes into a homeotropically aligned columnar architecture (Fig. 7-2), which is very important for OPV applications because such a molecular arrangement can provide an efficient path for electrons and/or holes along the columnar axis where the light-harvesting molecules are arranged with the largest area toward the incident light.

Bulk heterojunction OPV cells based on a structurally similar liquid crystalline porphyrin **2** and $PC_{61}BM$ exhibited V_{oc} of 0.510 V, J_{sc} of 1.340 mA/cm^2, and power conversion efficiency (PCE) of 0.222%, where V_{oc} is the voltage measured at open circuit and J_{sc} is the current density measured at short circuit. Importantly, after the postannealing, the device achieved a J_{sc} of 3.990 mA/cm^2 and a PCE of 0.712%, leading to a significant increase of $\sim$300% higher PCE than the counterpart device without postannealing, which might result from the alignment of the porphyrin-$PC_{61}BM$ blend by the thermal annealing [33]. The preliminary photovoltaic performance is encouraging, since a homeotropic alignment is difficult to achieve in such a blend film. We can anticipate dramatic improvements in efficiency by tailoring molecule structure, optimized cell

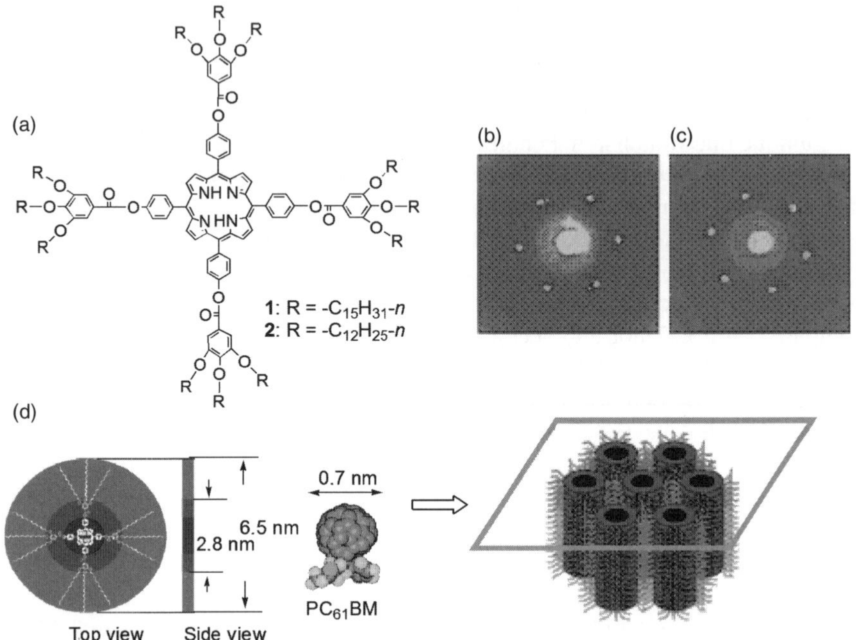

Figure 7-2   (a) Chemical structure of liquid crystalline porphyrin **1** and **2**. (b) Synchrotron XRD patterns from homeotropic monodomain of material **1** in a 8-μm-thick glass cell. (c) Synchrotron XRD patterns from homeotropic monodomain of the blend of **1** with PC$_{61}$BM in a 8-μm-thick glass cell. (d) Calculated geometric dimensions of porphyrin **1** and 3D ChemDraw spacing-filling model of fullerene derivative PC$_{61}$BM and the schematic representations of homeotropically aligned architecture of the blend of **1** and PC$_{61}$BM. Reproduced with permission from the American Chemical Society [32]. (A full color version of this figure appears in the color plate section.)

structure, and engineering if we can maintain the homeotropically aligned architecture in bulk, double, or multiheterojunction OPV cells.

Torres et al. reported a highly ordered mesophase formed by mixing the mesomorphic phthalocyanine (Pc) **3** with the nonmesomorphic Pc-C$_{60}$ dyad **4**, **5**, or **6** [34]. A differential scanning calorimetry (DSC) diagram of the blends exhibited a clear transition from the crystalline state to a mesophase, which was identified as a hexagonal columnar (Col$_h$) structure by XRD. Considering the fact that there is no LC property in the pure Pc-C$_{60}$ dyads **4** and **5** and the mismatch in the column diameter would result in no mesomorphism in domains of separated molecules of Pc-C$_{60}$ dyad and Pc **3**, a predominance of alternating stacking was proposed as shown in Figure 7-3.

Very recently, Han et al. improved the performance of a well-established P3HT:PC$_{61}$BM bulk heterojunction OPV cell by adding a small amount of discotic LC 2,3,6,7,10,11-hexaacetoxytriphenylene. For example, photovoltaic devices fabricated with an active layer containing 3 wt% of the discotic LC

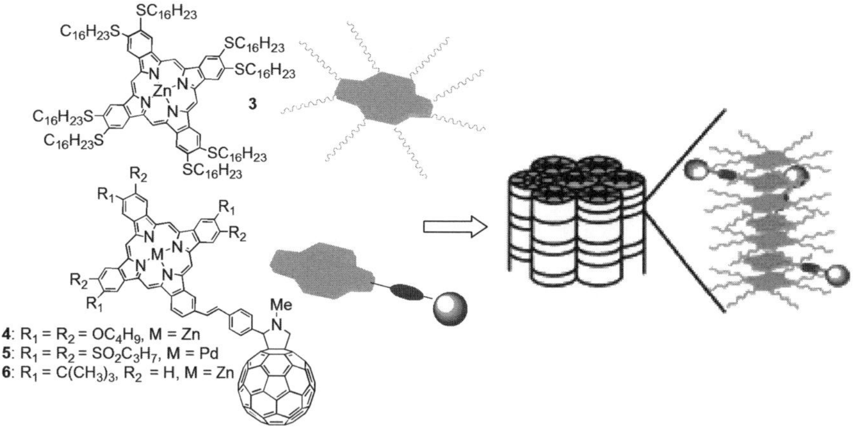

Figure 7-3 Schematic representation of the LC phase formed by the blend of mesomorphic phthalocyanine (Pc) **3** and the non-mesomorphic Pc-C_{60} dyads **4, 5** and **6**. Reproduced with permission from the American Chemical Society [34]. (A full color version of this figure appears in the color plate section.)

resulted in an average PCE of 3.97% after thermal annealing, compared to the reference cells without addition of the discostic LC with PCE of 3.03% [35].

7.3. FULLERENE-BASED LIQUID CRYSTALLINE COVALENTLY LINKED DONOR-ACCEPTOR DYADS

The fullerene-based covalently linked donor-acceptor system possessing the unique self-organizing property of LCs is an attractive molecular design strategy for OPV applications. Covalently linking an electron-donating mesogenic unit to a fullerene moiety imparts liquid crystallinity to the whole molecule. Such fullerene-containing liquid crystalline donor-acceptor dyads have shown better performance in OPV applications [36–38]. Liquid crystalline C_{60}-donor dyads where ferrocene (Fc), oligophenylenevinylene and tetrathiafulvalene (TTF) act as electron donors have been reported [39–42]. The fullerene-based covalently linked donor-acceptor system with the unique self-organizing property of liquid crystals could be used as a supramolecular platform in OPV technology.

The liquid crystalline ferrocene-fullerene dyad **7** prepared by a Bingel reaction between C_{60} and a ferrocene exhibited a smectic A phase (Fig. 7-4). It was found that the electron transfer occurred from the donor ferrocene to the acceptor fullerene. A long-lived radical pair with a lifetime of the order of several hundred nanoseconds was formed by the photoexcitation [43]. Taking into account the cross-sectional areas of the C_{60} and the mesogenic fragments as well as the possible interactions, the supramolecular organization of **7** was postulated as in Figure 7-5. The fullerene moieties formed a double sublayer inserted between aliphatic sublayers. Space filling was obtained through a microsegregation of the

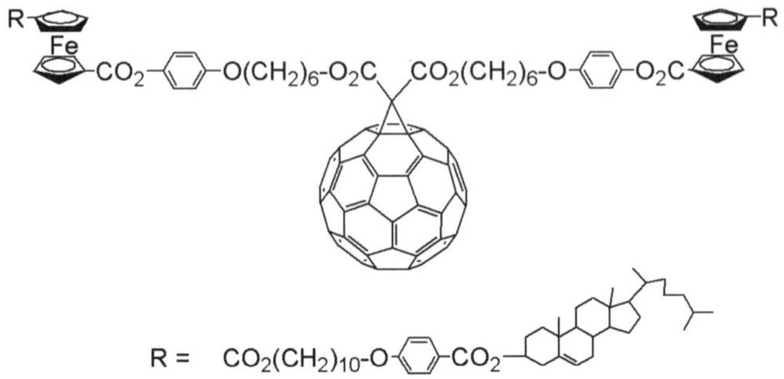

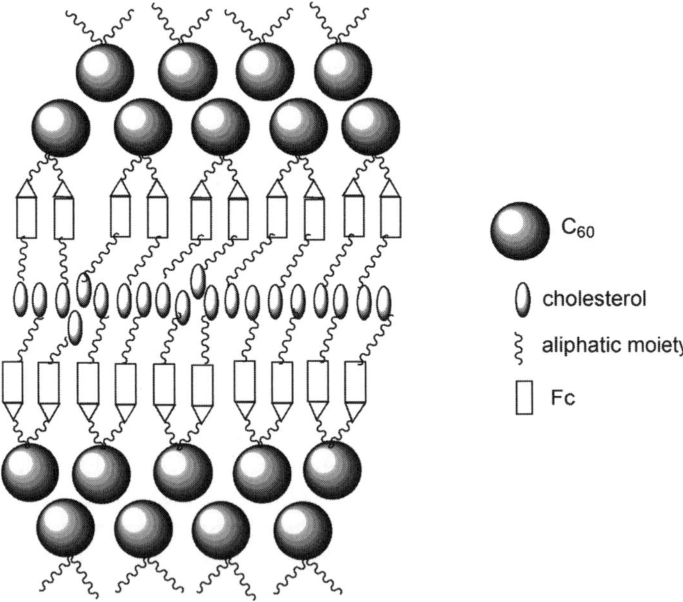

Figure 7-4   Chemical structure of the ferrocene-fullerene dyad **7**.

Figure 7-5   Proposed supermolecular organization of **7** in smectic A phase.

different molecular species together with good apposition between the interfaces of the different sublayers.

Liquid crystalline oligophenylenevinylene-fullerene dyads **8** and **9** were reported by Nierengarten et al. (Fig. 7-6) [44, 45]. Their preliminary luminescence measurements in CH_2Cl_2 showed a dramatic quenching of the oligophenylenevinylene fluorescence by the fullerene unit in both derivatives upon selective excitation at the oligophenylenevinylene band maximum (**8**: 367 nm; **9**: 389 nm), indicating the occurrence of an intramolecular photoinduced process that might be useful for OPV application.

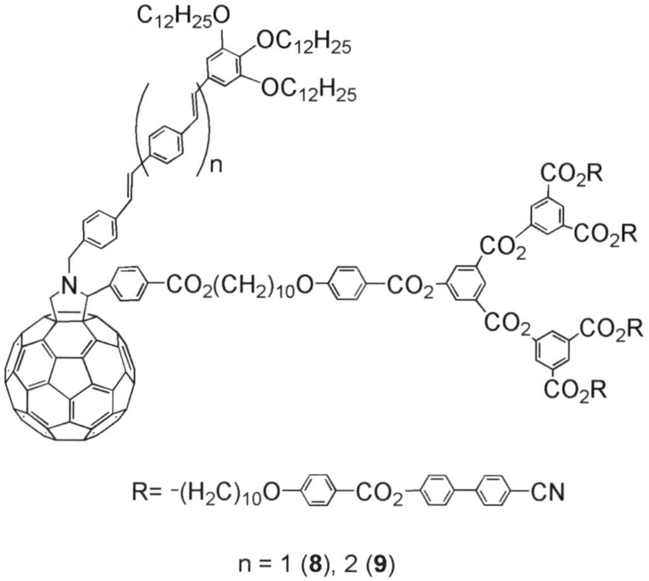

$R = -(H_2C)_{10}O-\langle\!\!\langle\rangle\!\!\rangle-CO_2-\langle\!\!\langle\rangle\!\!\rangle-\langle\!\!\langle\rangle\!\!\rangle-CN$

$n = 1\ (\mathbf{8}),\ 2\ (\mathbf{9})$

Figure 7-6 Chemical structures of the liquid crystalline oligophenylenevinylene-fullerene dyads **8** and **9**.

A second-generation ferrocene-fullerene dyad **10** prepared by 1,3-dipolar cycloaddition reaction exhibited a smectic A phase (Fig. 7-7) [46], which was identified based on the observation of focal-conic and homeotropic textures. Its cyclic voltammetric behavior was investigated in THF solution under strictly aprotic conditions. The experimental results showed that three redox waves at -0.46, -1.04, and -1.65 V $(E_{1/2})$ were associated with the reversible subsequent one-electron reduction of the fulleropyrrolidine. The anodic peak with $E_{1/2} = 0.70$ V was attributed to the reversible one-electron oxidation of the ferrocene moiety. Photoinduced electron transfer from ferrocene to fullerene was identified with lifetimes for the charge-separated state of 560 ns (THF) and 490 ns (benzonitrile), respectively.

The TTF-fullerene dyad **11** with a bis-mesogenic fragment showed smectic A and smectic B phases (Fig. 7-8) [47]. The supramolecular organization of **11** in the smectic layers was of the bilayer type in a head-to-tail fashion with the C_{60} units slightly shifted. Its electrochemical property was studied by cyclic voltammetry. The dyad **11** gave rise to two quasireversible, one-electron reduction waves at -0.60 and -0.99 V, corresponding to the reduction of C_{60}, and two reversible oxidation waves at 0.52 and 0.78 V due to the formation of the radical cation and dication species of the TTF fragment. The reduction potentials were a little more negatively shifted compared to C_{60} (-0.54 and -0.92 V) because of the saturation of a double bond in C_{60}, which raised the LUMO energy [48]. The oxidation potentials were similar to the TTF derivative **12** without fullerene unit (0.52 and 0.79 V). The electrochemical property of **12** indicated that both donor

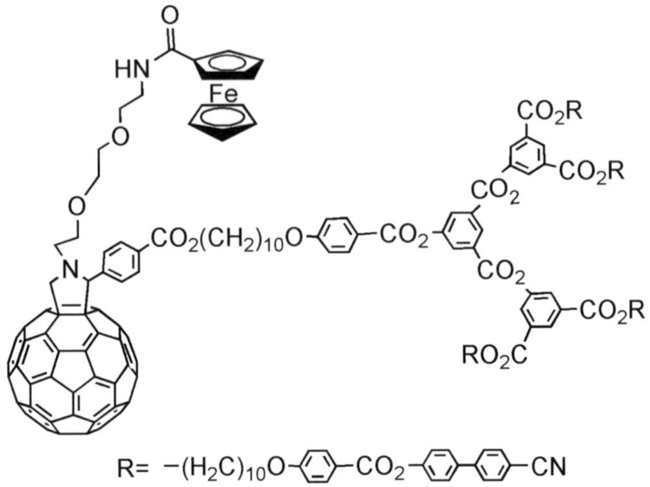

10

Figure 7-7   Chemical structure of the ferrocene-fullerene dyad **10**.

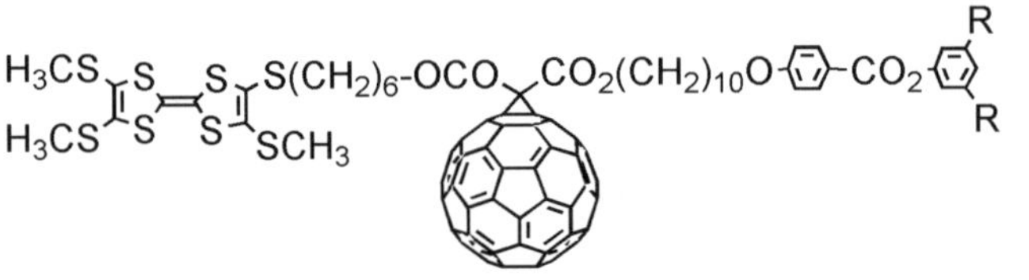

11

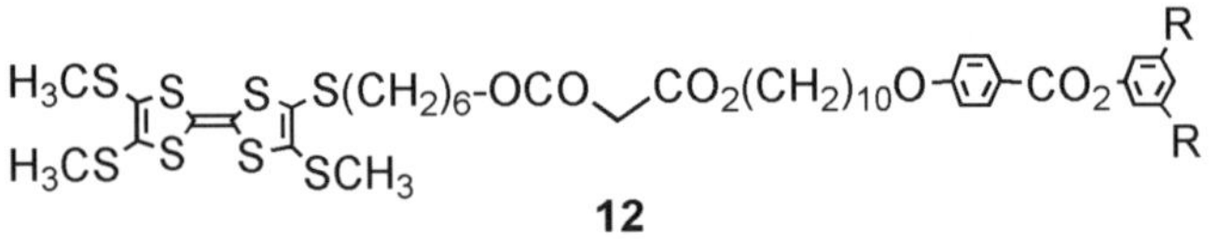

12

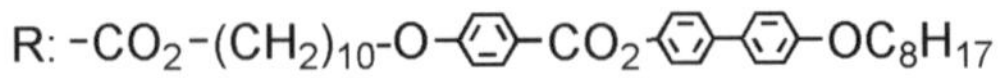

Figure 7-8   Chemical structures of TTF-fullerene dyad **11** and TTF derivative **12**.

(TTF fragment) and acceptor (C_{60} fragment) units preserved their individual electroactive identities.

The ferrocene-fullerene dyad **13** (Fig. 7-9) has an enantiotropic smectic A phase from 57 to 155°C. Detailed investigation of its photophysical property by steady-state and time-resolved fluorescence as well as transient absorption spectroscopy in polar and apolar solvents showed interesting photoinduced electron transfer phenomena. Its cyclic voltammetric (CV) behavior was investigated in THF solution under strictly aprotic conditions. The CV curve displayed a series of

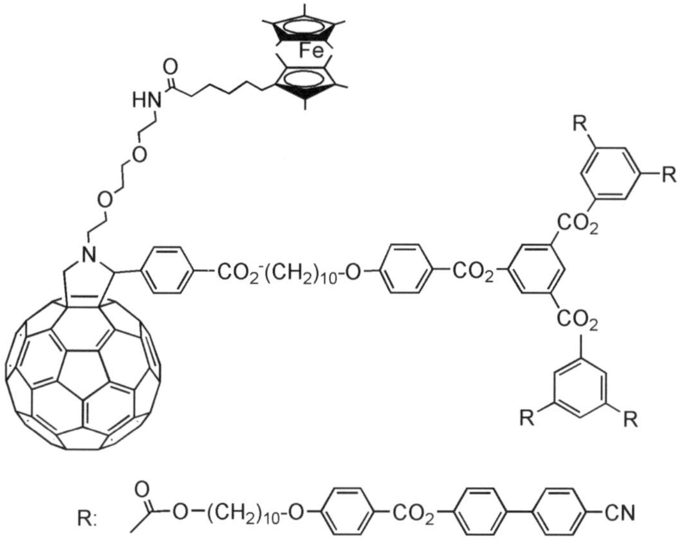

13

Figure 7-9 Chemical structure of the ferrocene-fullerene dyad **13**.

subsequent reduction peaks that were partially attributed to the fulleropyrrolidine moiety, while the other peaks were due to the dendrimer [49].

Matsuo et al. reported a series of conical molecules with a fullerene apex **14–19** (Fig. 7-10), in which a fullerene core was connected to many hydrocarbon groups by silyl acetylene tethers [50–53]. The shuttlecock molecules were found to be able to self-assemble into layered lamellar structures in crystal and LC states. For example, compound **14** showed a lamellar structure. The distance between the fullerene layers was 22.44 Å determined by single X-ray crystallographic technique. The crystal packing of compound **15** was a pseudolamellar system, in which two molecules were interdigitated and showed a shorter interlayer distance of 16.22 Å. With a longer hydrocarbon chain on the mesogenic molecule, the volume between the neighboring fullerene layers became larger, and the interlayer distance increased accordingly. Compounds **16–19** with long alkyl chains exhibited the liquid crystalline property as well. Typically, batonnet textures of a smectic phase were observed after cooling from the isotropic liquid phase. In addition, XRD analyses of compounds **16–19** showed that they did not form the interdigitated dimer structure but formed true layered structures with interlayer distances of 20.5, 22.6, 23.9, and 25.8 Å, respectively.

Nakanishi et al. reported long-range ordered lamellar mesophases of the fulleropyrrolidines bearing long alkyl chains, providing high C_{60} content and a comparably high electron mobility of $\sim 3 \times 10^{-3}$ $cm^2V^{-1}s^{-1}$ [54]. Fullerene derivatives **20**, **21**, and **23** with enough long alkyl chains exhibited ordered lamellar mesophase, whereas derivatives **22**, **24** and **25** were found

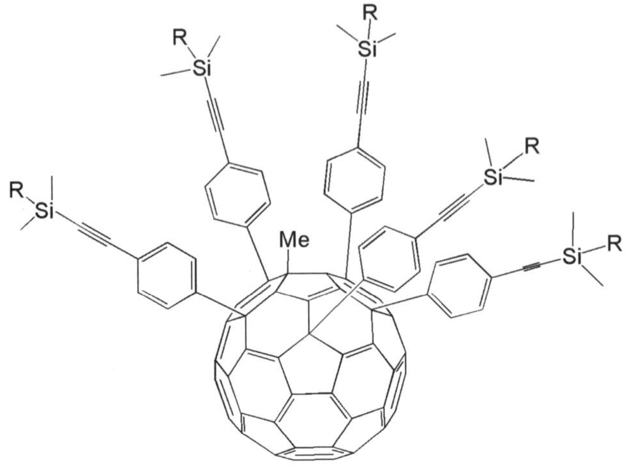

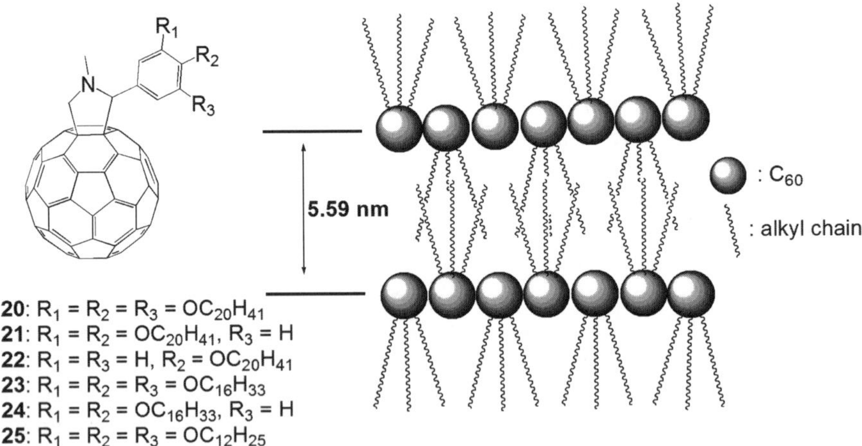

14: R = Me; **15**: R = n-C$_4$H$_9$

16: R = n-C$_8$H$_{17}$; **17**: R = n-C$_{10}$H$_{21}$; **18**: R = n-C$_{12}$H$_{25}$; **19**: R = n-C$_{14}$H$_{29}$

Figure 7-10   Chemical structures of compounds **14–19**.

20: R$_1$ = R$_2$ = R$_3$ = OC$_{20}$H$_{41}$
21: R$_1$ = R$_2$ = OC$_{20}$H$_{41}$, R$_3$ = H
22: R$_1$ = R$_3$ = H, R$_2$ = OC$_{20}$H$_{41}$
23: R$_1$ = R$_2$ = R$_3$ = OC$_{16}$H$_{33}$
24: R$_1$ = R$_2$ = OC$_{16}$H$_{33}$, R$_3$ = H
25: R$_1$ = R$_2$ = R$_3$ = OC$_{12}$H$_{25}$

Figure 7-11   Chemical structures of the fulleropyrrolidines **20–25** (left) and the proposed lamellar self-organization of **20** (right).

to be nonmesomorphic (Fig. 7-11). A schematic model of the supramolecular organization of **20** is presented in Figure 7-11. A cast film of **23** in 0.1 M aqueous n-Bu$_4$NCl solution at 60°C in the liquid crystalline state showed the first and second redox events corresponding to the generation of C$_{60}$ monoanion and dianion at potentials of $E_{red,1} = -0.70$ and $E_{red,2} = -0.87$ V respectively, whereas no redox response was observed at 15°C in the liquid crystalline state.

Liquid crystalline oligothiophene-C_{60} dyads 1_{Amphi} and 1_{Lipo} were reported by Aida et al. [55]. The smectic A phases of 1_{Amphi} (136.1 to 18.3°C) and 1_{Lipo} (111.4 to 12.5°C) were confirmed by POM and XRD measurements. The proposed molecular orientations in the smectic A phase are given in Figure 7-12. Both of the two dyads exhibited a distinct photoconductive character, whereas no photoconductivity was observed for the blends of the corresponding reference donor **26** and acceptor **27**.

Another example of Pc-C_{60} dyad **29** designed for heterojunction OPV cells was reported by Geerts et al. (Fig. 7-13) [56]. The dyad exhibited a LC phase at room temperature. The cyclic voltammetry result of **29** in CH_2Cl_2 corresponded to the superposition of those for $PC_{61}BM$ and Pc **28**, and no intramolecular charge transfer in the ground state was observed.

7.4. FULLERENE-BASED HYDROGEN-BONDED DONOR-ACCEPTOR ENSEMBLES

Compared to the covalently linked donor-acceptor dyads, noncovalent supramolecular interactions such as hydrogen-bonding donor-acceptor ensembles were found to be beneficial for the formation of long-lived photoinduced radical-ion pairs, although the hydrogen-bonded system was paid less attention until recently. Moreover, in OPVs, the use of hydrogen-bonding molecular recognition motifs was found to be able to enhance the photovoltaic response by the formation of an ordered supermolecular active layer [57, 58]. For instance, Bassani et al. reported that a 2.5-fold enhancement in light energy to electrical energy conversion was observed when comparing the photovoltaic device with hydrogen-bonding to that without hydrogen bonding [59]. The device was prepared by using oligothiophene **30** in the presence of two equivalents of fullerene-substituted barbituric acid **31**. It was found that the tapelike structure I is the most common among the three architectures observed in the melamine-barituric acid system, which is the only one that satisfies all the hydrogen-bonding requirements of both molecular constituents **30** and **31** (Fig. 7-14).

Guldi et al. observed a very strong electronic communication through hydrogen bonds between porphyrin and C_{60} (Fig. 7-15) [60]. The association constant of **32** was up to 10^7 M^{-1}. Moreover, it exhibited the longer-lived formation of radical-ion-pair state ($\sim$10 μs in THF) compared to that of the similar covalent C_{60} conjugates ($\sim$1 μs in THF) [61]. As for the hydrogen-bonding ensemble **33** in which the donor and acceptor units were linked by a three-point hydrogen-bonding (Fig. 7-16), it showed that the lifetime of the photogenerated radical-ion pair is 2.02 μs. This value is higher than that of the corresponding covalently linked porphyrin-C_{60} dyads because of the beneficial effect of the hydrogen bonds [62–65].

A series of hydrogen-bonding TTF-C_{60} ensembles **34–37** in which TTF (tetrathiafulvalene) acts as an electron donor were synthesized by Martín et al.

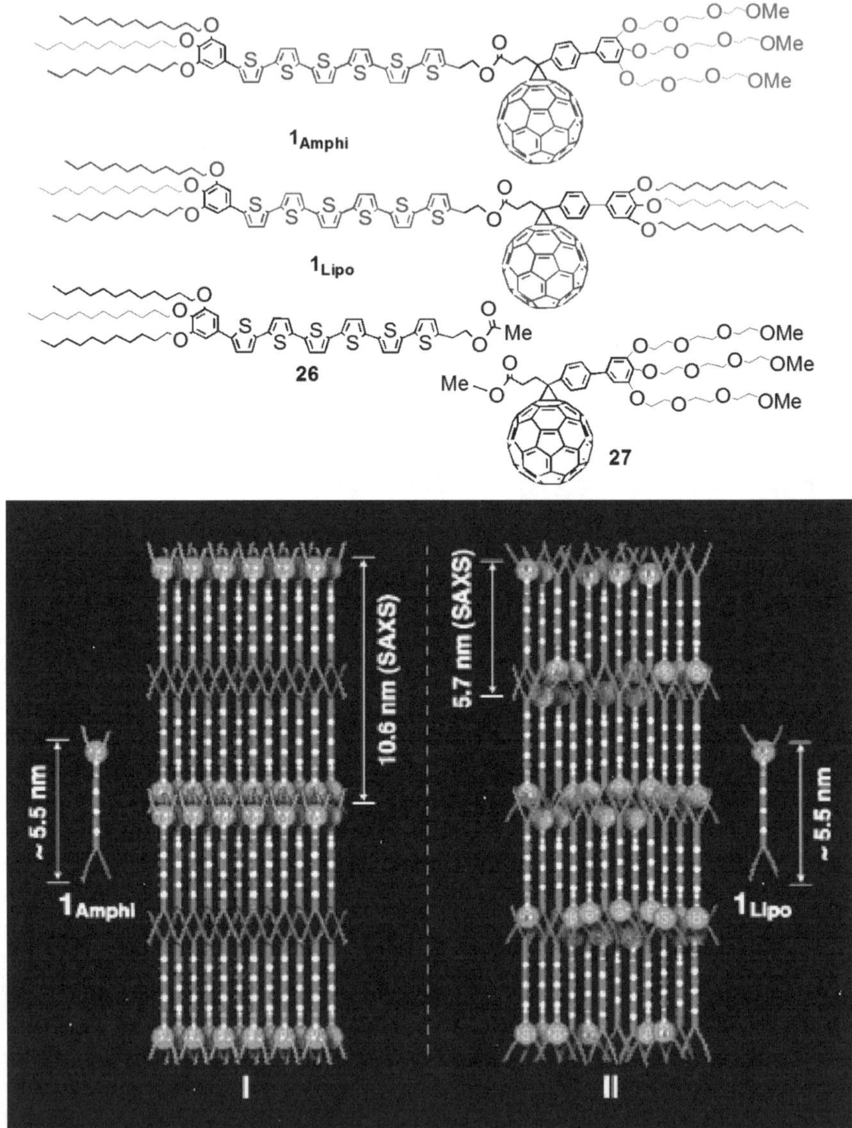

Figure 7-12 Chemical structures of **1**Amphi, **1**Lipo and the corresponding reference compounds **26** and **27** (top); schematic representations of molecular orientations in **1**Amphi (I) and **1**Lipo (II) in smectic A phase (bottom). Reproduced with permission from the American Chemical Society [55]. (A full color version of this figure appears in the color plate section.)

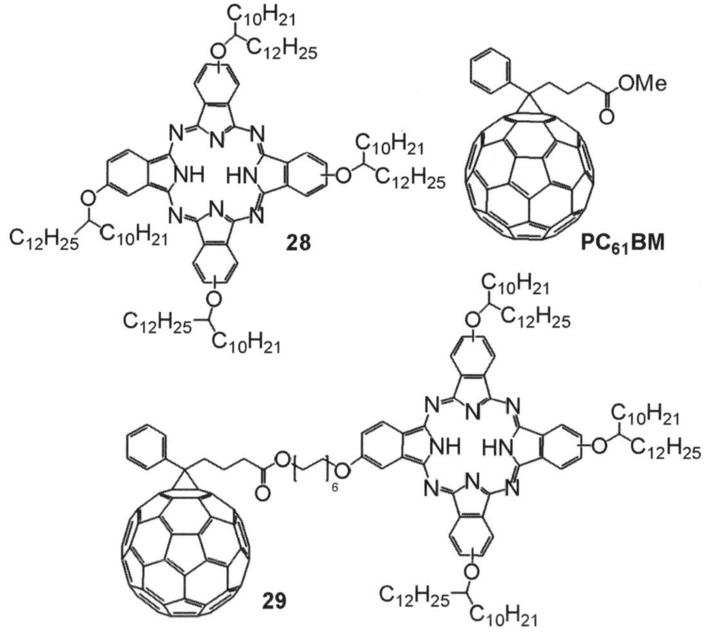

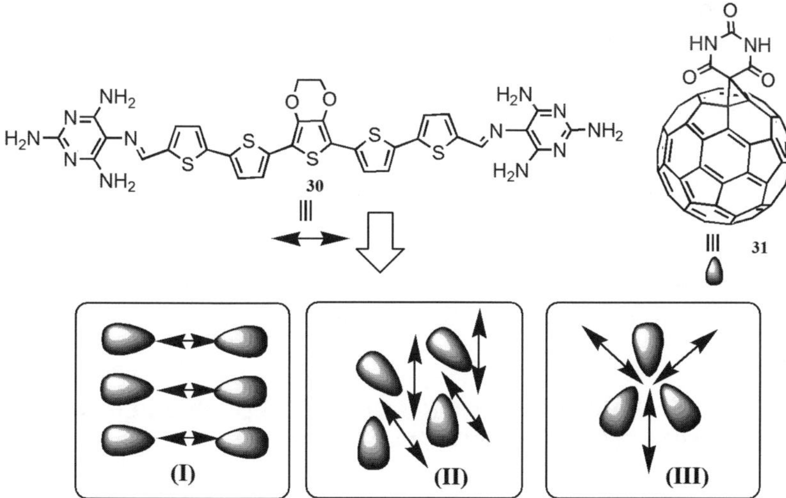

Figure 7-13 Chemical structures of phthalocyanine-C_{60} dyad **29** and the corresponding reference compounds **28** and **PC$_{61}$BM**.

Figure 7-14 Chemical structures of compounds **30** and **31** and the proposed arrangements of the self-assembly.

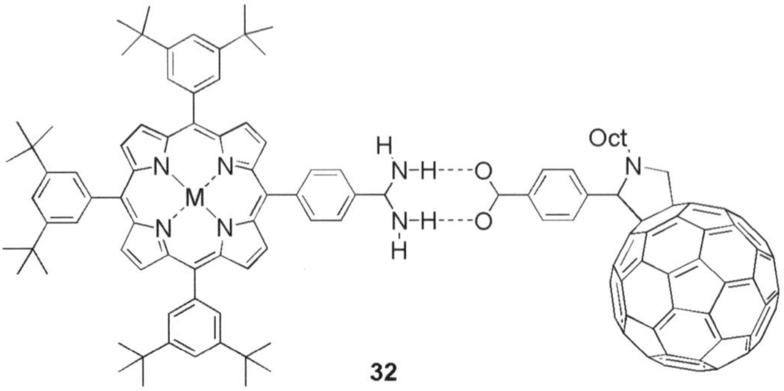

32

Figure 7-15   Chemical structure of the fullerene-based hydrogen-bonded donor-acceptor ensembles **32**.

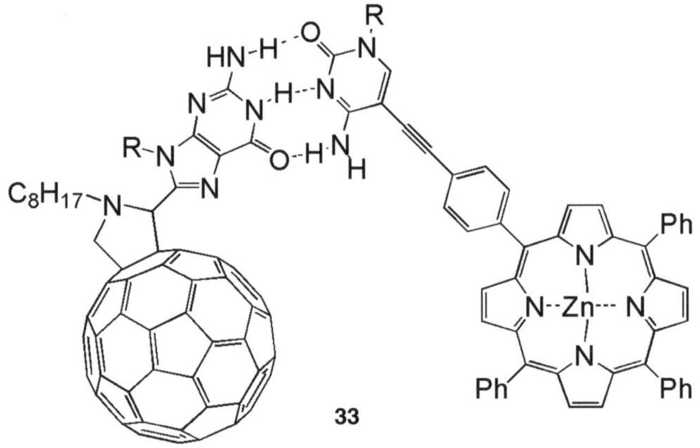

33

Figure 7-16   Chemical structure of the fullerene-based hydrogen-bonded donor-acceptor ensembles **33**.

(Fig. 7-17) [66, 67]. In these ensembles, both hydrogen bonding and electrostatic interactions are responsible for the collection of donor and acceptor units. The lifetime of the radical-ion pair state in those ensembles is in the range of hundreds of nanoseconds, several orders of magnitude higher than those corresponding covalently linked TTF-C_{60} dyads [68–70].

7.5. FULLERENE-BASED DONOR-ACCEPTOR BLENDS LINKED BY OTHER NONCOVALENT INTERACTIONS

Apart from hydrogen bonding, other noncovalent interactions such as π-π stacking, dipole-dipole, coordination, and formation of the donor-acceptor complex were also found to help to form the self-organized supermolecular structure in OPV cells.

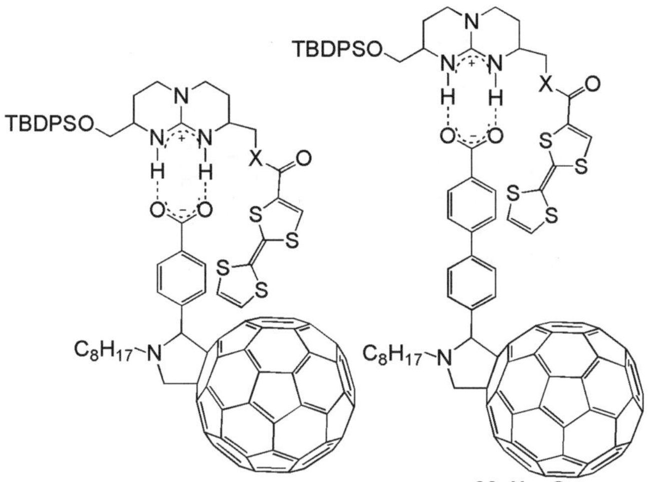

Figure 7-17 Chemical structures of the fullerene-based hydrogen-bonded donor-acceptor ensembles **34–37** (TBDPS: *tert*-butyldiphenylsilyl).

Recently, Rubin et al. successfully used a molecular approach to control both the length-scale of polymer/fullerene phase segregation and the electron mobility in the fullerene network [71]. Instead of using $PC_{61}BM$, they used the fullerene derivative **38** or **39** as electron acceptor in the fullerene-based OPV cells (Fig. 7-18). Fullerene derivative **38** with large *para tert*-butyl substitutions was found to self-organize into one-dimensional (1-D) domains within the active layer, which were confirmed by the X-ray structure, whereas its analog **39** with *para* methyl substitutions was found to hinder the formation of such 1-D stacks. Even with identical processing condition, the nonstacking fullerene **39** gave poor performance in OPV cells based on blends with regioregular P3HT, whereas the stacking fullerene **38** led to enhanced photovoltaic efficiency (Fig. 7-18). Under a thermal annealing process, the efficiency of the device based on P3HT-**38** was greatly enhanced compared to that of the device based on P3HT-**39**. This might result from the increased self-organization of the fullerene **38** system.

Intercalation between fullerene molecules and the side chains of the conjugated polymers by π-π interactions was also found to improve the efficiency of OPV cells. This new type of self-ordering was found in the blends of fullerene and polymer by McGehee and coworkers [72, 73]. A highly ordered two-phase film composed of the bimolecular crystal and a pure fullerene phase with approximately equal volumes was observed by synchrotron XRD measurements in a blend of $PC_{61}BM$ and poly(2,5-bis(3-tetradecylthiophen-2-yl)thieno[3,2-*b*]thiophene (PBTTT), in which the fullerene content is at an 80% weight ratio (Fig. 7-19). However, no such pure fullerene phase was observed at a ratio of 1:1. The former system exhibited much higher solar cell efficiency compared to the latter heterojunction system.

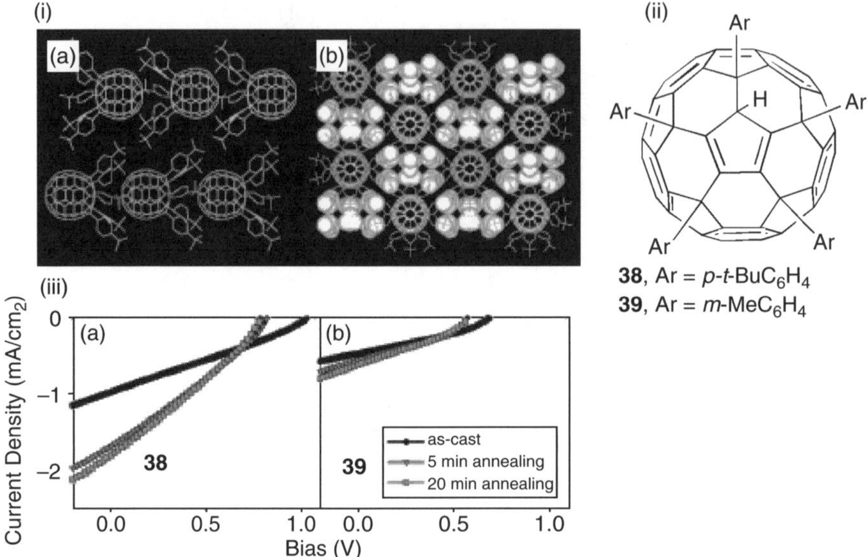

Figure 7-18 (i) Crystal structure of $38 \cdot (C_5H_{12})_3$. Antiparallel a-axis stacks, viewed along the bc plane (a) and a-axis (b). Hydrogens are removed for clarity; n-pentane solvent molecules are shown in space-filling mode. (ii) Chemical structures of compounds **38** and **39**. (iii) (a) I-V curves for as-cast and thermally annealed BHJ solar cells made from a 1:0.35 blend of **38** and P3HT. Annealing of these cells also occurs rapidly, with significantly improved device efficiencies compared to the data in (b). (b) I-V curves for as-cast and thermally annealed BHJ solar cells made from a 1:0.45 blend of **39** and **P3HT**. Annealing of these cells is complete in 5 min, but the overall device performance is low. Reproduced with permission from the American Chemical Society [71]. (A full color version of this figure appears in the color plate section.)

A Pd-mediated stepwise self-assembly of zinc porphyrin donor **40** with pyridylfullerene acceptor **41** was obtained by the coordination of the pyridyl moiety to the zinc atom together with π-π interactions between the pyridylfullerenes [74] (Fig. 7-20). After construction of a vertical alignment of the bicontinuous D-A arrays on a flat SnO_2 electrode modified with an additional C_{60}-acid component, the resulting device with five porphyrin layers showed the maximum IPCE value (21%), which is much higher than that without coordination and π-π interaction-based self-assemblies (the maximum IPCE of 13%). Notably, the IPCE value (21%) is the highest value reported for vertical arrangements of bicontinuous donor-acceptor arrays on electrodes [75, 76].

Another kind of self-assembled structure formed by a charge-transfer process was found between block copolymers and C_{60}. For example, Ikkala et al. reported that the spherical morphology was formed if an aged xylene solution of C_{60}

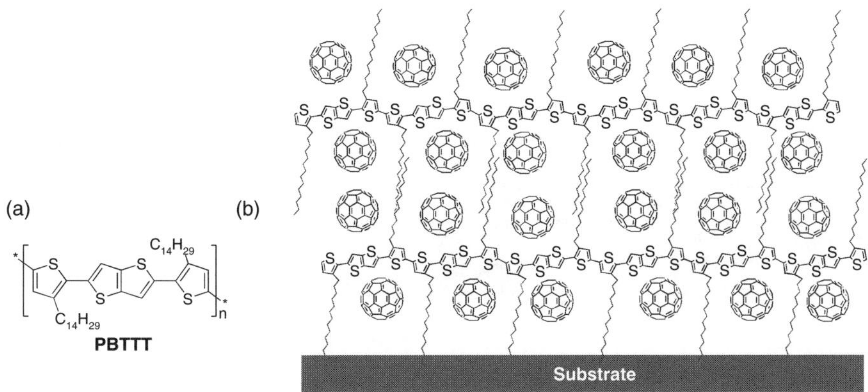

Figure 7-19 (a) Chemical structure of poly(2,5-bis(3-tetradecylthiophen-2-yl)thieno[3,2-b]thiophene) (PBTTT) and (b) Possible structure of a PBTTT-PC$_{61}$BM blend. The functional groups on C$_{60}$ were omitted for clarity. Intercalation of PC$_{61}$BM is associated with a lattice spacing increase of about 9 Å in the direction of the side chains. (A full color version of this figure appears in the color plate section.)

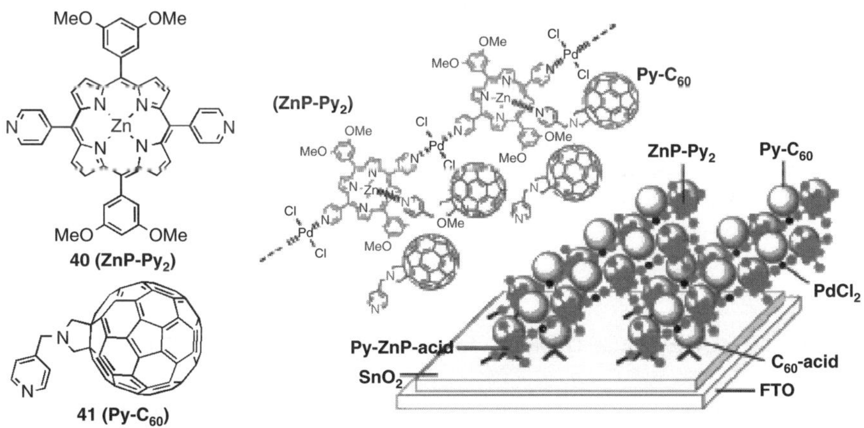

Figure 7-20 Chemical structures of zinc porphyrin donor **40** and pyridylfullerene acceptor **41**, and schematic illustration of the Pd-mediated self-assembly of coordinated donor-acceptor arrays on a C$_{60}$-acid modified SnO$_2$ electrode. Reproduced with permission from the American Chemical Society [74]. (A full color version of this figure appears in the color plate section.)

(acceptor)/PS-*block*-P4VP [donor, polystyrene-*block*-poly(4-vinylpyridine)] was used for film casting [77]. This is due to the formation of the charge-transfer complex between C$_{60}$ and the pyridine group. As shown in Figure 7-21a, a pure spherical morphology image was observed by transmission electron microscopy (TEM).

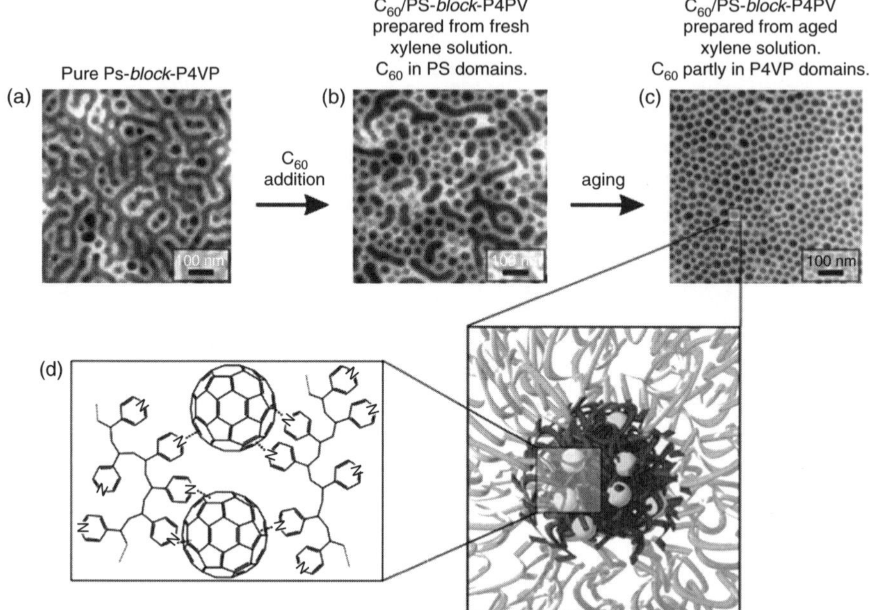

Figure 7-21 TEM images of (a) pure PS-*block*-P4VP, (b) C_{60}/PS-*block*-P4VP from freshly made xylene solution, and (c) C_{60}/PS-*block*-P4VP from aged xylene solution. (d) Schematic representation of the self-assembled PS-*block*-P4VP/C_{60} structures formed through charge-transfer complexation with the pyridines. The strong tendency of C_{60} molecules to aggregate combined with the possibility that each C_{60} molecule can bind multiple P4VP chains together through charge transfer is suggested to cause the morphological change from P4VP cylinders to C_{60}-containing P4VP spheres. Reproduced with permission from the American Chemical Society [77].

7.6. FULLERENE-BASED SELF-ASSEMBLED MONOLAYERS

Self-assembled monolayers (SAMs) are organic thin films that are formed onto a surface of metals, semiconductors, and insulators. Highly organized monolayers can be formed spontaneously by chemisorptions to yield robust, well-defined structures on substrates. It is well known that alkanethiols with a polymethylene chain (number of methylenes >9) are able to form densely packed SAMs on gold surfaces using S-Au covalent linkage [78]. The underlying densely packed monolayer of alkanethiols and alkylsulfanes on a metal surface would ensure the orientational array of the donor-acceptor unit over the SAMs to improve the corresponding device performance [79–81]. In OPV cells, the use of SAMs might result in much better electron and energy transfer efficiency compared to that without SAMs. SAMs are widely used in various organic semiconductors, leading to better performance of the corresponding devices [82–85]. The use of C_{60}-based self-assembled monolayer (C_{60}-SAM) in OPV cells would be able to significantly improve conversion efficiencies. Jen et al. fabricated a solar cell

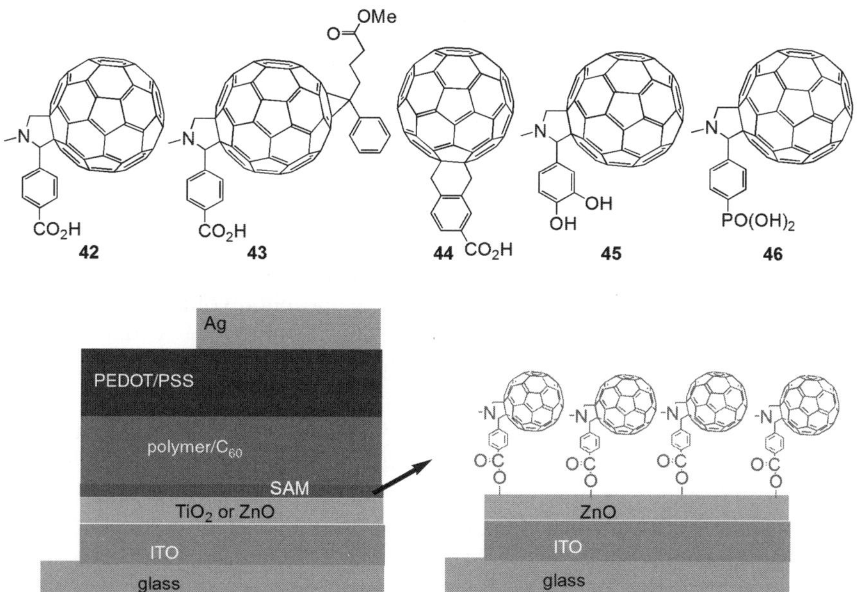

Figure 7-22 Chemical structures of the C_{60} derivatives **42–46** containing different anchoring groups (top); Architecture of polymer solar cells with self-assembled molecule mercaptoundecanoic acid (**42**) modified zinc oxide (ZnO) nanoparticle film/metal bilayer cathodes (bottom). TiO_2: Titanium dioxide. ITO: Indium tin oxide. PEDOT/PSS: Poly(3,4-ethylenedioxythiophene)/poly(styrenesulfonate). (A full color version of this figure appears in the color plate section.)

using the fullerene **42**-based SAM in an inert environment, showing efficiencies as high as 4.9% compared to the corresponding device without the C_{60}-SAM (3.8%) [86]. Here, the highly organized monolayers were formed by the interaction of the carbonic acid with zinc oxide (ZnO) (Figure 7-22). The insertion of the C_{60}-SAM enhanced the electronic coupling of the inorganic/organic interface, leading to improved fill factor and photocurrent. Other fullerene-based self-assembled monolayers containing different anchoring groups (carboxylic acid **43**, **44**, catechol **45**, and phosphonic acid **46**) were also found to be able to improve the performance of the inverted polymer solar cells. For example, using carboxylic acid, catechol, and phosphonic acid-based C_{60}-SAMs, the devices led to twofold, 75%, and 30% efficiency improvement, respectively, because of the open-circuit voltage affected by different anchoring groups and functionalization of the C_{60}-SAM. Moreover, an efficiency enhancement of $\sim$6–28% compared to devices without SAM modification was due to the improved photoinduced charge transfer from polymer to the C_{60}-SAM/ZnO. However, degradation of the ZnO surface occurred when immersing the ZnO substrate into a solution containing the C_{60}-phosphonic acid SAM for a longer time because of the strong acidic nature of the phosphonic acid anchoring group, leading to the corresponding devices without any photovoltaic activity. On the other hand, for weaker acids, such as carboxylic acid and catechol-based C_{60}-SAMs, there is no degradation between the anchoring acids with the ZnO layer. Thus, they will lead to devices with average efficiencies of 4.4% and 4.2%, respectively [87].

As mentioned above, because of the hydrogen bonding between compounds **30** and **31**, a 2.5-fold enhancement in light energy to electrical energy conversion was observed for the photovoltaic device compared to that without hydrogen bonds. As expected, with introduction of the SAM into the system by the alkanethiol-barbiturate **47** (Fig. 7-23), it was clearly found that resultant device showed much higher photocurrent [88]. The improvement of the photovoltaic response when using the SAM-modified surfaces might be due to either the variations in the surface properties and/or inhibition of the phonon quenching of the excited states generated in close proximity of the gold surface [89].

The formation of SAM of ferrocene-porphyrin-C_{60} triad **48** on a gold surface was reported by Imahori et al. [90]. The triad molecules were well packed with an almost perpendicular orientation on the gold surface, and the corresponding electron transfer process was found to be a long-lived, charge-separated state with a high quantum yield (Fig. 7-24). The artificial photosynthetic cells showed the quantum efficiency of 20–25% for photoinduced vectorial multistep electron transfer.

On the other hand, SAMs of C_{60} donors on semiconductor electrodes, for example, on indium-tin oxide (ITO) with Si-O-ITO linkage, also showed great promise in the solar cell application. Chukharev and coworkers investigated photoinduced electron transfer in SAMs of porphyrin-fullerene dyads on ITO [91]. Two porphyrin-fullerene dyads were prepared to form SAMs on an ITO electrode with either ITO-fullerene-porphyrin **49** or ITO-porphyrin-fullerene **50** orientations (Fig. 7-25). The transient photovoltage response amplitude for the ITO-fullerene-porphyrin **49**, in which the primary intramolecular electron-transfer direction coincides with the direction of the final electron transfer from the dyad to ITO, is 25 times stronger than that for ITO-porphyrin-fullerene **50**. As for ITO-fullerene-porphyrin **49**, the quantum yields of photocurrent generation for the cathodic and anodic currents were 0.45% and 2.8%, respectively, higher than that for ITO-porphyrin-fullerene **50**, which showed cathodic currents of 0.36% and anodic currents of 2.1%. The higher performance in ITO-fullerene-porphyrin

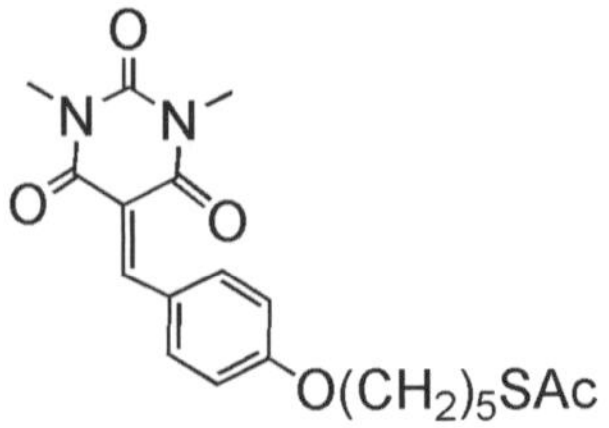

47

Figure 7-23   Chemical structure of compound **47**.

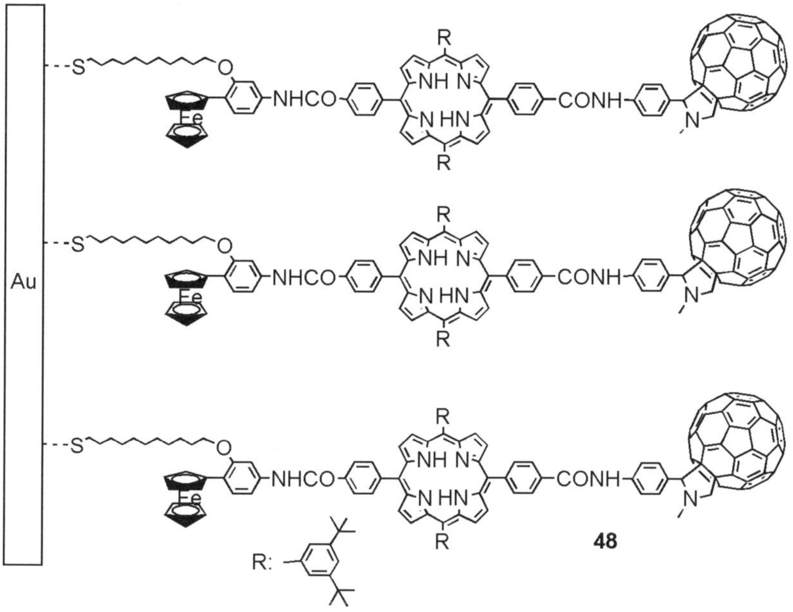

Figure 7-24   The proposed SAM organization of compound **48** on the gold surface.

49 is due to the fact that the primary intramolecular electron transfer is from porphyrin to fullerene, the direction toward ITO, while the primary electron transfer is opposite in the ITO-porphyrin-fullerene **50** structure. On the other hand, the quantum yields of the cell ITO-porphyrin **51** (Fig. 7-25) without fullerene structure are 0.25% and 1%, respectively, which means that the fullerene increased the photocurrent generation efficiency 2–3 times.

Park and coworkers fabricated a photovoltaic cell based on a fullerene metal cluster-porphyrin dyad SAM on an ITO electrode [92]. As seen in the C_{60}-porphyrin dyad, a triosmium carbonyl cluster moiety linking C_{60} and a porphyrin unit and 3-(triethoxysilyl)propyl isocyanide is employed as a surface anchoring ligand (Fig. 7-26). The dyad forms an ideal SAM (abbreviated as ZnP-C_{60}/ITO, **52**) with almost full-surface coverage on the ITO electrode in the presence of diazabicyclooctane. The cyclic voltammogram of ZnP-C_{60}/ITO in CH_2Cl_2 with 0.1 M Bu_4NPF_6 as an electrolyte at a scan rate of 0.5 V s^{-1} reveals three well-resolved redox waves at -1.17, -1.47, and -1.92 V ($E_{1/2}$). The first and second waves are successive one-electron reductions of C_{60}, and the third wave corresponds to an overlapped pattern of two-electron reduction of the C_{60}-triosmium cluster moiety and one-electron reduction of the porphyrin moiety. Two successive waves of 0.29 and 0.48 V ($E_{1/2}$) for ZnP-C_{60}/ITO belong to the first and second oxidations of the porphyrin moiety. A stable photocurrent appeared immediately during irradiation of the ITO electrode with 435-nm light (0.96 mWcm^{-2}) under a typical photocurrent measurement procedure [93, 94]. The quantum yield

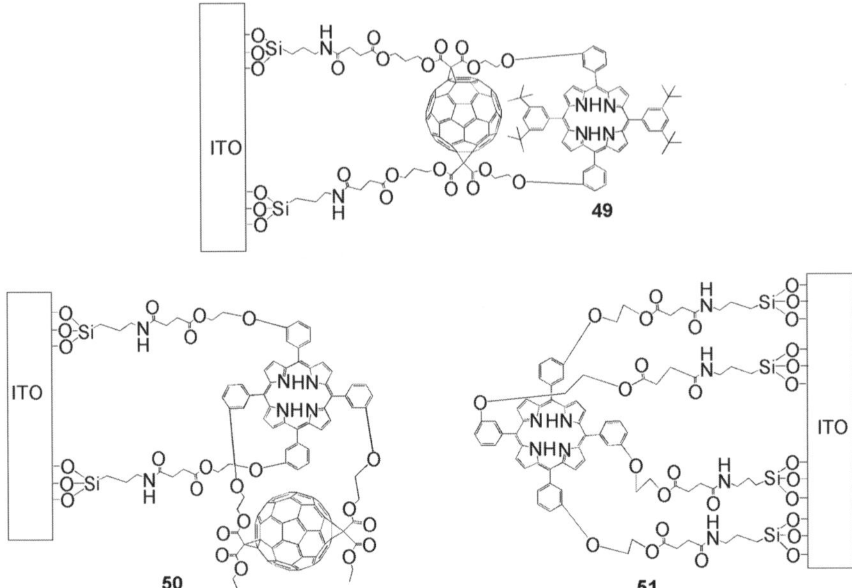

Figure 7-25   Schematic illustration of the SAMs **49**, **50**, and **51**.

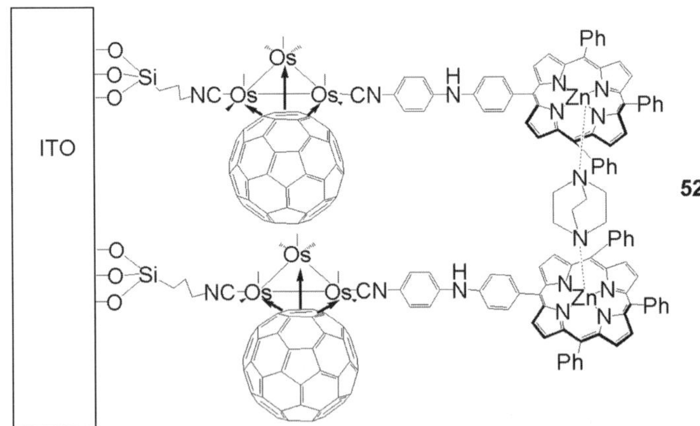

Figure 7-26   Schematic illustration of the SAM ZnP-C_{60}/ITO **52** (carbonyl ligands are omitted for clarity).

is 19.5%, which is pretty high among the cells made from the covalently linked D-A dyad structures on ITO.

An example of the high-degree (up to quaternary) self-organization from porphyrin (donor) and C_{60} (acceptor) was reported by Fukuzumi and coworkers [95, 96]. The electrode OTE/SnO_2/($H_2PC_nMPC + C_{60}$)$_m$ based on the quaternary self-organized porphyrin (donor) and fullerene (acceptor) dye units was obtained by using an electrophoretic deposition method as shown in Figure 7-27,

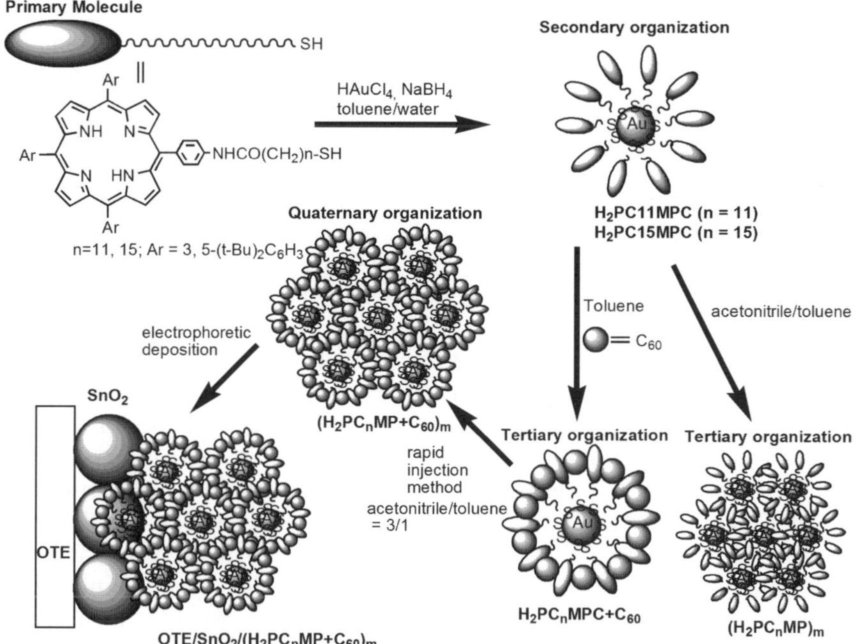

Figure 7-27 Schematic view of quaternary self-assembly of porphyrin and fullerene units by clusterization with gold nanoparticles on nanostructured SnO_2 electrodes.

in which the key step is the formation of tertiary organized nanoparticles based on the C_{60} molecules and the SAM of porphyrin. It was found that the device of $OTE/SnO_2/(H_2PC_{11}MPC + C_{60})_m$ showed an incident photon-to-electron conversion efficiency (IPCE) of 28% and overall power conversion efficiency of 0.61%. The performance of the device is much better than that of reference systems $OTE/SnO_2/(H_2PC_{11}MPC)_m$ in which no C_{60} was used. On the other hand, the length of the alkyl chain in the porphyrin was also found very important to improve the performance of the device. For example, the device of $OTE/SnO_2/(H_2PC_{15}MPC + C_{60})_m$ ($[H_2P] = 0.19$ mM, $[C_{60}] = 0.38$ mM) exhibits a maximum IPCE value of 54% which is 2.5 times better than that from the $OTE/SnO_2/(H_2PC_{11}MPC)_m$ system and a very broad photoresponse (up to ~ 1000 nm) that extends to the near-IR region. These results demonstrated that the quaternary organization approach of the porphyrin gold nanoparticles and fullerenes might provide a novel perspective for the development of efficient organic solar cells.

7.7. CONCLUSION AND OUTLOOK

In this chapter, we have presented a brief overview of various self-organizations involving fullerene [60] with a focus on their potential application in OPVs.

First, we discussed the fullerene-based self-assemblies formed by the addition of liquid crystal component or by the use of liquid crystalline dyads containing C_{60} moiety. Then the examples of fullerene-based self-assemblies formed by the noncovalent or weak supramolecular interactions such as hydrogen-bonds, π-π interactions, coordination bonds, charge-transfer, etc. were presented. Finally, the highly efficient organic solar cells using C_{60}-based SAMs were presented.

Improvement in the efficiency of OPV cells is a long-term challenge of commercialization, which is discussed in detail in Chapter 8. It will be possible to improve OPV efficiency by optimization at the molecular level, that is, exploring new electron donors and acceptors and fabricating highly ordered nanoscale thin films by various chemical and physical means. The use of noncovalent interactions between tailored p-conjugated systems and C_{60} derivatives specifically designed for molecular recognition and self-organization would be one of the most promising strategies for improving the performance of organic solar cells.

The fact that C_{60} exhibits novel properties in electron transfer stimulates the development of new electron transfer chemistry, and higher fullerenes such as C_{70}, heterofullerenes, and metallofullerenes, as well as highly "three-dimensional" molecules, such as nanotubes, also seem to be good candidates as electron acceptors and/or electron accumulators. In summary, the unique photophysical properties of fullerenes will extend a new stage for the design of fullerene-based self-assemblies in organic solar cells.

ACKNOWLEDGMENTS

The preparation of this chapter benefited from the support to Quan Li by the Ohio Board of Regents under its Research Challenge program, the Department of Energy (DOE DE-SC0001412), the Air Force Office of Scientific Research (AFOSR FA9950-09-1-0193 and FA9550-09-1-0254), and the National Science Foundation (NSF IIP 0750379).

REFERENCES

1. P. M. Allemond, A. Koch, F. Wudl, Y. Rubin, F. Diederich, M. M. Alvarez, S. J. Anz, and R. L. Whetten. Two different fullerenes have the same cyclic voltammetry. *J. Am. Chem. Soc.* **1991**, *113*, 1050–1051.

2. *Fullerenes: chemistry and reactions*, A. Hirsch, M. Brettreich. Eds., Wiley-VCH, Verlag GmbH & Co, 2005.

3. F. Wudl. The chemical properties of buckminsterfullerene (C_{60}) and the birth and infancy of fulleroids. *Acc. Chem. Res.* **1992**, *25*, 157–161.

4. H. Hoppe and N. S. Saricifci. Organic solar cells: an overview. *J. Mater. Chem.* **2006**, *16*, 45–61.

5. K. Matsuoka, T. Akiyama, and S. Yamada. Step-by-step fabrication of porphyrin-fullerene supramolecular assemblies and their photoelectrochemical properties. *J. Phys. Chem. C* **2008**, *112*, 7015–7020.

6. G. Decher, J. D. Hong and J. Schmitt. Buildup of ultrathin multilayer films by a self-assembly process: III. Consecutively alternating adsorption of anionic and cationic polyelectrolytes on charged surfaces. *Thin Solid Films* **1992**, *210/211*, 831–835.

7. G. Decher. Fuzzy nanoassemblies: toward layered polymeric multicomposites. *Science* **1997**, *277*, 1232–1237.

8. J. K. Mwaura, M. R. Pinto, D. Witker, N. Ananthakrishnan, K. S. Schanze, and J. R. Reynolds. Photovoltaic cells based on sequentially adsorbed multilayers of conjugated poly(p-phenylene ethynylene)s and a water-soluble fullerene derivative. *Langmuir* **2005**, *21*, 10119–10126.

9. P. Seta, E. Bienvenue, A. L. Moore, P. Mathis, R. V. Bensasson, P. A. Liddell, P. J. Pessiki, A. Joy, T. A. Moore, and D. Gust. Photodriven transmembrane charge separation and electron transfer by a carotenoporphyrin–quinone triad. *Nature* **1985**, *316*, 653–655.

10. G. Steinberg-Yfrach, P. A. Liddell, S.-C. Hung, A. L. Moore, D. Gust, and T. A. Moore. Artificial photosynthetic reaction centers in liposomes: photochemical generation of transmembrane proton potential. *Nature* **1997**, *385*, 239–241.

11. G. Steinberg-Yfrach, J.-L. Rigaud, E. N. Durantini, A. L. Moore, D. Gust, and T. A. Moore. Light-driven production of ATP catalysed by F0F1-ATP synthase in an artificial photosynthetic membrane. *Nature* **1998**, *392*, 479–482.

12. K. Liang, K.-Y. Law, and D. G. Whitten. Synthesis, characterization, photophysics, and photosensitization studies of squaraine-bipyridinium diads. *J. Phys. Chem. B*, **1997**, *101*, 540–546.

13. N. V. Tkachenko, A. Y. Tauber, P. H. Hynninen, A. Y. Sharonov, and H. Lemmetyinen. Light induced electron transfer in pyropheophytin-anthraquinone dyads: vectorial charge transfer in langmuir-blodgett films. *J. Phys. Chem. A* **1999**, *103*, 3657.

14. L. Sánchez, N. Martín, and D. M. Guldi. Hydrogen-bonding motifs in fullerene chemistry. *Angew. Chem. Int. Ed*. **2005**, *44*, 5374–5382.

15. P. J. de Rege, S. A. Williams, and M. J. Therien. Direct evaluation of electronic coupling mediated by hydrogen bonds: implications for biological electron transfer. *Science*, **1995**, *269*, 1409–1413.

16. A. M. van de Craats, and J. M. Warman. The core-size effect on the mobility of charge in discotic liquid crystalline materials. *Adv. Mater*. **2001**, *13*, 130–133.

17. D. Demus, J. Goodby, G. W. Gray, H.-W. Spiess, and V. Vill. *Handbook of Liquid Crystals* Wiley, Weinheim, 1998.

18. D. Adam, P. Schumacher, J. Simmerer, L. Haussling, K. Siemensmeyer, K. H. Etzbach, H. Ringsdor, and D. Haarer. Fast photoconduction in the highly ordered columnar phase of a discotic liquid crystal. *Nature* **1994**, *371*, 141–143.

19. Q. Li and L. Li. Photoconducting Discotic Liquid Crystals. *Thermotropic Liquid Crystals*; Ramamoorthy, A., Ed.; Springer: New York, **2007**; Chapter 11.

20. L. Schmidt-Mende, A. Fechtenkotter, K. Müllen, R. H. Friend, and J. D. MacKenzie. Self-organized discotic liquid crystals for high-efficiency organic photovoltaics. *Science* **2001**, *293*, 1119–1122.

21. B. A. Greg, B. A. Fox, and A. J. Bard. Photovoltaic effect in symmetrical cells of a liquid crystal porphyrin. *J. Phys. Chem*. **1990**, *94*, 1586–1598.

22. M. A. Fox, J. V. Grant, T. Torimoto, C.-Y. Liu, and A. J. Bard. Effect of structural variation on photocurrent efficiency in alkyl-substituted porphyrin solid-state thin layer photocells. *Chem. Mater*. **1998**, *10*, 1771–1776.

23. L. Schmidt-Mende, M. Watson, K. Müllen, and R. H. Friend. Organic thin film photovoltaic devices from discotic materials. *Mol. Cryst. Liq. Cryst*. **2003**, *396*, 73–90.

24. L. Li, S. Kang, J. Harden, Q. Sun, X. Zhou, L. Dai, A. Jakli, S. Kumar, and Q. Li, Nature inspired light-harvesting liquid crystalline porphyrins for organic photovoltaics. *Liq. Cryst*. **2008**, *35*, 233–239.

25. S. Sergeyev, W. Pisula, and Y. H. Geerts. Discotic liquid crystals: a new generation of organic semiconductors. *Chem. Soc. Rev*. **2007**, *36*, 1902–1929.

26. G. Zucchi, B. Donnio, and Y. H. Geerts. Remarkable miscibility between disk- and lathlike mesogens. *Chem. Mater*. **2005**, *17*, 4273–4277.

27. S. Kumar. Discotic liquid crystals for solar cells. *Current Science*. **2002**, *82*, 256–257.

28. L. Schmidt-Mende, A. Fechtenkotter, K. Müllen, E. Moons, R. H. Friend, and J. D. MacKenzie. Self-organized discotic liquid crystals for high-efficiency organic photovoltaics. *Science* **2001**, *293*, 1119–1122.

29. B. Kippelen, S. Yoo, B. Domercq, C. L. Donley, C. Carter, W. Xia, B. A. Minch, D. F. O'Brien, and N. R. Armstrong. Organic photovoltaics based on self-assembled mesophases. *NCPV and Solar Program Review Meeting* **2003**, 431–434.

30. S. Günes, H. Neugebauer, and N. S. Sariciftci. Conjugated polymer-based organic solar cells. *Chem. Rev*. **2007**, *107*, 1324–1338.

31. W. Cai, X. Gong, and Y. Cao. Polymer solar cells: Recent development and possible routes for improvement in the performance. *Solar Energy Materials & Solar Cells* **2010**, *94*, 114–127.

32. X. Zhou, S.-W. Kang, S. Kumar, R. R. Kulkarni, S. Z. D. Cheng, and Q. Li. Self-assembly of porphyrin and fullerene supramolecular complex into highly ordered nanostructure by thermal annealing. *Chem. Mater*. **2008**, *20*, 3551–3553.

33. Q. Sun, L. Dai, X. Zhou, L. Li, and Q. Li. Bilayer- and bulk-heterojunction solar cells using liquid crystalline porphyrins as donors by solution processing. *Appl. Phys. Lett*. **2007**, *91*, 253505.

34. A. Escosura, M. V. Martínez-Díaz, J. Barberá, and T. Torres. Self-organization of phthalocyanine-[60]fullerene dyads in liquid crystals *J. Org. Chem*., **2008**, *73* 1475–1480.

35. S. Jeong, Y. Kwon, B.-D. Choi, H. Ade, and Y. S. Han. Improved efficiency of bulk heterojunction poly(3-hexylthiophene):[6,6]-phenyl-C_{61}-butyric acid methyl ester photovoltaic devices using discotic liquid crystal additives. *Appl. Phys. Lett*. **2010**, *96*, 183305.

36. J. L. Segura, N. Martín, and D. M. Guldi. Materials for organic solar cells: the C60/π-conjugated oligomer approach. *Chem. Soc. Rev*. **2005**, *34*, 31–47.

37. S. Günes, H. Neugebauer, and N. S. Sariciftci. Conjugated polymer-based organic solar cells. *Chem. Rev*. **2007**, *107*, 1324–1338.

38. B. C. Thompson and J. M. J. Frechet. Polymer-fullerene composite solar cells. *Angew. Chem., Int. Ed*. **2008**, *47*, 58–77.

39. S. Campidelli, C. Eng, I. M. Saez, J. W. Goodby, and R. Deschenaux. Functional polypedes—chiral nematic fullerenes. *Chem. Commun*. **2003**, 1520–1521.

40. T. Chuard and R. Deschenaux. Design, mesomorphic properties, and supramolecular organization of [60]fullerene-containing thermotropic liquid crystals. *J. Mater. Chem*. **2002**, *12*, 1944–1951.

41. B. Dardel, D. Guillon, B. Heinrich, and R. Deschenaux. Fullerene-containing liquid-crystalline dendrimers. *J. Mater. Chem.* **2001**, *11*, 2814–2831.

42. S. Campidelli, L. Pérez, J. Rodríguez-López, J. Barberá, F. Langa, and R. Deschenaux. Dendritic liquid-crystalline fullerene–ferrocene dyads. *Tetrahedron* **2006**, *62*, 2115–2122.

43. M. Even, B. Heinrich, D. Guillon, D. M. Guldi, M. Prato, and R. Deschenaux. A mixed fullerene-ferrocene thermotropic liquid crystal: synthesis, liquid-crystalline properties, supramolecular organization and photoinduced electron transfer. *Chem. Eur. J.* **2001**, *7*, 2595–2604.

44. S. Campidelli, R. Deschenaux, J.-F. Eckert, D. Guillon, and J.-F. Nierengarten. Liquid-crystalline fullerene–oligophenylenevinylene conjugates. *Chem. Commun.* **2002**, 656–657.

45. J.-F. Eckert, J.-F. Nicoud, J.-F. Nierengarten, S.-G. Liu, L. Echegoyen, F. Barigelletti, N. Armaroli, L. Ouali, V. Krasnikov, and G. Hadziioannou. Fullerene-oligophenylenevinylene hybrids: synthesis, electronic properties, and incorporation in photovoltaic devices. *J. Am. Chem. Soc.* **2000**, *122*, 7467–7479.

46. S. Campidelli, E. Vázquez, D. Milic, M. Prato, J. Barberá, D. M. Guldi, M. Marcaccio, D. Paolucci, F. Paolucci, and R. Deschenaux. Liquid-crystalline fullerene–ferrocene dyads. *J. Mater. Chem.* **2004**, *14*, 1266–1272.

47. E. Allard, F. Oswald, B. Donnio, D. Guillon, J. L. Delgado, F. Langa, and R. Deschenaux. Liquid-crystalline [60]fullerene-TTF dyads. *Org. Lett.* **2005**, *7*, 383–386.

48. L. Echegoyen and L. E. Echegoyen. Electrochemistry of fullerenes and their derivatives. *Acc. Chem. Res.* **1998**, *31*, 593–601.

49. S. Campidelli, M. Séverac, D. Scanu, R. Deschenaux, E. Vázquez, D. Milic, M. Prato, M. Carano, M. Marcaccio, F. Paolucci, G. M. A. Rahman, and D. M. Guldi. Photophysical, electrochemical, and mesomorphic properties of a liquid-crystalline fullerene–peralkylated ferrocene dyad. *J. Mater. Chem.* **2008**, *18*, 1504–1509.

50. Y.-W. Zhong, Y. Matsuo, and E. Nakamura. Lamellar assembly of conical molecules possessing a fullerene apex in crystals and liquid crystals. *J. Am. Chem. Soc.* **2007**, *129*, 3052–3053.

51. M. Sawamura, K. Kawai, Y. Matsuo, K. Kanie, T. Kato, and E. Nakamura. Stacking of conical molecules with a fullerene apex into polar columns in crystals and liquid crystals. *Nature* **2002**, *419*, 702–705.

52. Y. Matsuo, A. Muramatsu, R. Hamasaki, N. Mizoshita, T. Kato, and E. Nakamura. Stacking of molecules possessing a fullerene apex and a cup-shaped cavity connected by a silicon connection. *J. Am. Chem. Soc.* **2004**, *126*, 432–433.

53. Y. Matsuo, A. Muramatsu, Y. Kamikawa, T. Kato, and E. Nakamura. Synthesis and structural, electrochemical, and stacking properties of conical molecules possessing buckyferrocene on the apex. *J. Am. Chem. Soc.* **2006**, *128*, 9586–9587.

54. T. Nakanishi, Y. Shen, J. Wang, S. Yagai, M. Funahashi, T. Kato, P. Fernandes, H. Möhwald, and D. G. Kurth. Electron transport and electrochemistry of mesomorphic fullerenes with long-range ordered lamellae. *J. Am. Chem. Soc.* **2008**, *130*, 9236–9237.

55. W.-S. Li, Y. Yamamoto, T. Fukushima, A. Saeki, S. Seki, S. Tagawa, H. Masunaga, S. Sasaki, M. Takata, and T. Aida. Amphiphilic molecular design as a rational strategy for tailoring bicontinuous electron donor and acceptor arrays: photoconductive liquid crystalline oligothiophene-C60 dyads. *J. Am. Chem. Soc.* **2008**, *130*, 8886–8887.

56. Y. H. Geerts, O. Debever, C. Amato, and S. Sergeyev. Synthesis of mesogenic phthalocyanine-C_{60} donor–acceptor dyads designed for molecular heterojunction photovoltaic devices. *Beilstein J. Org. Chem*. **2009**, *5*, doi: 10.3762/bjoc.5.49.

57. M. Gutiérrez-Nava, H. Nierengarten, P. Masson, A. V. Dorsselaer, and J.-F. Nierengarten. A supramolecular oligophenylenevinylene–C_{60} conjugate. *Tetrahedron Lett*. **2003**, *44*, 3043–3046.

58. E. H. A. Beckers, A. P. H. J. Schenning, P. A. van Hal, A. El-ghayoury, L. Sánchez, J. C. Hummelen, E. W. Meijer, and R. J. Janssen. Preferential heterodimer formation and equilibrium dynamics of self-complementary bifunctional oligo (*p*-phenylenevinylene) and C60 ureido-pyrimidinone derivatives in solution. *Chem. Commun*. **2002**, 2888–2889.

59. C.-H. Huang, N. D. McClenaghan, A. Kuhn, J. W. Hofstraat, and D. M. Bassani. Enhanced photovoltaic response in hydrogen-bonded all-organic devices. *Org. Lett*. **2005**, *7*, 3409–3412.

60. L. Sánchez, M. Sierra, N. Martín, A. J. Myles, T. J. Dale, J. Rebek, W. Seitz, and D. M. Guldi. Exceptionally strong electronic communication through hydrogen bonds in porphyrin-C_{60} pairs. *Angew. Chem. Int. Ed*. **2006**, *45*, 4637–4641.

61. H. Imahori, H. Yamada, D. M. Guldi, Y. Endo, A. Shimomura, S. Kundu, K. Yamada, T. Okada, Y. Sakata, and S. Fukuzumi. Comparison of reorganization energies for intra- and inter-molecular electron transfer. *Angew. Chem. Int. Ed*. **2002**, *41*, 2344–2347.

62. H. Yamada, H. Imahori, Y. Nishimura, I. Yamazaki, T. K. Ahn, S. K. Kim, D. Kim, and S. Fukuzumi. Photovoltaic properties of self-assembled monolayers of porphyrins and porphyrin-fullerene dyads on ITO and gold surfaces. *J. Am. Chem. Soc*. **2003**, *125*, 9129–9139.

63. H. Imahori and Y. Sakata. Donor-linked fullerenes: Photoinduced electron transfer and its potential application. *Adv. Mater*. **1997**, *9*, 537–546.

64. N. Watanabe, N. Kihara, Y. Forusho, T. Takata, Y. Araki, and O. Ito. Photoinduced intrarotaxane electron transfer between zinc porphyrin and [60]fullerene in benzonitrile. *Angew. Chem. Int. Ed*. **2003**, *42*, 681–683.

65. H. Imahori, M. E. El-Khouly, M. Fujitsuka, O. Ito, Y. Sakata, and S. Fukuzumi. Solvent dependence of charge separation and charge recombination rates in porphyrin-fullerene dyad. *J. Phys. Chem. A* **2001**, *105*, 325–332.

66. M. Segura, L. Sánchez, J. de Mendoza, N. Martín, and D. M. Guldi. Hydrogen bonding interfaces in fullerene-TTF ensembles. *J. Am. Chem. Soc*. **2003**, *125*, 15093–15100.

67. J. L. Segura and N. Martín. New concepts in tetrathiafulvalene chemistry. *Angew. Chem. Int. Ed*. **2001**, *40*, 1372–1409.

68. D. M. Guldi, S. González, N. Martín, A. Antón, J. Garín, and J. Orduna. Efficient charge separation in C_{60}-based dyads: triazolino[4′, 5′:1,2][60]fullerenes. *J. Org. Chem*. **2000**, *65*, 1978–1983.

69. J. L. Segura, E. M. Priego, N. Martín, C. Luo, and D. M. Guldi. A new photoactive and highly soluble C_{60}-TTF-C_{60} dimer: charge separation and recombination. *Org. Lett*. **2000**, *2*, 4021–4024.

70. N. Martín, L. Sánchez, M. A. Herranz, and D. M. Guldi. Evidence for two separate one-electron transfer events in excited fulleropyrrolidine dyads containing tetrathiafulvalene (TTF). *J. Phys. Chem. A* **2000**, *104*, 4648–4657.

71. R. D. Kennedy, A. L. Ayzner, D. D. Wanger, C. T. Day, M. Halim, S. I. Khan, S. H. Tolbert, B. J. Schwartz, and Y. Rubin. Self-assembling fullerenes for improved bulk-heterojunction photovoltaic devices. *J. Am. Chem. Soc.* **2008**, *130*, 17290–17292.

72. R. Gysel, N. Cates, M. McGehee, A. Mayer, and M. Toney. Molecular self-ordering in organic solar cell materials. *Solar & Alternative Energy* **2009**, SPIE Newsroom. DOI:10.1117/2.1200908.1769.

73. A. C. Mayer, M. F. Toney, S. R. Scully, J. Rivnay, C. J. Brabec, M. Scharber, M. Koppe, M. Heeney, I. McCulloch, and M. D. McGehee. Bimolecular crystals of fullerenes in conjugated polymers and the implications of molecular mixing for solar cells. *Adv. Func. Mater.* **2009**, *19*, 1173–1179.

74. A. Kira, T. Umeyama, Y. Matano, K. Yoshida, S. Isoda, J. K. Park, D. Kim, and H. Imahori. Supramolecular donor-acceptor heterojunctions by vectorial stepwise assembly of porphyrins and coordination-bonded fullerene arrays for photocurrent generation *J. Am. Chem. Soc.* **2009**, *131*, 3198–3200.

75. H. J. Snaith, G. L. Whiting, B. Sun, N. C. Greenham, W. T. S. Huck, and R. H. Friend. Self-organization of nanocrystals in polymer brushes. Application in heterojunction photovoltaic diodes. *Nano Lett.* **2005**, *5*, 1653–1657.

76. A. L. Sisson, N. Sakai, N. Banerji, A. Fürstenberg, E. Vauthey, and S. Matile. Zipper assembly of vectorial rigid-rod π-stack architectures with red and blue naphthalenedi-imides: toward supramolecular cascade n/p-heterojunctions. *Angew. Chem., Int. Ed.* **2008**, *47*, 3727–3729.

77. A. Laiho, R. H. A. Ras, S. Valkama, J. Ruokolainen, R. Österbacka, and O. Ikkala. Control of self-assembly by charge-transfer complexation between C_{60} fullerene and electron donating units of block copolymers. *Macromolecules* **2006**, *39*, 7648–7653.

78. M. D. Porter, T. B. Bright, D. L. Allara, and C. E. D. Chidsey. Spontaneously organized molecular assemblies. 4. Structural characterization of n-alkyl thiol monolayers on gold by optical ellipsometry, infrared spectroscopy, and electrochemistry. *J. Am. Chem. Soc.* **1987**, *109*, 3559–3568.

79. J. Roncali. Linear π-conjugated systems derivatized with C_{60}-fullerene as molecular heterojunctions for organic photovoltaics. *Chem. Soc. Rev.* **2005**, *34*, 483–495.

80. H. Imahori and S. Fukuzumi. Porphyrin- and fullerene-based molecular photovoltaic devices. *Adv. Funct. Mater.* **2004**, *14*, 525–536.

81. T. Umeyama and H. Imahori. Self-organization of porphyrins and fullerenes for molecular photoelectrochemical devices. *Photosynthesis Research* **2006**, *87*, 63–71.

82. M. F. Calhoun, J. Sanchez, D. Olaya, M. E. Gershenson, and V. Podzorov. Electronic functionalization of the surface of organic semiconductors with self-assembled monolayers. *Nature Materials* **2008**, *7*, 84–89.

83. H.-L. Yip, S. K. Hau, N. S. Baek, H. Ma, and A. K.-Y. Jen. Polymer solar cells that use self assembled-monolayer-modified ZnO/metals as cathodes, *Adv. Mater.* **2008**, *20*, 2376–2382.

84. H.-L. Yip, S. K. Hau, N. S. Baek, and A. K.-Y. Jen. Self-assembled monolayer modified ZnO/metal bilayer cathodes for polymer:fullerene bulk-heterojunction solar cells, *Appl. Phys. Lett.* **2008**, *92*, 193313.

85. S. K. Hau, H.-L. Yip, H. Ma, and A. K.-Y. Jen. High performance ambient processed inverted polymer solar cells through interfacial modification with a fullerene self-assembled monolayer. *Appl. Phys. Lett.*, **2008**, *93*, 233304.

86. H.-L. Yip, S. K. Hau, and A. K.-Y. Jen. Interface engineering of stable, efficient polymer solar cells. *Solar & Alternative Energy* **2009**, SPIE Newsroom. DOI: 10.1117/2.1200905.1666.

87. S. K. Hau, Y.-J. Cheng, H.-L. Yip, Y. Zhang, H. Ma, and A. K.-Y. Jen. Effect of chemical modification of fullerene-based self-assembled monolayers on the performance of inverted polymer solar cells. *Applied Materials & Interfaces* **2010**, DOI: 10.1021/am100238e.

88. C.-H. Huang, N. D. McClenaghan, A. Kuhn, G. Bravic, and D. M. Bassani. Hierarchical self-assembly of all-organic photovoltaic devices. *Tetrahedron*, **2006**, *62*, 2050–2059.

89. S. Ekgasit, F. Yu, and W. Knoll. Displacement of molecules near a metal surface as seen by an SPR-SPFS biosensor. *Langmuir*, **2005**, *21*, 4077–4082.

90. H. Imahori, H. Yamada, and Y. Nishimura. Vectorial multistep electron transfer at the gold electrodes modified with self-assembled monolayers of ferrocene-porphyrin-fullerene triads. *J. Phys. Chem. B* **2000**, *104*, 2099–2108.

91. V. Chukharev, T. Vuorinen, A. Efimov, N. V. Tkachenko, M. Kimura, S. Fukuzumi, H. Imahori, and H. Lemmetyinen. Photoinduced electron transfer in self-assembled monolayers of porphyrin-fullerene dyads on ITO. *Langmuir* **2005**, *21*, 6385–6391.

92. Y.-J. Cho, T. K. Ahn, H. Song, K. S. Kim, C. Y. Lee, W. S. Seo, K. Lee, S. K. Kim, D. Kim, and J. T. Park. Unusually high performance photovoltaic cell based on a [60]fullerene metal cluster-porphyrin dyad SAM on an ITO electrode. *J. Am. Chem. Soc.* **2005**, *127*, 2380–2381.

93. A. Ikeda, T. Hatano, S. Shinkai, T. Akiyama, and S. Yamada. Efficient photocurrent generation in novel self-assembled multilayers comprised of [60]fullerene-cationic homooxacalix[3]arene inclusion complex and anionic porphyrin polymer. *J. Am. Chem. Soc.* **2001**, *123*, 4855–4856.

94. H. Imahori, M. Kimura, K. Hosomizu, T. Sato, T. K. Ahn, S. K. Kim, D. Kim, Y. Nishimura, I. Yamazaki, Y. Araki, O. Ito, and S. Fukuzumi. Vectorial electron relay at ITO electrodes modified with self-assembled monolayers of ferrocene-porphyrin-fullerene triads and porphyrin-fullerene dyads for molecular photovoltaic devices. *Chem. Eur. J.* **2004**, *10*, 5111–5122.

95. T. Hasobe, H. Imahori, P. V. Kamat, and S. Fukuzumi. Quaternary self-organization of porphyrin and fullerene units by clusterization with gold nanoparticles on SnO_2 electrodes for organic solar cells. *J. Am. Chem. Soc.* **2003**, *125*, 14962–14963.

96. T. Hasobe, H. Imahori, P. V. Kamat, T. K. Ahn, S. K. Kim, D. Kim, A. Fujimoto, T. Hirakawa, and S. Fukuzumi. Photovoltaic cells using composite nanoclusters of porphyrins and fullerenes with gold nanoparticles. *J. Am. Chem. Soc.* **2005**, *127*, 1216–1228.

High-Efficiency Organic Solar Cells Using Self-Organized Materials

PAUL A. LANE

U.S. Naval Research Laboratory, Washington DC

8.1. INTRODUCTION

8.1.1. Background

Understanding structure-property relationships is fundamental to the research and application of the organic semiconductors. The interplay between chemical structure, intermolecular self-organization, and morphology in films and at interfaces determines important electronic and spectroscopic properties. As an example, many molecular semiconductors have a planar structure with efficient charge transport perpendicular to the plane of the molecule due to pi-stacking between adjacent chromophores [1, 2]. The self-organizing properties of organic semiconductors have been exploited by researchers to develop organic solar cells with high efficiency. The primary challenge of these materials lies in the excitonic nature of organic semiconductors that is a consequence of their chemical structure. Figure 8-1 shows the chemical structure of molecular semiconductors used in organic solar cells. The basic building block of these materials is sp- and sp^2-hybridized carbon. The hybridized σ orbitals form the structural backbone of the molecule, and the delocalized π orbitals determine its optical and electronic properties. Group V and VI elements are often incorporated into the chemical structure, particularly nitrogen with a nonbonding pair that influences the electronic structure via n-π transitions. Macrocyclic molecules such as porphyrins and phthalocyanines as well as fused aromatic acenes such as pentacene and perylene have particular importance for organic photovoltaics.

Molecular orbital theory provides a useful framework to describe organic semiconductors, with the frontier orbitals playing a determining role in charge transfer.

Self-Organized Organic Semiconductors: From Materials to Device Applications, First Edition.
Edited by Quan Li.
© 2011 John Wiley & Sons, Inc. Published 2011 by John Wiley & Sons, Inc.

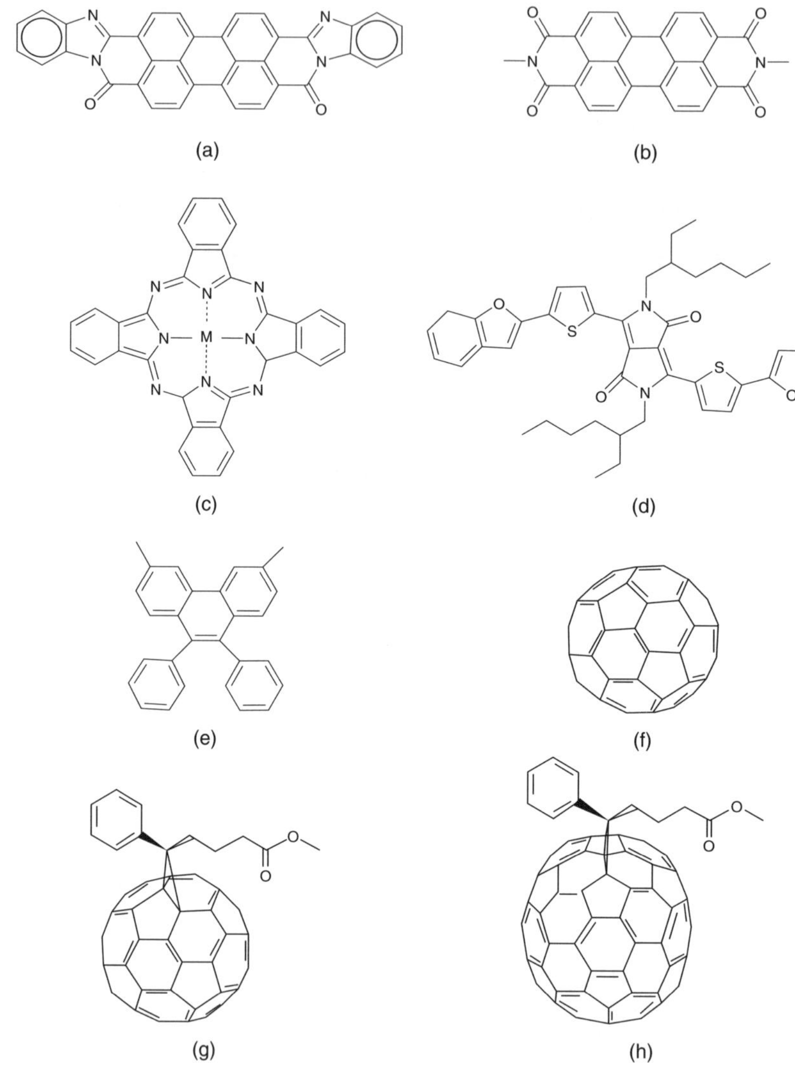

Figure 8-1 Chemical structure of molecules used in organic solar cells. (a) PTCBI. (b) Me-PTCDI. (c) MPc (where the most widely used metal is copper or zinc). (d) DPP(TBFu)$_2$ (e) BCP. (f) C$_{60}$. (g) PC$_{61}$BM. (h) PC$_{71}$BM.

The primary photoexcitations are bound states that require a significant amount of energy to break and have often been compared to Frenkel excitons of inorganic semiconductors. Although there has been vigorous debate as to the magnitude of the exciton binding energy [3], there is general agreement that at least several tenths of an electron volt are required to split an exciton into free carriers [4, 5]. As a consequence, the intrinsic photocarrier yield for an organic semiconductor is poor and primarily due to defects and impurities [6]. This makes charge

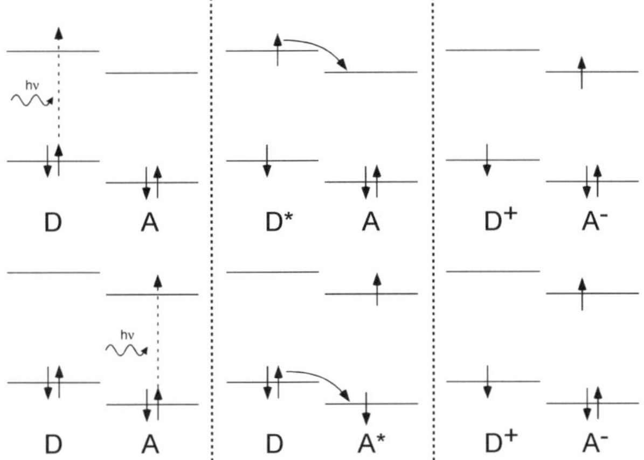

Figure 8-2 Illustration of charge transfer reaction for excitation of the electron donor (top) or electron acceptor (bottom).

photogeneration in an organic solar cell fundamentally different from that of cells based on inorganic semiconductors. In a conventional *p-i-n* solar cell, charge carriers are directly created by absorption of light within the intrinsic layer and are swept to the electrodes by a built-in electric field due to the difference in the Fermi energy relative to the band edges of the *p*-type and *n*-type layers. The built-in electric field ($\sim 10^5$ V/cm) in an organic cell is insufficient to ionize the excited state of an organic semiconductor. Given typical values for the exciton binding energy (≥ 0.2 eV) and exciton radius (1 nm), a field of 10^6 V/cm is required for excitons to dissociate [7].

The photocurrent in organic solar cells must therefore be generated by a photoinduced charge transfer reaction between adjacent chromophores—the electron donor and electron acceptor. An energetic offset between either the highest occupied molecular orbital (HOMO) or the lowest unoccupied molecular orbital (LUMO) or both of the two molecules drives the charge transfer reaction. Figure 8-2 illustrates possible charge transfer reactions. Charge transfer can follow excitation of either the donor or the acceptor, with the electron transferring into the LUMO of the acceptor following excitation of the donor or from the HOMO of the donor following excitation of the acceptor. There is also evidence that Förster resonance energy transfer (FRET) can play an important role in charge photogeneration [8]. Förster transfer is a long-range resonant dipole-dipole interaction that is contingent upon strong spectral overlap between the emission spectrum of one chromophore and the absorption spectrum of the other [9, 10]. As the acceptor and donor generally do not have the same optical gaps, it is therefore energetically favorable for energy transfer to occur. Finally, it is also important for photocurrent generation that the back transfer rate of charged excitations be relatively slow so that carriers can escape their mutual coulomb attraction.

The interface between the electron donor and acceptor, referred to as the organic heterojunction, is fundamental to the operation of organic solar cells. This also presents the greatest challenge to obtaining a device with efficient charge transfer and charge transport. Optimizing the former has often meant sacrificing the latter. The most efficient way to maximize charge transfer is to completely mix the electron donor and acceptor, creating a bulk heterojunction. Intimate contact between the donor and acceptor results in complete charge transfer—every photon absorbed results in the creation of an electron-hole pair. Complete mixing, however, interferes with charge transport. With few exceptions, organic semiconductors preferentially transport electrons or holes. It should be noted that charge carriers in organic semiconductors are often referred to as polarons, because of the associated deformation of the molecular structure. In a mixed system, charge trapping and recombination negatively impact device performance, increasing the series resistance of the cell and reducing the photocurrent density fill factor. The alternative approach is to prepare a multilayered cell with a planar heterojunction in which neat layers of the electron donor and acceptor are deposited sequentially, permitting optimized charge transport in each layer. The limitation of this approach is that charge transfer occurs only at the heterojunction and energy from photons absorbed away from this interface can be lost by recombination or quenching of the excited state before it migrates to the heterojunction.

The ideal bulk heterojunction consists of two interpenetrating networks that provide a large interfacial area between donor and acceptor domains and per-colation pathways for charge carriers to be transported within separate domains. Self-organization has proved key to the development of high-efficiency organic solar cells that have such a morphology, both from semiconducting, π-conjugated polymers and from molecular semiconductors. The drive for the optimized bulk heterojunction morphology, in combination with advances in materials, has led to the development of organic solar cells with a power conversion efficiency of nearly 8% in late 2009 [11]. Calculations show that cell efficiencies as high as 15% might be attainable from stacked (tandem) cells [12]. Self-organization of organic semiconductors has proven essential to advances in organic solar cell performance.

8.1.2. Solar Cell Characterization

It is useful to overview solar cell characterization before a detailed discussion of organic solar cells. Figure 8-3 shows a schematic diagram of a system designed to characterize solar cells. Two types of measurements are needed to provide a complete set of data. The current density-voltage (JV) characteristics of the cell are measured in the dark and under simulated solar illumination. The solar simulator (Fig. 8-3a) consists of an arc lamp light source that is filtered to eliminate atomic lines of the arc lamp spectrum and shaped to approximate the solar spectrum [13]. It is particularly important that diffuse illumination be provided to prevent hot spots that can result in an incorrect measurement of the illumination intensity and thus unreliable measurements of device efficiency. It is also necessary that the device area and illumination spot size be matched to avoid artifacts such as current spreading that alter the measured device efficiency [14].

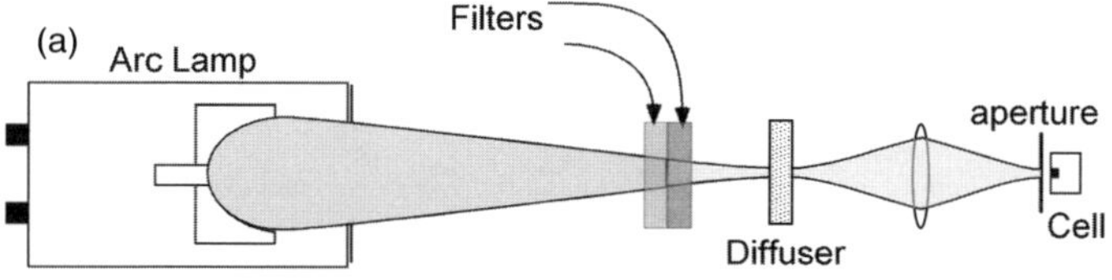

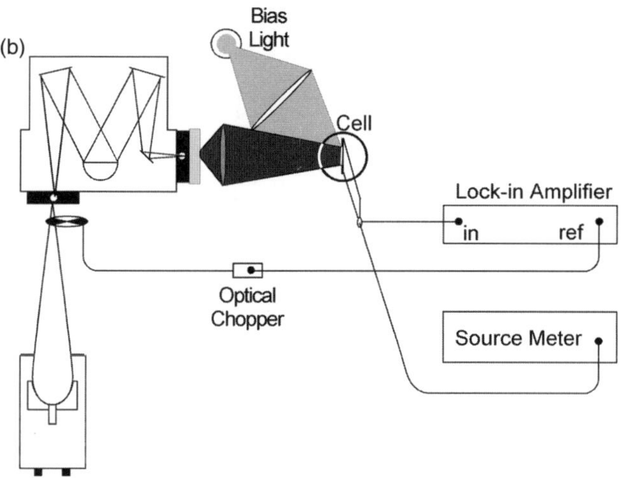

Figure 8-3   (a) Schematic diagram of solar simulator used to measure current density-voltage characteristics of solar cells. (b) Spectrometer used to measure IPCE spectra of solar cells.

Figure 8-4 shows the JV characteristics of an organic solar cell, measured in the under $\sim$1 sun illumination [15]. The most important device parameters are the short-circuit current density (J_{sc}), the open-circuit voltage (V_{oc}), and the fill factor (ff). The fill factor of a device, defined as the ratio between the maximum power delivered to an external circuit and the product of V_{oc} and J_{sc}, is the ratio of the darkly shaded to lightly shaded regions in Figure 8-4. The power conversion efficiency of a cell, defined as the ratio between the maximum electrical power generated and the incident optical power P_0, is related to the device parameters by Eq. 8.1:

$$\eta_{\text{power}} = \frac{V_{oc} \cdot J_{sc} \cdot ff}{P_{in}} m \tag{8.1}$$

where m is a spectral mismatch factor that corrects for the differences between the filtered output spectrum of the Xe lamp and the true AM1.5D (ASTMG173) solar reference spectrum.

The electrical behavior of a solar cell is often modeled in terms of discrete components. A solar cell can be treated as a current source in parallel with a diode; series resistance and shunt (parallel) resistance are added to the model.

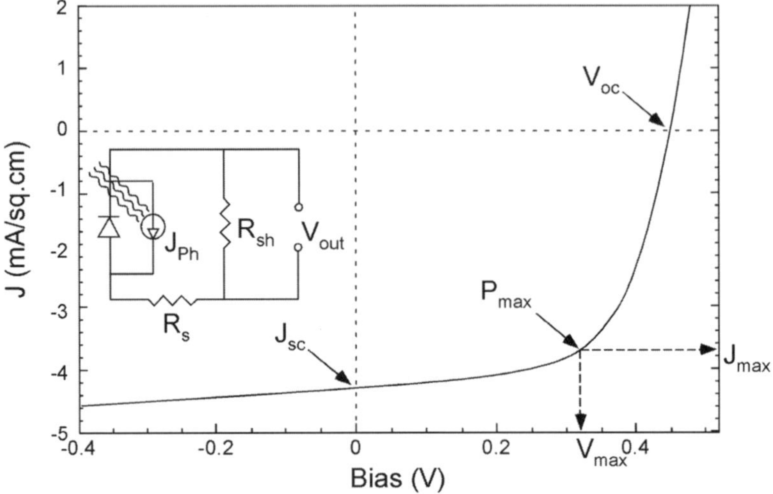

Figure 8-4   Sample *JV* characteristics of an organic solar cell under simulated AM1.5 solar illumination [15]. Inset: diode model used for solar cells.

The modified Schottky equation is given below:

$$J(V) = J_0\{\exp[q(V - JR_S)/nkT] - 1\} + (V - JR_S)/R_P + J_{SC} \qquad (8.2)$$

where R_S is the series resistance, R_P is the shunt resistance, and n is the diode ideality factor. The current density appears on both sides of this equation, and it must be solved iteratively. An ideal cell has zero series resistance and infinite shunt resistance. In practice, deviations from ideality can affect the fill factor dramatically and reduce both V_{oc} and J_{sc} to a degree. Sources of series resistance include the intrinsic mobility of the organic semiconductor, contact resistance at interfaces, and electrode resistance. Partial electric shorts as well as photoenhanced conductivity combine to reduce the shunt resistance. It is worth noting that significant deviations from Schottky-like behavior have been observed for organic solar cells. For example, many organic semiconductors show photoconductivity that is dependent upon the illumination intensity and electrical bias. Thus, Eq. (8.2) is only a qualitative guide to device characterization.

The spectral response is necessary for a complete characterization of characterize solar cells. This measurement serves as a check against possible errors in the *JV* characterization and is particularly important for organic solar cells, as one can determine the relative conversion efficiencies of the chromophores. The external quantum efficiency spectrum, also known as the incident photon conversion efficiency (IPCE), is given by the number of electrons generated per incident photon of a given wavelength. The IPCE spectrum, integrated over the solar spectrum, should yield the short-circuit current density. The charge generation efficiency of a solar cell can be dramatically affected by the light intensity [16], and so it is important that the IPCE spectrum be measured while the device is illuminated, a condition known as bias illumination. In the schematic setup shown in Figure 8-3b, the probe consists of a weak monochromatic light source that is optically chopped and dispersed through a monochromator while the cell is illuminated. Lock-in amplification is used to detect changes in the photocurrent due to the probe. In practice, it is not necessary to match the solar spectrum, only the short-circuit current density, and so an inexpensive incandescent light source may be used for bias illumination.

8.2. SMALL-MOLECULE SOLAR CELLS

8.2.1. Planar Heterojunction Cells

Photovoltaic effects have long been observed in molecular semiconductors [17]. Molecular semiconductors are particularly attractive as purification by vacuum train sublimation can reduce contaminant levels in molecular semiconductors below 0.1%, much lower than can be obtained from solution-processed compounds. As a consequence, carrier mobilities exceeding $10 \ cm^2/V \cdot s$ have been observed from molecular semiconductors [18], several orders of magnitude higher than for any conjugated polymer [19–21]. Molecular semiconductors such as phthalocyanines and perylenes are widely used for xerographic applications, resulting in widespread availability of pure materials at moderate cost. Furthermore, these organic dyes have linear absorption coefficients exceeding $10^5 \ cm^{-1}$, permitting the use of thin films to absorb light. Figure 8-5 shows the extinction spectra of zinc phthalocyanine (ZnPc) and N,N'-dimethyl-3,4:9,10-perylene tetracarboxylic acid diimide (Me-PTCDI), two molecular materials widely used in organic solar cells. These materials have complementary absorption spectra covering a wide range of the solar spectrum and high extinction coefficients in the visible.

Early planar heterojunction cells were made from combinations of electron-transporting dyes such as rhodamines or triphenylmethane with hole-transporting dyes such as phthalocyanines or merocyanines [22]. Harima et al. reported efficient charge photogeneration at an organic heterojunction between ZnPc and 5,10,15,20-tetra(3-pyridyl) porphyrin in 1984 [23]. The internal quantum efficiency was calculated to be as high as 17% and open-circuit voltages of $\sim$1.0 V could be obtained. The first breakthrough in organic solar cells came in 1985, when C. W. Tang reported a bilayer structure that achieved a power efficiency of nearly 1% under simulated solar illumination [24]. The device is conceptually similar to that of Harima et al., but used a perylene derivative as the electron acceptor. The cell was fabricated by successively evaporating layers of copper phthalocyanine (CuPc), 3,4,9,10-perylene tetracarboxylic-bis-benzimidazole (PTCBI) and silver onto an ITO-coated glass substrate. The device parameters under simulated AM2 solar illumination at an intensity of 75 mW/cm^2 were $J_{sc} = 2.3 \pm 0.1$ mA/cm^2, $V_{oc} = 0.45 \pm 0.02$ V, and $ff = 0.69$. The overall power conversion efficiency was 0.95%, roughly an order of magnitude better than had

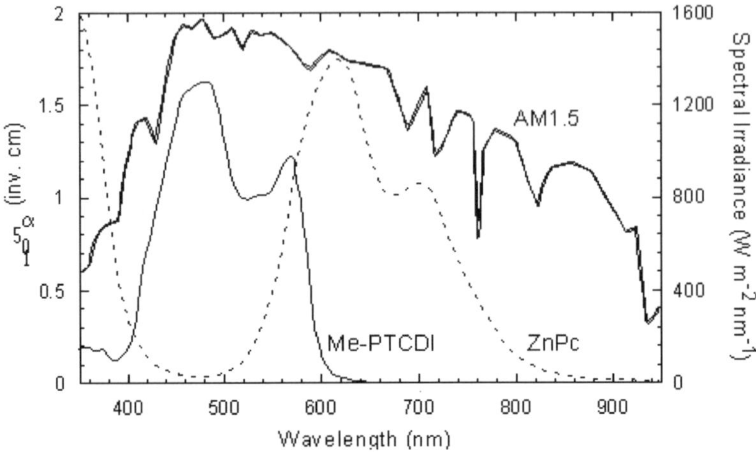

Figure 8-5   Extinction spectra of (dashed line) ZnPc and (solid line) Me-PTCDI. The AM1.5 solar spectrum is shown as a bold line for comparison.

been previously achieved. The IPCE spectrum roughly follows the absorption spectrum of the bilayer, demonstrating that charge photogeneration occurs in both layers. The peak collection efficiency was about 15% at $\lambda = 620$ nm, and the corresponding internal quantum efficiency is roughly 25%.

Improvements in materials and cell design have led to the development of planar heterojunction cells with power conversion efficiencies in the range of 2–4%. Buckminsterfullerene (C_{60}) has attracted considerable attention since its discovery [25] and synthesis in bulk form [26]. Its attractive electronic proper-ties for photovoltaics include being a strong electron acceptor [27] and having a high electron mobility for an organic semiconductor [28, 29]. Furthermore, intersystem crossing in C_{60} results in a nearly unit triplet yield that in turn yields long excited-state lifetimes [30] and long exciton diffusion lengths (7.7 ± 1.0nm) [31]. Electron transfer from an organic electron donor to C_{60} occurs on a sub-picosecond timescale [32], and the back transfer rate is many orders of magnitude slower [33]. Despite the attractive photophysics of C_{60} that make it an ideal electron acceptor, the first C_{60}-based planar heterojunction devices had relatively modest efficiencies [34, 35]. One particular problem when using materials with long exciton diffusion lengths is that excitons diffuse to the electrode where they are quenched rather than contributing to the photocurrent.

Much higher OPV efficiencies were obtained by inserting a layer between the electron acceptor and the cathode to prevent photogenerated excitons from being quenched at the cathode [36]. The first such device was a modified Tang structure with a 15-nm-thick CuPc film as the electron donor, 6 nm of PTCBI as the electron acceptor, and 15 nm of bathocuproine (BCP) as an exciton blocking layer (EBL). The BCP layer was doped with 10% PTCBI to inhibit crystallization. Although the energy of the lowest unoccupied molecular orbital (LUMO) of BCP is well above the LUMO of PTCBI, the insertion of a thin layer of BCP did not

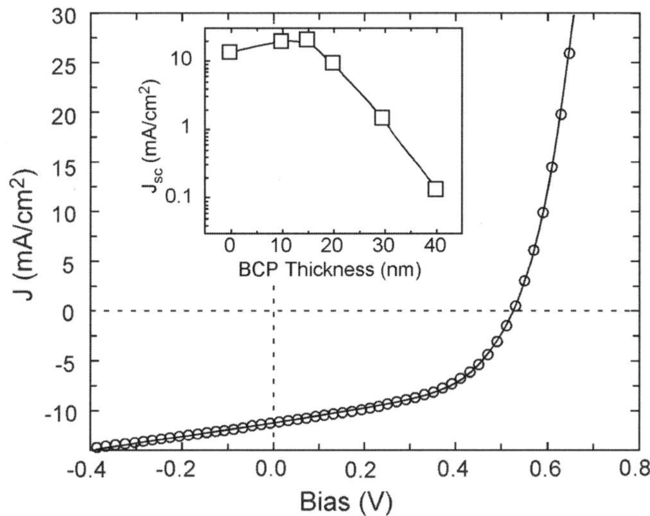

Figure 8-6 *JV* characteristics of planar heterojunction cell with the structure ITO/PEDOT:PSS/CuPc/C_{60}/BCP/Ag. Inset: dependence of J_{sc} on the thickness of the exciton blocking layer consisting of BCP. From Ref. 37 [P. Peumans et al. *J. Appl. Phys.* **2003**, *93(7)*, 3693–3723.] Copyright 2003 American Institute of Physics.

appear to interfere with electron transport. BCP blocked excitons created in the PTCBI layer from being quenched at the Ag electrode and permitted photocarriers to exit the structure. Even though extremely thin organic layers were used, devices show a power conversion efficiency of $\sim$1% under a wide range of illumination intensities. The effectiveness of such a structure was demonstrated when C_{60} was used as the electron acceptor in a device with a BCP exciton blocking layer [37]. This device also used poly(3,4-ethylenedioxythiophene):poly(styrenesulfonate) (PEDOT:PSS) as a hole injection layer. Although not a focus of this chapter, the development of hole injection layers such as PEDOT:PSS has played an important role in improved device efficiencies. Figure 8-6 shows the *JV* characteristics of a cell with the structure ITO/PEDOT:PSS/(20 nm) CuPc/(40 nm) C_{60}/(12 nm) BCP/Al under 150 mW/cm^2 simulated AM1.5 illumination. A power conversion efficiency of 3.6% was obtained from a circular device with a 1-mm diameter ($\sim$0.01 cm^2). There have since been a number reports of planar heterojunction cells with efficiencies of several percent that use C_{60} as an electron donor and have an exciton blocking layer [38–41].

8.2.2. Self-Organized Bulk Heterojunction Molecular Solar Cells

The performance of a planar heterojunction cell is ultimately determined by the efficiency of charge photogeneration, transport, and extraction. Although tremendous strides in performance have been made, planar heterojunction cells are ultimately limited by the fact that charge transfer occurs only at the organic heterojunction. Absorption must therefore occur within the exciton diffusion length in the respective materials. In principle, the exciton diffusion length can be calculated by analyzing the spectral response of a device [41] and is typically estimated to be 5–10 nm [31, 42–44]. Such thin layers can only absorb a fraction of incident photons, and thus recombination losses will limit cell performance. There is also a significant risk of electrical shorts in devices made with organic layers less than 50 nm thick.

Bulk heterojunction devices based on evaporated molecular semiconductors have had disappointing performance. For example, coevaporated films of H_2Pc or ZnPc with C_{60} result in devices with low J_{sc} and fill factors of 0.25 [34]. It is difficult to obtain the desired morphology by cosublimation of the electron donor and acceptor because of the high energy of reorganization [45]. Many molecular semiconductors, however, are solution processible and self-organize to form supramolecular assemblies. A number of research groups have exploited the fact that many organic semiconductors also have liquid crystalline phases. Supramolecular interactions such as van der Waals forces, multipolar interactions, and hydrogen bonding play a crucial role in the formation of liquid crystals and determine their mesomorphic properties. Molecules in these phases combine the fluidity of liquids with the anisotropy of crystals. In the context of organic photovoltaics, these self-organizing materials hold out the possibility of a segregated molecular bulk heterojunction suitable for use in organic photovoltaics. For example, porphyrins are promising materials for photovoltaic applications that have been shown to form nanoparticles and nanorods [46]. An early study by Gregg et al. demonstrated that unusual photovoltaic effects could be obtained from ordered arrays of zinc octakis (β-octyloxyethyl)porphyrin, a liquid crystalline porphyrin [47, 48]. The porphyrin was infiltrated by capillary action between two ITO-coated glass plates separated by a spacer film. The cell was heated above the isotropic liquid phase until full and then cooled slowly through the liquid crystalline (LC) phase to the solid. Time spent in the LC phase was used to order the film. Even though the devices were microns thick and only had a single component, reasonable performance was obtained. A short-circuit current density of 250 $\mu A/cm^2$ was obtained from a 1.5-μm-thick cell under 150 mW/cm^2 illumination, falling to 15 $\mu A/cm^2$ from a 5-μm-thick cell.

More recently, the properties of discotic liquid crystalline organic semiconductors have been used to form bulk heterojunction solar cells. These materials have attracted interest because of their fundamental importance as model systems for the study of charge and energy transport and because of the possible applications in organic electronics [49]. Discotic liquid crystals are formed by disclike molecules typically comprising a rigid aromatic core and flexible peripheral chains that stack into columns with a high degree of order and having efficient charge transport. For example, the carrier mobility of hexahexylthiotriphenylene (HHTT) was measured by time of flight [50], and time-resolved microwave conductivity [51] was found to be of order 10^{-1} $cm^2/V \cdot s$ in the helical mesophase. Figure 8-7 shows the dependence of the mobility on the temperature as HHTT is cooled through phase transitions. A peak mobility

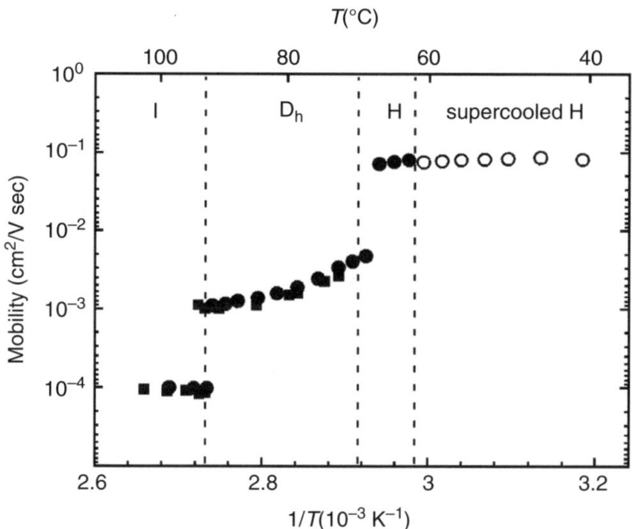

Figure 8-7 Mobility dependence of HHTT on temperature, measured by time-of-flight method. Squares show the mobility as the sample is being heated, and circles show the mobility as the sample is being cooled [51]. Used with permission from Nature Publishing Group.

of 0.4 $cm^2/V \cdot s$ was measured [50]. For photovoltaic applications, a bulk heterojunction can be formed from an electron donor and an electron acceptor that will preferentially order into separated columns (Fig. 8-8). Photoinduced charge transfer occurs between molecules in neighboring columns, followed by charge transport within columns and perpendicular to the plane of the device.

The first photovoltaic cell constructed from discotic liquid crystalline materials with reasonable efficiency used hexaphenyl-substituted hexabenzocoronene (HBC-PhC$_{12}$) as the electron donor and a soluble perylene derivative, N,N'-bis(1-ethylpropyl)-3,4,9,10-perylene tetracarboxylic diimide (EP-PTCDI), as the electron acceptor [52]. Intracolumn hole mobilities as high as 0.22 $cm^2/V \cdot s$ have been measured for HBC-PhC$_{12}$ [53], and perylenes are well-known electron transport materials. Cyclic voltammetry was used to measure the ionization potentials of the two materials (5.25 eV for HBC-PHC$_{12}$ and 5.32 eV for EP-PTCDI). As the optical gap of the coronene is $\sim$0.7 eV larger than that of the perylene, the coronene should be a suitable electron donor. Both the hexabenzocoronene and the perylene exhibit characteristic morphologies upon spin coating from solution. Tapping-mode atomic force microscopy (AFM) images of HBC-PhC$_{12}$ showed a smooth-textured surface with an rms surface roughness of $\sim$10 nm. Polarized light microscopy indicates the presence of large domains with typical in-plane length scales of hundreds of microns. The perylene surface has a surface roughness of 40 nm with crystals protruding from the surface. Spin coating of blend solutions of 40% HBC-PhC$_{12}$ and 60% perylene resulted in a markedly different surface morphology. The surface roughness of the composite film is much less

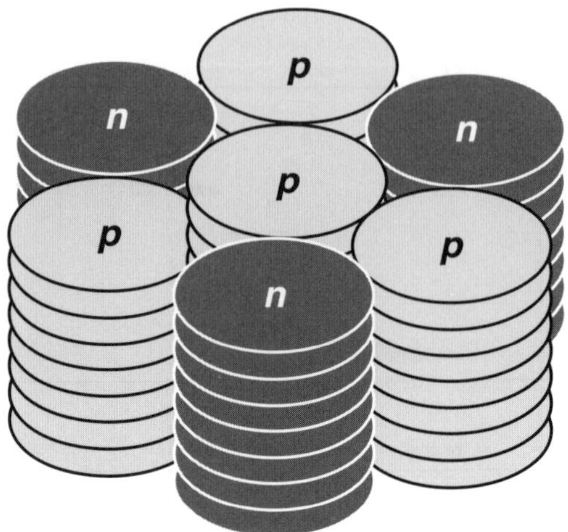

Figure 8-8   Schematic diagram of desired discotic LC morphology for use in an organic solar cell.

than either of the single component films, and there is nearly continuous surface coverage of shortened perylene crystallites. The cross section of the composite film was imaged by a field-emission scanning electron microscope (SEM) capable of high-resolution, low-accelerating-voltage operation. Perylene crystallites with a cross section resembling that of the long crystallites found in the pure perylene film are dispersed through the intermediate layer in the SEM image. Stratification is clearly visible in the cross section, with a polycrystalline perylene sheet extending down the top surface of the film. These results suggest a phase-segregated film suitable for photovoltaic applications.

Devices were made by spin-casting a blend of 40% HBC-PhC$_{12}$ and 60% EP-PTCDI (15 mg/ml) onto an ITO-coated substrate and then evaporating an aluminum electrode on top. The IPCE spectrum of the cell shows that both materials contribute to the photocurrent, with the EQE of EP-PTCDI being somewhat higher than HBC-PhC$_{12}$. The performance of these devices compared favorably to that of bilayer devices made by evaporation of the perylene onto a spin-cast HBC-PhC$_{12}$ film. The EQE of the self-organized cells was an order of magnitude higher, demonstrating that a large interfacial area between the electron donor and acceptor was achieved. The measured JV characteristics of the device under 0.47 mW/cm^2 illumination at 490 nm resulted in a power conversion efficiency of ~2%. Although not competitive with planar heterojunction devices, these results demonstrate that phase segregation and improved ordering can be obtained from liquid crystalline materials.

Work continues on discotic liquid crystals for organic solar cells. Schmidtke et al. followed up on the above work, using a slightly different coronene derivative [54]. The quantum efficiency of the cell could be improved by annealing following deposition of the cathodem, though the efficiency of the annealed device was comparable to the previous work. This is a significant step in that LC-based cells have the potential to be self-healing structures to remove degradation-related defects. Oukachmich et al. made cells by using a triphenylene ether in combination with one of several different perylenes [55]. These cells were prepared by vacuum sublimation of layered or bulk-heterojunction devices, although annealing was not attempted. External quantum efficiencies of several percent under monochromatic illumination were obtained. Feng et al. fabricated cells with dicotic graphenes and obtained peak monochromatic quantum efficiencies of $\sim$20% at 500 nm [56]. Palermo and coworkers recently reported work on a rigid polymer that can be used as a macromolecular scaffold to which liquid crystalline sidegroups are attached [57]. The nonconjugated polymer defines the geometery, and then functional groups are attached to exploit it. Although LC solar cells have yet to challenge the efficiency of more convential structures, this remains a promising avenue of development.

This section concludes with a recent report of a high-efficiency bulk heterojunction cell based on solution-processed small molecules. Polymer-based cells have dominated recent work in organic solar cells, primarily because of the better film forming properties of polymers and the superior morphology of self-organized polymer:fullerene blends. Nguyen and coworkers synthesized a new class of small molecules based on diketopyrrolopyrrole and oligothiophene moieties [58]. Introducing solubilizing groups onto the aryl groups bound to the DPP core leads to films that self-assemble into ordered domains. The electron donor in this study contained 3,6-bis(5-(benzofuran-2-yl)thiophen-2-yl)-2,5-bis(2-ethylhexy)pyrrolo[3, 4-c]pyrrole-1,4-dione or DPP(TBFu)$_2$ (see Fig. 8-1) [59]. Solution-processed films of DPP(TBFu)$_2$ have broad absorption across the visible spectrum and when combined with a functionalized fullerene, self-assemble into a bulk heterojunction with bicontinuous networks of donor and acceptor molecules. The device parameters of an optimized cell based on a 3:2 ratio of DPP(TBFu)$_2$ to fullerene were $J_{sc} = 10.0$ mA/cm^2, $V_{oc} = 0.92$ V, and $ff = 0.48$, yielding a power conversion efficiency of 4.4%. Given that this fill factor of this device is significantly below that of the best cells, efficiencies of 6% or greater may be attainable from this materials system.

8.2.3. *P-I-N* Molecular Solar Cells

An intermediate approach between planar and bulk heterojunction cells is to prepare a multilayered cell with a charge generation layer (CGL) formed by codeposition of the electron donor and acceptor. This increases the optically active area of the cell but does not entirely sacrifice the attractive properties of neat films of molecular semiconductors. Electrons and holes are photogenerated in the composite layer and are swept to the transport layers by built-in chemical and electric potentials. As with bulk heterojunction cells, the morphology of the CGL plays a key role in determining device efficiency. We refer to these cells as *p-i-n* cells in analogy with cells based on inorganic semiconductors. The first *p-i-n* cells were reported by Hiramoto et al. [60], who made cells from perylene and phthalocyanine derivatives with and without a codeposited film between the electron-transport layer (ETL) and the hole-transport layer (HTL). CuPc or a free base phthalocyanine (H_2Pc) was used as the electron donor, and a perylene derivative was used as the electron donor—either PTCBI or Me-PTCDI. Measured under 100 mW/cm^2 white light illumination, the current density of the Me-PTCDI/H_2Pc device increased from 0.94 mA/cm^2 to 2.14 mA/cm^2 without a negative effect on V_{oc} (0.51–0.54 V) or fill factor (0.48). The cell efficiency increased from 0.29% to 0.63%. For a PTCBI/CuPc-based device, the increase in J_{sc} was about 60% (1.61 mA/cm^2 to 2.56 mA/cm^2), but the fill factor dropped from 0.42 to 0.25 and there was no improvement in device efficiency. The poor fill factor of the CuPc/CuPc:PTCBI/PTCBI device was attributed to a reduced shunt resistance, which in turn was due to poor film quality of the codeposited layer. This highlights the importance of morphology in bulk heterojunctions.

Rostalski and Meissner used a codeposited layer of C_{60} and ZnPc (1:1 by weight) as the CGL in a cell with a ZnPc HTL and an Me-PTCDI ETL [61]. The device parameters were $J_{sc} = 5.26$ mA/cm^2, $V_{oc} = 0.39$ V, and $ff = 0.45$, yielding a power conversion efficiency of 1.05%. The photocurrent action spectrum has a maximum quantum efficiency of 37.5%, more than twice that of the Tang device, but the fill factor is somewhat lower. Overall device performance, therefore, did not improve significantly. Both Harimoto [60] and Rostalski [61] suggested that the morphology of the bulk heterojunction could limit the power efficiency of molecular OPVs. The crystallinity and phase segregation of a film can be enhanced by evaporating onto a substrate at an elevated temperature. This, however, comes at the cost of increased surface roughness. For example, Geens et al. [62] studied coevaporated films of C_{60} and a five-ring phenylene-vinylene oligomer. Low substrate temperatures result in amorphous films with a surface roughness of only 5 nm. Increasing the substrate temperature to 100°C resulted in the formation of crystals and increased the surface roughness up to 200 nm. Such a rough film is unsuitable for device applications, as holes reaching down to the substrate will short the device. Postdeposition annealing results in similar problems with increased surface roughness [63].

The Leo group at Dresden has made remarkable progress in the development of high-efficiency organic p-i-n solar cells. A unique aspect to their approach is to dope the electron (n-type) and hole (p-type) layers [64]. Tetrafluoro-tetracyano-quinodimethane (F_4-TCNQ) was used as an electron acceptor to dope the p-type layer (ZnPc or an amorphous amine), and rhodamine B was used as an electron acceptor to dope the n-type layer (C_{60} or a perylene derivative). The dopants are sufficiently strong donors and acceptors that there is a ground state charge transfer, increasing the conductivity of the transport layer by several orders of magnitude and reducing the series resistance of the cell. The first efficient p-i-n cells using this concept were reported in 2004 [65]. Figure 8-9 shows the cell structure and its JV characteristics under 1 sun illumination. A composite layer of ZnPc and C_{60} was used as the CGL. The cell parameters are $V_{oc} = 0.44$ V, $J_{sc} = 8.44$ mA/cm^2, ff of 0.45 and a power conversion efficiency of 1.67%. An even higher efficiency of 1.9% was obtained from a device using Me-PTCDI as

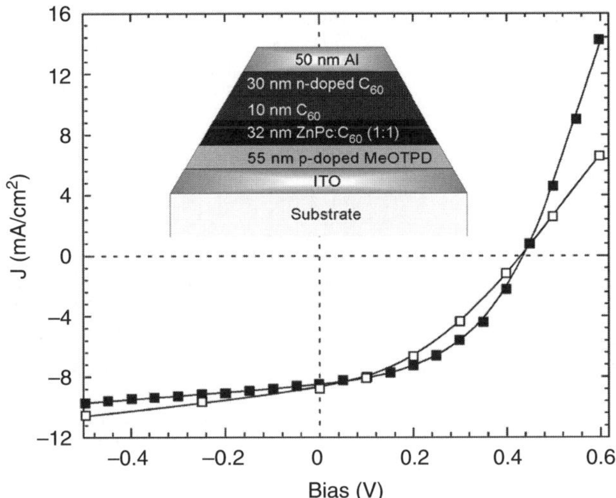

Figure 8-9 *JV* characteristics of an optimized *p-i-n* solar cell with device areas of 1.1 cm^2 (open squares) and 2.4 mm^2 (filled squares). The lower fill factor of the larger area device is due to the series resistance of ITO [65]. Used with permission from Springer-Verlag.

the ETL. Me-PTCDI has stronger absorption than C_{60} in the visit, contributing to a higher J_{sc}.

The morphology of the CGL has proven critical to the development of high performance *p-i-n* cells. Figure 8-10 shows electron bright-field and electron diffraction images of films of ZnPc, C_{60}, and a ZnPc:C_{60} composite [65]. The crystal lattice of ZnPc can be imaged with lattice planes having a spacing of 6.94 Å, and the crystalline domains have an average size of 20–30 nm. High-resolution images of C_{60} layers display small crystallites of 5–10 nm size, and the corresponding electron diffraction images show Debye–Scherrer rings resulting from a polycrystalline arrangement. In contrast, the images of the mixed ZnPc:C_{60} layer (1:1 by weight) have no visible structural order. Although this layer is relatively thin compared to bulk junction cells, the mixed layer will have a deleterious effect on charge transport.

The essential problem is how to exploit the natural tendency of molecular crystals to self-order while obtaining intimate contact between the electron donor and acceptor. Self-ordering is inhibited by codeposition of materials with very different crystal structures. Hong et al. exploited the natural tendency of organic crystals to self-order by alternating deposition of ultrathin layers (10–30 Å) of the electron donor and acceptor [66]. In comparison with planar heterojunction cells ($V_{oc} = 0.48$ V, $J_{sc} = 6.33$ mA/cm^2, $ff = 0.51$, $\eta = 1.3\%$) or *p-i-n* cells with a codeposited CGL ($V_{oc} = 0.47$ V, $J_{sc} = 9.52$ mA/cm^2, $ff = 0.44$, $\eta = 1.7\%$), the cell with an alternately deposited CGL equaled the high V_{oc} (0.49 V) and ff (0.51) of the planar heterojunction cell and had an even higher current density ($J_{sc} = 10.13$ mA/cm^2) than the cell with a codeposited CGL. The morphologies of the

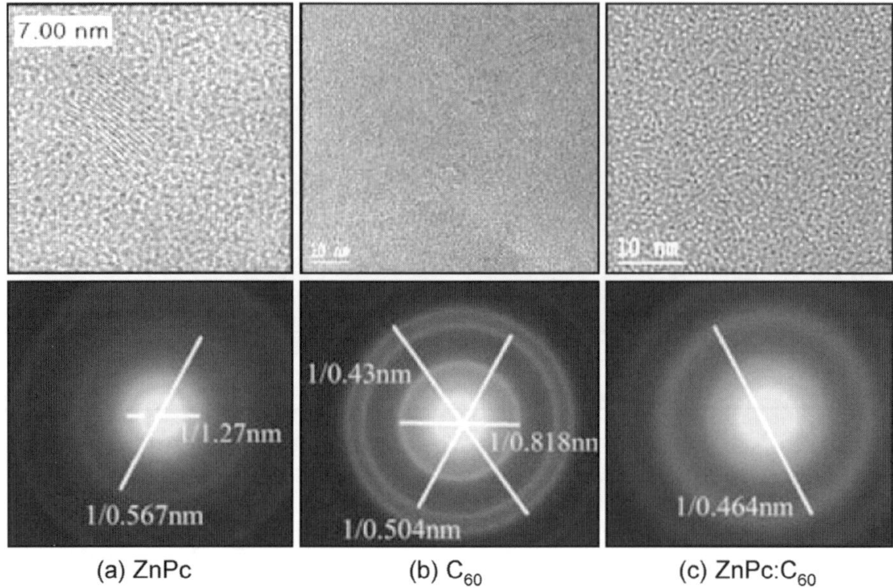

Figure 8-10   High-resolution images (top) and electron diffraction patterns (bottom) of films of ZnPc (a), C_{60} (b), and a ZnPc:C_{60} composite (c) [65]. Used with permission from Springer-Verlag.

codeposited and alternately deposited layers were compared by analyzing high-resolution transmission electron micrographs (Fig. 8-11). Fourier transforms of these micrographs show enhanced crystallinity in the alternately deposited films.

Observation of crystallinity in a nanolayered ZnPc/C_{60} film suggests that a degree of phase segregation has been achieved. Optical spectroscopy can provide insight into the intrinsic charge transfer dynamics. Esenturk et al. investigated the intrinsic properties of composite and layered ZnPc/C_{60} films by time-resolved THz spectroscopy [67, 68]. An optical pump induces changes in the transmission amplitude of a THz probe, and these are recorded as a function of the relative time delay between the pump pulse and the THz probe pulse. The decay dynamics were recorded by measuring the pump-induced modulation of the THz peak field strength as a function of delay time of the incident UV excitation pulse with respect to the THz pulse. The differential transmission signal yields information about the ultrafast carrier dynamics during generation, evolution and recombination. At any given time delay the measured differential THz transmission is proportional to the effective mobility and carrier density, which in turn, is proportional to the carrier generation efficiency. The early time dynamic investigation of these films offers a unique opportunity to measure the intrinsic photon-to-carrier efficiency of these films and improve the photoactive characteristics of the active layers.

The differential transmission of composite films of ZnPc and C_{60} with varying weight ratios is shown in Figure 8-12, and those of superlattice films with the

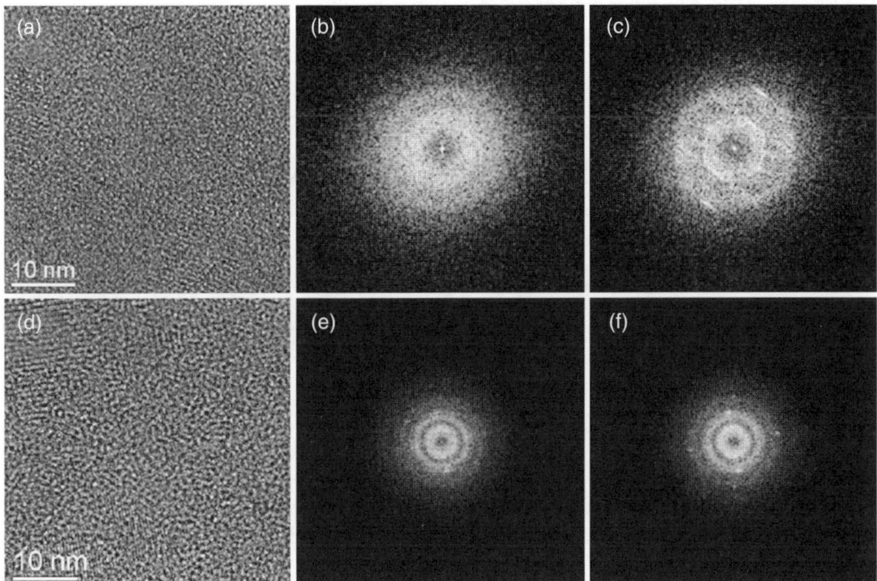

Figure 8-11   (a) HRTEM image of 1:1 mixture of C_{60} and ZnPc. It shows partial crystallinity as visualized by the FFT images (b) and (c). (d) HRTEM image of the C_{60}/ZnPc multilayered system, showing highly mosaic crystal domains of C_{60}. (e) FFT shows strong (202) C_{60} crystal reflections. (f) FFT of the left upper part of (d) indicates a single crystalline domain [66]. Copyright 2007 American Institute of Physics.

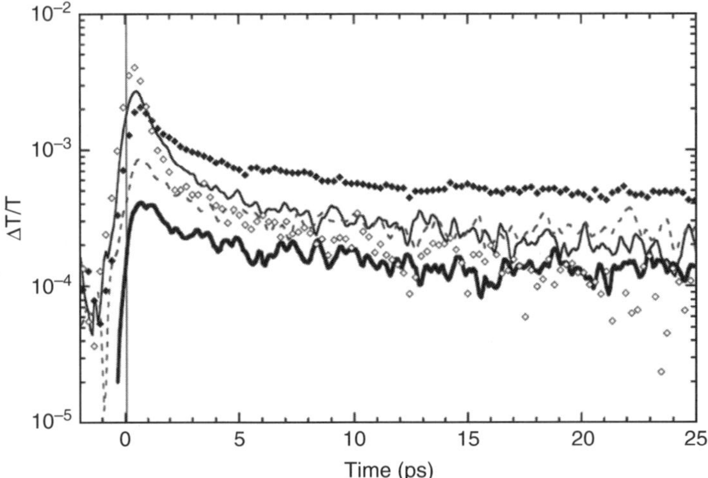

Figure 8-12   THz absorption transients for films of neat C_{60} (open symbols), 75% C_{60} (dashed line), 50% C_{60} (filled symbols), 25% C_{60} (thin solid line), and 10% C_{60} (bold solid line) [68]. Copyright 2009 SPIE.

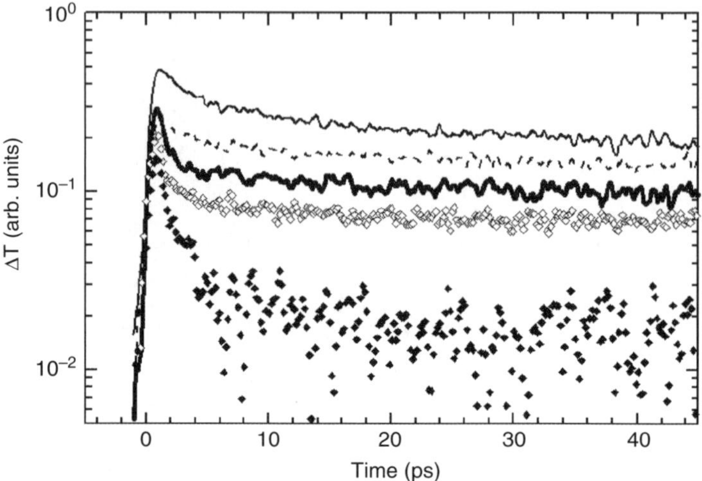

Figure 8-13   THz absorption transients for multilayer films of C_{60} and ZnPc with thicknesses of 2 nm (line), 5 nm (dashed line), 10 nm (bold line), 20 nm (open symbols), and 40 nm (filled symbols) [68]. Copyright 2009 SPIE.

varying layer thicknesses are shown in Figure 8-13. The transients of all samples have a sharp rise in THz absorption upon excitation to a peak followed by a rapid decay within a few picoseconds and then a much slower decay. The rise time dynamics were virtually independent of composition, 1.0 ± 0.1 ps from the onset of absorption to the peak. The neat C_{60} film has the strongest peak, but this signal decays exponentially with a lifetime $\tau = 0.65 \pm 0.12$ ps and there is no evidence for significant photogeneration of long-lived carriers. The peak amplitude in composite films is proportional to the C_{60} content, suggesting that this is a charge transfer state associated with C_{60}. It is worth noting that the lifetime of the fast decay increases from 0.65 ps for neat C_{60} to 1.4–1.7 ps for the samples containing 50% or more ZnPc, showing that the dynamics of the CT state of C_{60} are affected by the presence of neighboring ZnPc molecules. The slow decay takes place on a longer timescale than the pump-probe delay, indicating the formation of long-lived carriers. There is a clear dependence on the composition of the sample, with the strongest signal at longer pump-probe delays occurring for the sample with a 50:50 composition of C_{60} and ZnPc.

The decay of nanolayered films also has two components, with a fast component having exponential decay with a lifetime less than 2 ps and a second slowly decaying component that occurs on a timescale longer than that of the experiment. Amplitudes of the differential transmission at long pump-probe delay ($t > 10$ ps) suggest that the relative efficiencies of these model active layers increase as the total number of layers increases. The intensity of the slowly decaying component also increases with decreasing layer thickness, as a greater fraction of the sample will be available for charge transfer. Exciton diffusion will be minimal during the short timescale of this experiment, and thus photocarriers

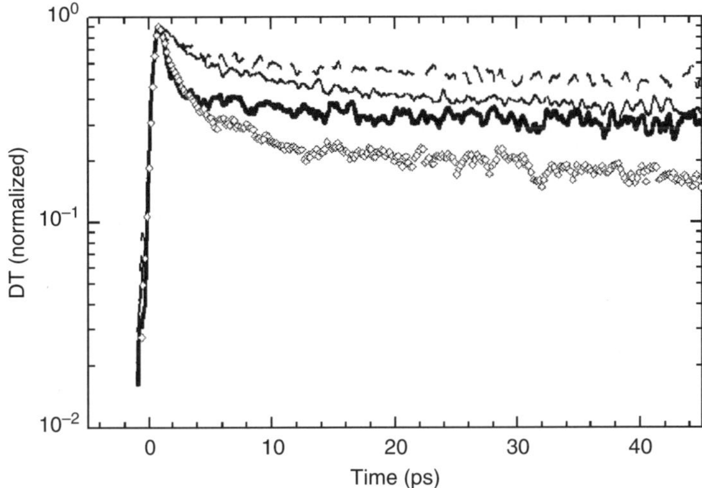

Figure 8-14 Normalized THz absorption transients for multi-layer films of C_{60} and ZnPc with thicknesses of 2 nm (line), 5 nm (dashed line), 10 nm (bold line), and a 1:1 ZnPc:C_{60} blend (open symbols) [68]. Copyright 2009 SPIE.

are generated only from excitations formed close to the heterojunction. The signal from the nanolayered sample with 10-nm-thick layers is comparable to that of the best composite film, even though the timescale is too short for exciton diffusion. Most important is that the multilayer films retain their superior mobility even when the layer thickness corresponds to only a few monolayers. The THz absorption of the layered 2-nm structure is several times that of the composite film, even though the two films contain the same C_{60} fraction. Of potentially greater significance for device applications is that the decay of the THz signal is slower in the multilayer films than in composites. Figure 8-14 compares the THz absorption of different multilayered samples to that of the 1:1 blend, normalized to the initial peak amplitude. These data indicate that photocarriers have separated into the distinct domains and can move under the influence of the built-in field.

Recent work at Heliatek, a start-up working with the Leo group, has focused on improvement of the bulk heterojunction morphology and incorporation of such junctions in tandem organic solar cells. A tandem cell consists of two cells stacked vertically with an intermediate recombination layer [69, 70]. In April of 2009, a Heliatek cell was certified with a power conversion efficiency of 5.87% at the National Renewable Energy Laboratory (NREL). Figure 8-15 shows the *JV* characteristics of the cell with a device area of 2.0 cm^2 under 100 mW/cm^2 simulated solar illumination, conducted at NREL [71]. The composition of the CGL and processing method used have not been published as of the date of this work, but this achievement illustrates that self-organization processes in molecular semiconductors have proven critical to advances in cell design and performance.

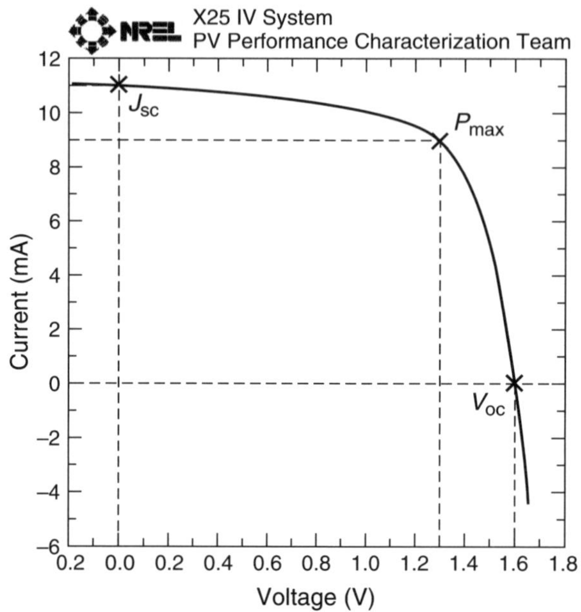

Figure 8-15 *JV* characteristics of an optimized *p-i-n* tandem cell device measured at NREL under 1 sun simulated solar illumination. The device area was 2 cm^2 [71]. Copyright 2009 SPIE.

8.3. POLYMER SOLAR CELLS

8.3.1. Polymer and Polymer Blend Cells

Conjugated polymers have drawn widespread interest as a novel class of materials with a variety of applications in optoelectronic devices such as light-emitting diodes, thin-film transistors, and solar cells. Reports of photovoltaic effects in π-conjugated polymers date back nearly 30 years [72–75]. The earliest devices were based upon trans-poly(acetylene), which has an absorption spectrum well-matched to the peak of the solar spectrum. Thin film sandwich structure devices with active layers of trans-poly(acetylene) exhibited photovoltaic effects, though both the photocurrent and fill factor were quite low [73]. In addition to the problem of bound excited states discussed above, most conjugated polymers are primarily hole transporters, sometimes being referred to as p-type materials. For example, poly(*para*-phenylene vinylene) (PPV) has a much higher hole than electron mobility, and the electron current is strongly reduced by trapping [76]. A time-of-flight study of PPV has shown that the mobility-lifetime product and, hence, the range of holes is three orders of magnitude larger than that of electrons [77]. Thus, these devices suffered from low photocurrent density and high series resistance. The dual problems of charge photogeneration and transport have been addressed by doping polymers with a material having a higher electron affinity and better electron transport. Both conjugated polymers and molecules have been used to sensitize charge photogeneration, and composites of conjugated polymers and functionalized fullerenes show the greatest promise.

One of the most attractive properties of conjugated polymers is that they are amenable to solution processing, promising significant cost savings compared with other deposition methods. Early polymer solar cells used a planar hetero-junction structure. A hole-transporting polymer film is spin-coated onto the anode, followed by evaporating a molecular sensitizer or spin-coating a second polymer. It is possible to avoid dissolving the first layer by thermal converting the first film to an insoluble form [79] or using a solvent for the second layer in which the already spun polymer is insolvent. Planar heterojunction polymer cells suffer both the limitation that charge photogeneration only occurs near the organic heterojunction as well as the reduced mobility associated with their lower purity vis-à-vis molecular semiconductors.

Solution processing naturally lends itself to preparation of bulk heterojunction cells, and the efficiency of bulk heterojunction polymer cells has advanced rapidly in the last five years. Self-organization of the bulk heterojunction coupled with the synthesis of new low-gap conjugated polymers has led to the development of high-efficiency organic solar cells. Figure 8-16 shows repeat units of various conjugated polymers used in active layers of organic solar cells. Derivatives of PPV with different side-chain substitution came under initial investigation. More recently, polythiophene and other low-gap polymers have been the focus of efforts. The advantage of these materials is better coverage of the solar spectrum and, in the case of polythiophene, superior hole mobility. Most recently, a number of low-gap polymers have been synthesized specifically for photovoltaic applications. These are discussed in detail below.

Polymer blends have the potential to form a bulk heterojunction with the desired morphology—efficient charge photogeneration at interfaces between the two polymers and separate pathways for transport of electrons and holes along the polymer chains. Conjugated polymers with high electron affinity have been developed by side-chain cyano-substitution [79] or by polymerization of model compounds with high electron affinity such as pyridine [80]. The first efficient solar cells based upon polymer blends combined poly(2-methoxy-5-ethyl(2′-hexyloxy) *para*-phenylene vinylene) (MEH-PPV) and a dialkoxy-substituted PPV with cyano side groups on the vinyl bond (CN-PPV) [81, 82]. Cyano-substitution modifies the position of the HOMO and LUMO levels, resulting in charge transfer of CN-PPV when blended with MEH-PPV. The PL quantum yield of a 1:1 blend of CN-PPV and MEH-PPV is less than 5% as compared to 32% for neat CN-PPV. Photoluminescence quenching in polymer blends is much weaker than that observed in C_{60} doping, suggesting that the film consists of phase-segregated domains of the two polymers. Recent investigations therefore concentrated on improving the polymer morphology [83] Polymer blends tend to phase separate into separate domains because of a low entropy of mixing [84]. The morphology of polymer blends can be modified either during deposition by varying the polymer concentration, the solvent, or deposition conditions or by postdeposition

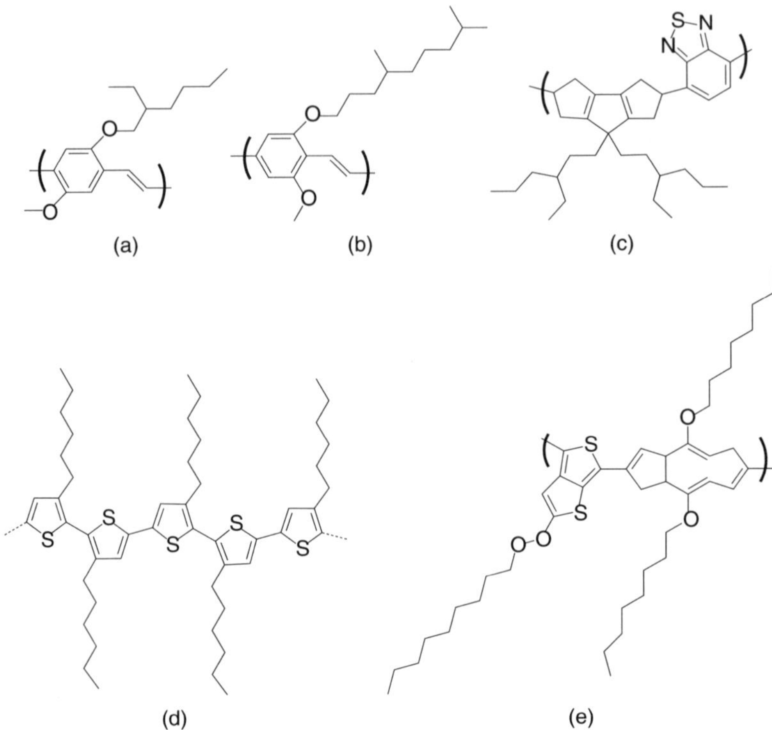

Figure 8-16 Chemical structure of polymers used in organic solar cells. (a) MEH-PPV. (b) MDMO-PPV. (c) PTB1. (d) Regioregular P3HT. (e) PCPDTBT.

treatments such as annealing at elevated temperatures. We note a particularly interesting theoretical study on the potential efficiency of various bicontinuous structures for use in polymer blend solar cells [85].

Several groups have recently reported solar cells based on polymer blends with power conversion efficiency >1.5% [86–88]. Kietzke et al. [88] exploited self-organization in a polymer blend to obtain cells with a power conversion efficiency of 1.7%. The cells were fabricated from an alkoxy- and a cyano-substituted PPV derivative: poly[2,5-dimethoxy-1,4-phenylene-1,2-ethenylene-2-methoxy-5-(2-ethylhexyloxy)-(1,4-phenylene-1,2-ethenylene)]) and CN-ether-PPV (poly[oxa-1,4-phenylene-1,2-(1-cyano)ethenylene-2,5-dioctyloxy-1,4-phenylene-1,2-(2-cyano)ethenylene-1,4-phenyene]). The polymer blend layer was spin-cast onto the substrate from a chlorobenzene solution and then annealed at 110°C in a nitrogen atmosphere for 40 min. Comparison of the IPCE spectrum of devices and the corresponding absorption spectra of films led to the conclusion that the film has a vertical composition gradient with an M3EH-PPV-rich region close to the anode where the light intensity and thus the exciton generation is largest. M3EH-PPV is less soluble in chlorobenzene than CN-ether-PPV, and thus M3EH-PPV should precipitate first during evaporation of the solvent in the spin-coating process. Through sidechain substitution of the electroactive polymers, it should be possible in principle to optimize the gradient blend. The efficiency of the devices was believed to be limited by the formation of exciplexes in the polymer blend.

8.3.2. Polymer Fullerene Blend Solar Cells

The development of functionalized fullerenes has enabled the development of polymer solar cells with high power conversion efficiency. A strong photoinduced electron transfer reaction between C_{60} and MEH-PPV was reported by Sariciftci et al. in 1992 [89]. The PL intensity of MEH-PPV was quenched by almost three orders of magnitude in a 1:1 by weight blend of MEH-PPV and C_{60}. Light-induced electron spin resonance (LESR) spectra contained distinct signals from positive polarons on MEH-PPV chains and the second from C_{60} anions. Transient PL measurements show a decay within the time resolution of the experiment (60 ps). Later work, including the recent THz study, showed that charge transfer occurs on a sub-picosecond timescale [67, 90]. Polaron lifetimes in MEH-PPV were measured to be on the order of milliseconds [91]. C_{60} therefore meets the criteria of ultrafast charge transfer and slow back transfer. Morita and coworkers showed evidence of photoinduced charge transfer in other polymer/fullerene composites [92, 93]. The term fullerene is generally used to comprise the family of cagelike carbon molecules such as C_{60} and C_{70} as well as their functionalized derivatives.

The combination of efficient, ultrafast charge transfer-producing metastable photocarriers has resulted in fullerene/polymer composites being one of the most promising light-harvesting classes of materials for organic solar cell technologies. The first photovoltaic devices using C_{60} were prepared by spin-coating a polymer film onto an ITO-coated glass substrate, followed by vacuum evaporation of C_{60} to form a p-n heterojunction [94]. This device had poor performance under monochromatic illumination ($V_{oc} = 0.44$ V, $J_{sc} = 2.1 \mu A/cm^2$, $ff = 0.20$, $\eta = 0.02\%$). Bulk heterojunction devices made from a blend of MEH-PPV and C_{60} showed significantly better photoresponse than the first bilayer devices [95]. Devices fabricated from a xylene solution of MEH-PPV and C_{60} (10:1 by weight) showed improved performance, though poor by current standards. Under 2.8 mW/cm^2 illumination at 500 nm, the current density $J \approx$ 15.3 μA/cm^2 corresponds to an EQE of roughly 1.3%.

The poor performance of bulk heterojunction devices made from C_{60} and PPV derivatives was a consequence of the morphology of spin-cast C_{60}:polymer films. C_{60} has a tendency to crystallize during film formation, and its solubility is lower than that of the electron donor polymers in organic solvents used for spin-casting polymer films [96]. The solution to this problem lies in the field of fullerene chemistry. Fullerenes have a rich chemistry, resulting in the synthesis of a variety of fullerene derivatives [97] shortly after Krätschmer and Huffman [98] discovered how to synthesize and purify large quantities of fullerenes. The electron-withdrawing nature of fullerenes makes them ideal dienophiles for Diels–Alder cycloaddition [99]. This was exploited by Wudl and coworkers to synthesize methanofullerenes [100]. These functionalized fullerenes have high solubility in the same solvents used for conjugated polymers. Synthesis of these highly soluble fullerene derivatives was necessary to the formation of fullerene:polymer bulk heterojunctions with the desired morphology for use in organic solar cells and has enabled the development of high-efficiency cells.

The first bulk heterojunction solar cells using a methanofullerene combined MEH-PPV and 1-(3-methoxycarbonyl)-propyl-1-phenyl-(6,6)C_{61} or [6, 6]-PC_{61} BM [101]. Yu et al. found that the carrier collection and power conversion efficiency of their cell was critically dependent on the morphology of the blend and its chemical composition. By changing the solvent from xylene to 1,2-dichlorobenzene, they were able to prepare films with up to 80% $PC_{61}BM$ by weight, roughly one fullerene per polymer repeat unit. Devices made with a low concentration of C_{60} or $PC_{61}BM$ have quantum efficiencies of about 15% at low intensities, falling to less than 10% at 20 mW/cm^2. The device containing 80% $PC_{61}BM$, in contrast, reaches an EQE of 45% at low intensities, falling to 29% at 20 mW/cm^2. An open-circuit voltage of 0.82 V was obtained for a device with the structure ITO/PC_{61}BM:MEH-PPV/Ca. Because of rise in the photovoltage with intensity, the power efficiency remained relatively constant: $\eta_e = 3.2\%$ at 10 μW/cm^2, falling to 2.9% at 20 mW/cm^2.

Although improved device efficiencies did not immediately follow the early use of $PC_{61}BM$ in polymer solar cells, work continued on optimizing the morphology of $PC_{61}BM$/polymer blends [102, 103]. This work led to a breakthrough in 2001 with the announcement of a solar cells with 2.5% power conversion efficiency based on a blend of poly[(2-methyl,5-(3′,7″-dimethyl-octyloxy)]-1,4-phenylenevinylene) (MDMO-PPV) and $PC_{61}BM$ [104]. The choice of the proper casting solvent allowed a nearly threefold improvement of the device power conversion efficiency up to that date. Devices with a BHJ cast from a chlorobenzene solution were compared to devices with a BHJ cast from toluene, representative for other solvents reported in literature. Figure 8-17 shows AFM images of the surfaces of MDMO-PPV:PC_{61}BM films spin cast from toluene and chlorobenzene as well as the *JV* characteristics of cells using films cast from these solutions. The surface of the toluene cast film contains features with horizontal dimensions on the order of 0.5 mm. Measurements of the mechanical stiffness and adhesion properties of the surface showed that the vertical features have a chemical composition different than the surrounding valleys. As these features were not observed in films of neat MDMO-PPV, they were assigned to phase-segregated domains of PC_{61}BM-rich material. In contrast, the film cast from chlorobenzene-cast film contains structures with horizontal dimensions on the order of only 0.1 mm, indicating more uniform mixing of the constituents. Furthermore, the surface roughness of the toluene-cast films has height variations

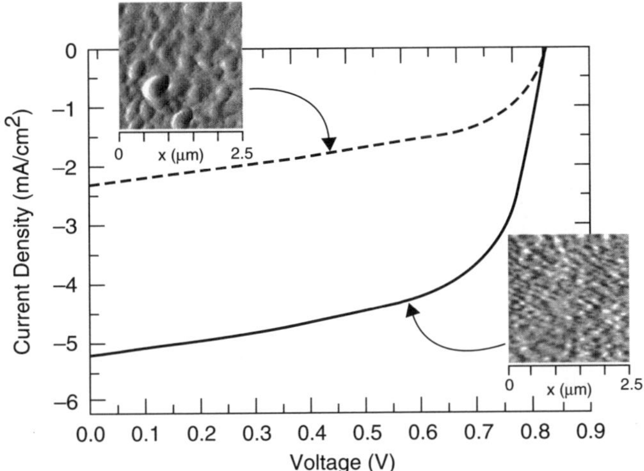

Figure 8-17 *JV* characteristics for devices with an active layer that is spin coated from a toluene solution (dashed line) and from a chlorobenzene solution (solid line). Data are for devices illuminated with an intensity of 80 mW/cm^2 with an AM1.5 spectral mismatch factor of 0.753. Inset: AFM images showing the surface morphology of MDMOPPV:PC$_{61}$BM (1:4 by wt.) blend films with a thickness of approximately 100 nm spin-coated from toluene and chlorobenzene solutions [104]. Copyright 2001 American Institute of Physics.

an order of magnitude larger than those of the chlorobenzene-cast film. The difference in film morphology was attributed to the higher solubility of PC$_{61}$BM in chlorobenzene.

Devices using films cast from toluene or chlorobenzene demonstrate the impact of film morphology on photovoltaic device performance. The device structure consists of ITO/PEDOT:PSS/MDMO:PC$_{61}$BM(1:4)/LiF/Al. The devices have identical open circuit voltages, but the current density of the device with a chlorobenzene-cast film is more than twice that of the device with a film cast from toluene, 5.25 vs. 2.33 mA/cm^2. The absorption spectra of the devices were nearly identical; thus the difference in quantum efficiency cannot be due to absorbance. The fill factor of the device with a chlorobenzene-cast film was also higher, resulting in a power conversion efficiency of 2.5%, nearly three times that of the device with a toluene-cast film. The tendency of the PC$_{61}$BM molecules to phase segregate into clusters is suppressed when chlorobenzene is used as the solvent, resulting in higher current density and fill factor. The much smoother surface of the chlorobenzene-cast active layer also leads to better interfacial contact with the cathode. These results show that spin coating the active layer blend from a chlorobenzene solution has enhanced the morphological microstructures of both components that form the bulk heterojunction. In a follow-up study, optimizing the thickness of the LiF layer between the anode and the polymer:PC$_{61}$BM film increased the power conversion efficiency of the devices to 3.3% [105].

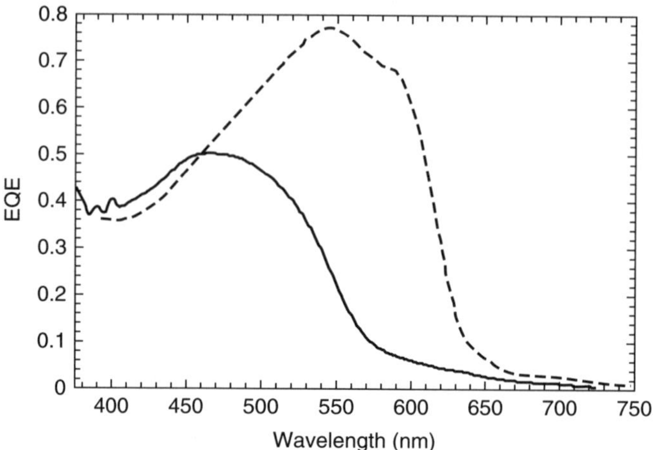

Figure 8-18   Comparison of the IPCE spectra of bulk heterojunction solar cells based on blends of $PC_{61}BM$ with MDMO-PPV [104] and P3HT [106]. Copyright 2001, 2002 American Institute of Physics.

The efficiency of devices based on PPV derivatives is limited by the absorption spectrum of the polymer. The absorption edge of MDMO-PPV, for example, is at approximately 550 nm. Although fullerene derivatives have a lower gap, they are weakly absorbing in the red. Lower gap materials therefore merit attention. Polyalkylthiophenes (P3ATs) have drawn interest, particularly regioregular poly(3-hexylthiophene) or rr-P3HT. Schilinsky et al. [106] reported bulk heterojunction devices from blends of P3HT and [6, 6]-$PC_{61}BM$ that achieve levels of performance comparable to the aforementioned MDMO-PPV:$PC_{61}BM$ device. The device construction is similar, that is a layer of PEDOT:PSS is spin-cast onto ITO-coated glass, followed by a blend of P3HT and $PC_{61}BM$ and a metal cathode is evaporated on top (calcium capped by silver). The photocurrent of the P3HT:$PC_{61}BM$ device is higher than that of PPV-based devices (8.7 mA/cm^2), because of greater absorption of the solar spectrum by the polymer layer. Figure 8-18 shows IPCE spectra of the cells based on blends of $PC_{61}BM$ with MDMO-PPV and P3HT. The P3HT-based cell has a higher peak quantum efficiency, 70% at ~530 nm vs. 50% at ~480 nm for the MDMO-PPV-based cell. These spectra also demonstrate improved coverage of the solar spectrum. The MDMO-PPV cell has only a weak response for $\lambda > 550$ nm, mainly due to absorbance by $PC_{61}BM$. Replacing MDMO-PPV by P3HT extends the strong spectral response out to 650 nm, with some additional contributions from $PC_{61}BM$ at longer wavelengths. The open-circuit voltage ($V_{oc} = 0.58$ V) and fill factor ($ff = 0.55$) of the P3HT-based cell were somewhat lower than the optimized MDMO-PPV device, and the power conversion efficiency was 2.8%, close to the best PPV:$PC_{61}BM$ device.

Control of the morphology of the polymer/fullerene bulk heterojunction through self-organization has led to the rapid improvement in organic solar cell efficiency. Several studies were published in 2005 and 2006[1] in which P3HT-based cells with power conversion efficiencies greater than 4% were demonstrated [107–109]. Each group used a distinct approach to obtaining a bulk heterojunction with an optimized morphology. Ma et al. focused on postdeposition annealing conditions [107]. P3HT:PC$_{61}$BM films were spin-cast at 700 rpm onto a PEDOT:PSS-coated ITO substrate from a chlorobenzene solution containing 10 mg/ml P3HT and 8 mg/ml PC$_{61}$BM. The devices were subsequently annealed at temperatures ranging from 25°C to 210°C either before or after an aluminum electrode was deposited to complete the device structure. X-ray diffraction spectra of devices before and after annealing showed increased intensity at peaks corresponding to interchain spacing in P3HT associated with interdigitated alkyl chains ($\theta \approx 5°$) and face-to-face packing of the thiophene rings ($\theta \approx 23°$) [110]. A recent study used grazing incidence small-angle X-ray scattering to study phase segregation of P3HT:PC$_{61}$BM thin films [111]. Locally phase-separated P3HT lamella was found to be intercalated with PC$_{61}$BM aggregates. When the P3HT crystallites and PC$_{61}$BM aggregates were of comparable size and density, the interpercolated networks for bipolar carrier transport were optimized. Figure 8-19 shows the dependence of the device efficiency on annealing conditions. Devices annealed before aluminum deposition have a maximum efficiency when annealed at 125°C (3.5%), whereas devices annealed after deposition could be annealed at higher temperatures and efficiencies as high as 5% were obtained. The parameters of devices annealed at 150°C under 80 mW/cm^2 AM1.5 illumination were $V_{oc} = 0.62$ V, $J_{sc} = 9.5$ mA/cm^2, and $ff = 0.68$.

The films used in the study discussed above are too thin to completely absorb incident light. Li et al. [108] addressed this essential problem of organic solar cells—complete absorption of solar energy requires such thick films that devices have a high series resistance and a low fill factor. Controlling the rate at which solvent is eliminated from the polymer:fullerene film enabled the use of a relatively thick films (>200 nm) in cells with low series resistance. The photoactive layer was spin-coated at 600 rpm from a 1,2-dichlorobenzene solution of P3HT at a concentration of 17 mg/ml to which PC$_{61}$BM had been added in equal weight. A glass substrate coated by ITO and PEDOT:PSS was used. Spin-coating at 600 rpm left the films wet. Films were dried either by baking at 70°C immediately after deposition (fast growth) or allowed to dry slowly in a covered petri dish (slow growth). Films were annealed at 110°C before the deposition of an aluminum cathode. AFM images of the P3HT:PC$_{61}$BM films showed a quite different surface texture. The surface roughness of the slow-grown film was 11.5 nm after deposition, decreasing to 9.5 nm after annealing. The fast-grown films were much smoother, with rms roughness of 0.9 nm after deposition, increasing to 1.9 nm after annealing. Although the roughness of the

---

[1]These studies are discussed in the order of publication, although it must be noted that the submission dates indicate that all three studies were conducted in parallel. This is an indication of rapid progress in the field over the last five years.

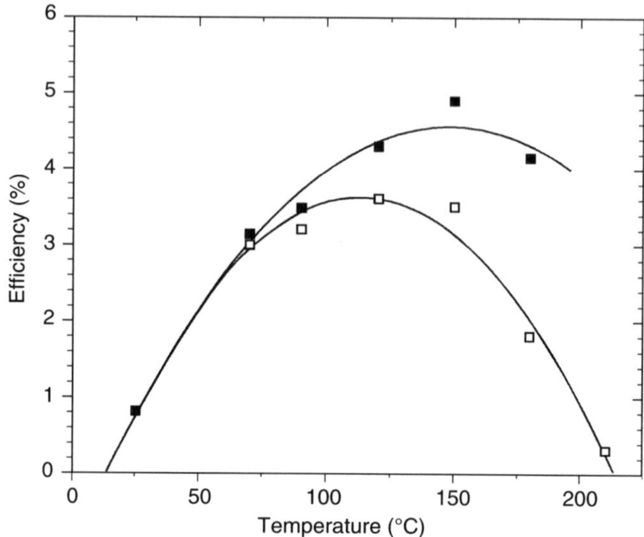

Figure 8-19   Solar cell efficiency vs. annealing temperature for devices annealed before (open squares) or after (filled squares) aluminum deposition. Adapted from Ref. 107. Copyright 2005 Wiley-VCH.

slow-grown film is problematic for devices based on films thinner than 100 nm [62], problems with device shorting were not experienced in this study.

The film morphology had a remarkable effect on the absorbance and charge transport capabilities of the different films. Figure 8-20 shows the absorption spectra of both types of film, as-cast and after thermal annealing. The absorbance of the slow-grown film is much stronger than that of the fast-grown film, particularly in the red. The vibronic replica from 500 to 600 nm is more pronounced in the slow-grown film, indicating a higher degree of order in the film [112]. Annealing also had quite different effects on the two films. The absorption spectrum of the slow-grown film is unaffected by annealing, whereas there is a substantial increase in the visible absorption of the fast-grown film. The greater absorption of the slow-grown film contributed to an increased photocurrent in devices, generally two to three times that of devices made from fast-grown films. Improved absorption is, however, insufficient to explain the difference in the photocurrent from devices made from fast- and slow-grown films. The absorption spectra of the two films are similar below 450 nm, yet IPCE spectra show that the slow-grown device is more than two times as efficient at these wavelengths.

To better understand the impact of film morphology on charge transport, time-of-flight (TOF) measurements were performed on slow-grown and fast-grown films. In a TOF experiment, a sheet of charge carriers is photogenerated by pulsed illumination through one electrode. The drift of carriers under an external bias to the collecting electrode results in a time-dependent current that is monitored across an external load resistor. A digital storage oscilloscope is used to average

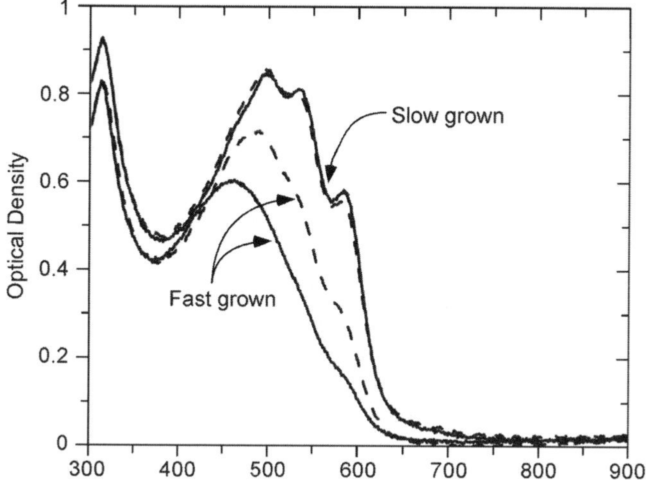

Figure 8-20   Absorption spectra for P3HT:PC$_{61}$BM (1:1 by weight) films for slow-grown and fast-grown films, before (solid line) and after (dashed line) annealing. The films were spun cast at 600 r.p.m. for 60 s (film thickness ~210 nm) and the annealing was done at 110°C for 20 min [108]. Used with permission from Nature Publishing Group.

multiple traces and generate the final TOF transient. The carrier mobility (μ) can then be calculated from the transit time (τ) by the relation $\mu = d/\tau F$, where d is the film thickness and F is the electric field magnitude. Electron or hole mobility is measured by varying the polarity of the electric field. Figure 8-21 shows TOF transients for fast- and slow-grown P3HT:PC$_{61}$BM films under an electric field of 200 kV/cm. Both current transients of the slow-grown film as well as the negative bias (electron) TOF transient of the fast-grown all exhibit characteristics of nondispersive charge transport. The transient consists of an initial current spike, followed by a constant current plateau and a drop in current due to the arrival of holes at the counterelectrode, where they are discharged. The observed nondispersive carrier transport is an indication of excellent purity and chemical regularity. The forward-biased transient (hole current) of the fast-grown film is characteristic of dispersive charge transport. No clear plateau is observed, and the mobility is an order of magnitude lower. Even more important than the nondispersive nature of charge transport in the slow-grown film are the similar electron and hole mobilities ($\mu_e = 7.7 \times 10^{-5}$ cm^2/V · s; $\mu_h = 5.1 \times 10^{-5}$ cm^2/V · s). Although these mobilities are relatively low compared to neat films of C$_{60}$ and P3HT, the most important aspect is that charge transport is balanced [113]. When charge transport is unbalanced as in the fast-grown film ($\mu_e/\mu_h = 21$), hole accumulation occurs and the photocurrent becomes limited by space charge effects [114]. For the devices prepared by Li et al., the mobility ratio of 1.5 results in significantly higher fill factors. The device parameters of the optimized device with a slow-grown film were $J_{sc} = 10.6 \pm 0.1$ mA/cm^2, $V_{oc} = 0.61 \pm 0.02$ V, and $ff = 0.67$, with a power conversion efficiency of 4.4%. The

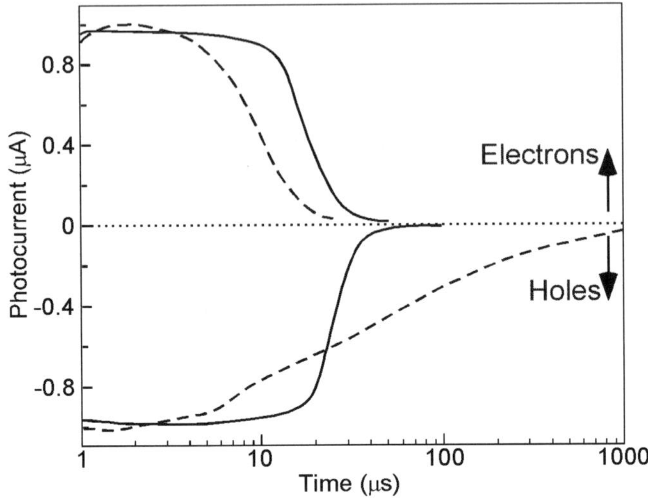

Figure 8-21   Time-of-flight transients for slow-grown (solid line) and fast-grown (dashed line) P3HT:PC$_{61}$BM films. Films in TOF devices were prepared the same way as those used in solar cells, and their thickness is 1 micron [108]. Used with permission from Nature Publishing Group.

greater thickness of the slow-grown film vis-à-vis typical active layer thickness of ~100 nm resulted in very high shunt resistance (>200 MΩ), which when coupled with the low series resistivity of the self-organized bulk heterojunction ($< 2\Omega$-cm^2), resulted in devices with high fill factors.

Kim et al. studied the demonstrated that the regioregularity of P3HT plays an important role in determining film morphology and device performance [109]. The transport properties of pristine P3HT polymer films are known to improve with the degree of regioregularity, defined as the percentage of monomers adopting a head-to-tail configuration rather than head-to-head. Sirringhaus et al. showed that self-organization in regioregular P3HT results in a lamella structure with two-dimensional sheets oriented perpendicular to the substrate formed by interchain stacking [20]. The mobility of rr-P3HT is consequently quite high for a conjugated polymer, of order 10^{-3} cm^2/V · s. Higher regioregularity enables closer packing of these lamellae such that excitations adopt some interchain character. This results in additional absorbance in the red due to a distinct shoulder observed on the long-wavelength side of the absorption maximum. The influence of regioregularity and annealing on in-plane chain stacking was demonstrated by grazing incidence X-ray diffraction (GIXRD) using a synchrotron X-ray beam (Fig. 8-22). Films of rr-P3HT with 90.7%, 93%, and 95.2% regioregularity were studied along with corresponding 1:1 blends with PC$_{61}$BM. The out-of-plane diffraction patterns (2θ = 5.3°, 10.7° and 15.9° for the primary, secondary and tertiary peaks, respectively) show a well-organized intraplane structure with lamallae oriented normal to the substrate. From the in-plane patterns for as-cast rr-P3HT (inset to Fig. 8-22a), the higher the degree of regioregularity, the weaker

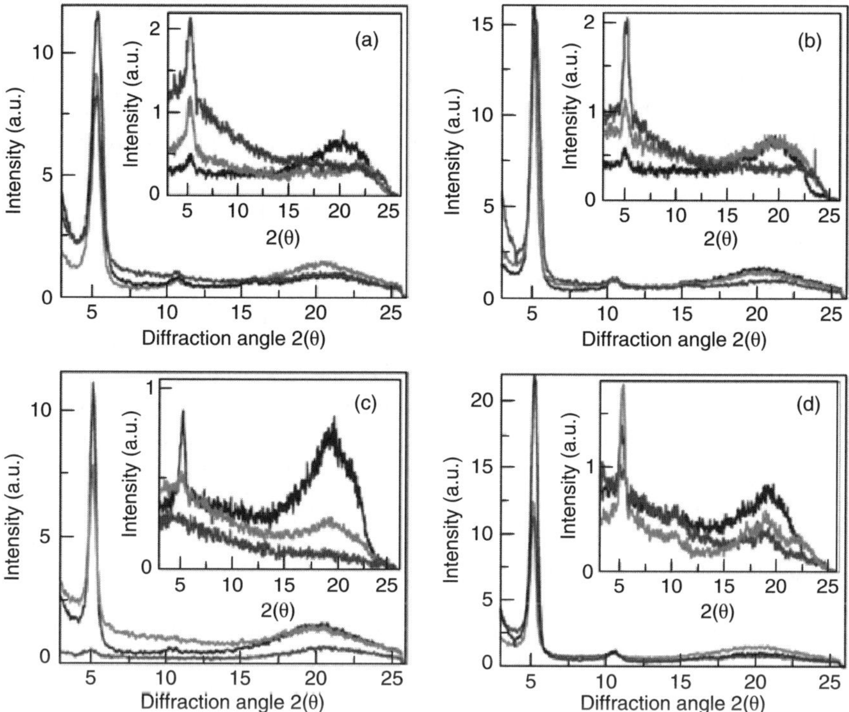

Figure 8-22 Grazing incidence X-ray diffractograms: (a) P3HT as cast, (b) annealed P3HT, (c) P3HT:PC$_{61}$BM as cast, and (d) annealed P3HT:PC$_{61}$BM. Samples were prepared with 90.7%, 93%, and 95.2% regioregularity [109]. Used with permission from Nature Publishing Group. (A full color version of this figure appears in the color plate section.)

the intraplane stacking in the plane of the substrate. Thus, intraplane stacking perpendicular to the substrate is correlated with the degree of regioregularity. After thermal annealing (Fig. 8-22b), the trend with regioregularity is preserved and the contrast in peak intensity for intraplane and out-of-plane geometries becomes more pronounced. For blend films of rr-P3HT and PC$_{61}$BM (Fig. 8-22c), the peaks almost entirely disappear for the sample with the least regioregularity, indicating that stacking almost disappears entirely upon blending with PC$_{61}$BM. Annealing the blend films (Fig. 8-22d) restores the intraplane stacking peaks, showing clear primary and secondary peaks for both in-plane and out-of-plane geometries. It was concluded that thermal annealing films at 140°C develops the intraplane chain stacking perpendicular to the film substrate and stimulates the segregation of crystalline P3HT domains. In a comparison of devices made from P3HT:PC$_{61}$BM blend films, both the current density and fill factor increased with increasing regularity with only slight influence on the open-circuit voltage. When the film thickness (175 nm) and annealing conditions were optimized, an efficiency of $\eta = 4.4\%$ was achieved.

Although P3HT is a better absorber of solar energy than MDMO-PPV and other PPV derivatives used in organic solar cells, there is further potential to optimize the energetic offset between molecular orbitals of the electron donor and acceptor. Scharber et al. [115] examined bulk heterojunction cells using a variety of polymers in combination with PC_{61}BM and calculated the following relation:

$$eV_{oc} = E_{HOMO}^{D} - E_{LUMO}^{A} - X_{B} \qquad (8.3)$$

where E_{HOMO}^{D} is the energy level of the HOMO of the electron donor, E_{LUMO}^{A} is the energy level of the LUMO of the electron acceptor, and X_{B} is an empirically determined value (0.3 eV) representing the energy lost during exciton dissociation. The ionization potential of P3HT has been measured by ultraviolet photoemission spectroscopy (UPS) to be 5.1 eV [115], and the electron affinity of PC_{61}BM has been estimated at 4.3 eV. Given an optical gap of 1.9 eV, the LUMO level of P3HT is roughly half an eV above what is needed for efficient charge transfer. As the HOMO level of P3HT determines V_{oc}, the larger optical gap of P3HT simply leads to energy loss in the cell. The development of conjugated polymers with lower optical gaps has driven advances in device efficiency.

New electron-donating polymers, however, bring their own set of challenges. Processing methods that encourage self-organization of one polymer:fullerene system may fail entirely for another. A good example of this issue is a polymer that was designed for use in photovoltaic applications—poly[(4,4-bis(2-ethylheyl)-cyclopenta-[2,1-*b*;3,4-*b'*]dithio–phene)-2,6-diyl-*alt*-2, 1,3-benzothiadiazole-4,7-diyl] or PCPDTBT (see Fig. 8-16 for the repeat unit) [116]. PCPDTBT is a push-pull material based on alternating electron-withdrawing (carbazole) and-donating (benzothiadiazole) moieties to stabilize the quinoidal form of the polymer and lower the optical gap. Because of the lower-lying LUMO, the optical gap of PCPDTBT is only 1.3 eV, providing much better coverage of the solar spectrum than P3HT without a corresponding drop in V_{oc}. Mühlbacher et al. initially reported solar cells made from a combination of PCPDTBT and a fullerene [117]. A fullerene based on C_{70} was used as the electron acceptor—phenyl C_{71}-butyric acid methyl ester (PC_{71}BM). Reasonable efficiencies were obtained from a PCPDTBT:PC_{71}BM-based cell (3.2%), although the fill factor was relatively low (0.45) compared to optimized P3HT:PC_{61}BM cells. The lower efficiency of the PCPDTBT-based cells persisted despite varying the solvent, annealing conditions, concentration, and molecular weight of the polymer [118].

The key to taking full advantage of the energetics of PCPDTBT was in using additives to give self-organization an assist. Alkanethiols have been used to stabilize gold nanoparicles [119] and were an alternative in this case to thermal annealing of the film. Peet et al. increased the efficiency of PCPDTBT:PC_{71}BM cells to 5.5% by the addition of alkanedithiols to the polymer:PC_{61}BM solution [120]. In this study, 24 mg/ml of 1,8-octanedithiol was added to a chlorobenzene solution of PCPDTBT and PC_{71}BM. Analysis by Fourier-transform infrared and Raman spectroscopy indicated that no thiol remained after a spin-cast film was dried in vacuum for 10 min. Optimization of the solution concentration, composition, and casting conditions led to devices with η varying from 5.2% to 5.8%, with $\eta = 5.5\%$ being obtained from most repeatable series of devices. These devices had a remarkably high short-circuit current density $J = 16.2$ mA/cm^2 and reasonable fill factor $ff = 0.55$. The IPCE spectra of devices prepared with additives showed a broad band increase in photocurrent generation, with a quantum efficiency >40% from 400 to 800 nm.

The last several years have seen rapid advances in efficiency of organic solar cells based on polymer:fullerene blends. In a number of cases, these have been independently certified by testing at national labs such as the National Renewable Energy Laboratory (US), the Fraunhofer Institute (Germany), or the National Institute of Advanced Industrial Science and Technology (AIST, Japan). Park et al. reported a polymer:PC$_{61}$BM cell with $\eta = 6\%$ using a dithienyl push-pull polymer—poly[N-9''-hepta-decanyl-2,7-carbazole-*alt*-5,5-(4',7'-di-2-thienyl-2',1',3'-benzothiadazole) or PCDTBT [121]. In addition to using a different electron donor, a titanium dioxide layer was used as an optical spacer to obtain a cell with an internal quantum efficiency of nearly 100%. Hoven et al. obtained similar efficiencies ($\eta = 5.4\%$) using a polymer with a silole moiety—poly[4,4-didodecyldithieno[3,2-*b*:2',3'-*d*]silole)-2,6-diyl-*alt*-(2,1,3-benzooxadiazole)-4,7-diyl] [122]. Plextronics reported a certified efficiency of 5.4% in 2007 using a proprietary ink as the active layer [123]. Although the development of new electron donors has received the bulk of attention, early work on PCDTBT demonstrates the potential for morphological problems associated with new materials. One strategy has been to alter the orbital energy levels of the fullerene acceptor through the development of endofullerenes. Contained within the fullerene cage is a rare earth trinitride that changes the position of the frontier orbitals without otherwise altering the fullerene solubility [124, 125]. These devices have a higher open-circuit voltage without a deleterious effect on the film morphology or the device current density or fill factor.

Low-gap polymers synthesized by Luping Yu's group at the University of Chicago and used in a collaboration with Solarmer have led to record cell efficiencies. The initial report used a polymer synthesized via the Stille polycondensation between an ester-substituted 2,5-dibromothieno[3,4-*b*]thiophene and dialkoxyl benzodithiophene distannane monomers, referred to as PTB1 [126]. The thieno[3,4-*b*]thiophene repeat unit stabilizes the quinoidal structure of the backbone that narrows the energy gap of the resulting polymers. The ester-substituted thieno[3,4-*b*]thiophene renders it both soluble and oxidatively stable. Devices made using PTB1 had an efficiency of $\eta = 4.4\%$ using PC$_{61}$BM as the acceptor and $\eta = 5.6\%$ using PC$_{71}$BM as the acceptor. The efficiency of these devices was limited by the open-circuit voltage, $V_{oc} = 0.58$ V. Through fine-tuning the sidechain substitution, the position of the HOMO level of PTB1 was modified, increasing V_{oc} and yielding higher efficiency [127]. In late 2009, a certified device efficiency of 6.8% was reported from devices using a derivative of PTB1 in which the alkoxyl chains of PTB1 were replaced with alkyl chains and an atom of high

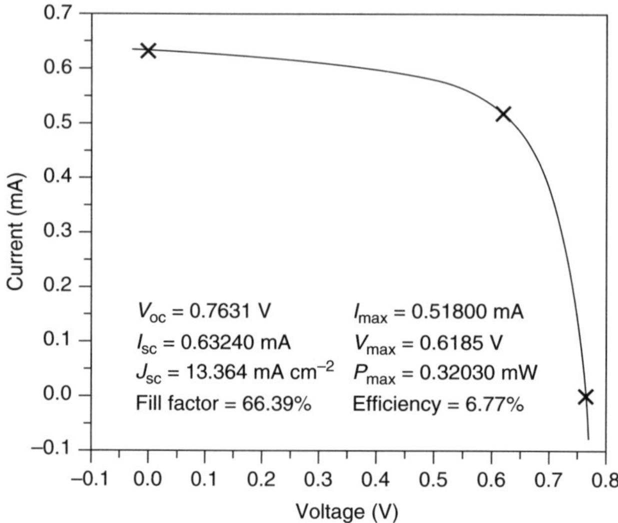

Figure 8-23 *JV* characteristics of an organic solar cell based on a derivative of PBT1 with alklyl sidechains and fluorine substitution [128]. Used with permission from Nature Publishing Group.

electron affinity was introduced to the thieno[3,4-*b*]thiophene unit [128]. The *JV* characteristics of this cell under 1 sun illumination are shown in Figure 8-23, and the repeat unit of the polymer, referred to as PBDTTT–CF, is inset. Continuing optimization of the electron donor led to a report in late 2009 of a cell with a power conversion efficiency of 7.9% [11]. This work has not been published as of the date of this writing, and thus the device structure and electron donor are not yet available.

8.4. CONCLUDING REMARKS

Taking advantage of the self-organizing properties of both molecular and semi-conductors has proven crucial to the development of efficient organic solar cells. Remarkable progress has been made, and organic solar cells are now close to the performance level of other thin-film technologies. The fabrication of devices with 10% power conversion efficiency, long seen as an unachievable goal, may soon be a reality. Coupled with the prospect of inexpensive processing, organic semiconductor solar cells are poised to take their place alongside other solar technologies.

REFERENCES

1. Garnier F, Yassar A, Hajlaoui R, Horowitz G, Deloffre F, Servet B, Ries S, and Alnot P. Molecular engineering of organic semiconductors—design of self-assembly

properties in conjugated thiophene oligomers. *J. Am. Chem. Soc.* **1993**, *115(19)*, 8716–8721.

2. Mattheus CC, Dros AB, Baas J, Oostergetel GT, Meetsma A, de Boer JL, and Palstra TTM. Identification of polymorphs of pentacene. *Synth. Met.* **2003**, *138(3)*, 475–481.

3. Sariciftci NS, Ed. *Primary Photoexcitations in Conjugated Polymers*. Singapore: World Scientific Publications, **1998**.

4. Campbell IH, Hagler TW, Smith DL, and Ferraris JP. Direct measurement of conjugated polymer electronic excitation energies using metal/polymer/metal structures. *Phys. Rev. Lett.* **1996**, *76(11)*, 1900–1903.

5. Alvarado SF, Seidler PF, Lidzey DG, and Bradley DDC. Direct determination of the exciton binding energy of conjugated polymers using a scanning tunneling microscope. *Phys. Rev. Lett.* **1998**, *81(5)*, 1082–1085.

6. Harrison MG, Grüner J, and Spencer GCW. Analysis of the photocurrent action spectra of MEH-PPV polymer photodiodes. *Phys. Rev. B* **1997**, *55(12)*, 7831–7849.

7. Gregg BA and Hanna MC. Comparing organic to inorganic photovoltaic cells: Theory, experiment, and simulation. *J. Appl. Phys.* **2003**, *93(6)*, 3605–3614.

8. Halls JJM, Cornil J, dos Santos DA, Silbey R, Hwang DH, Holmes AB, Bredas JL, and Friend RH. Charge- and energy-transfer processes at polymer/polymer interfaces: A joint experimental and theoretical study. *Phys. Rev. B* **1999**, *60(8)*, 5721–5727.

9. Forster T. 10th Spiers Memorial Lecture—Transfer mechanisms of electronic excitation. *Discuss. Faraday Soc.* **1959**, *27*, 7.

10. Scholes GD. Long-range resonance energy transfer in molecular systems. *Ann. Rev. Phys. Chem.* **2003**, *54*, 57–87.

11. Cheyney T. **2009**. Solarmer breaks organic solar PV cell conversion efficiency record, hits NREL-certified 7.9%. Available at http://www.pv-tech.org/news_c/opv_dssc/. Accessed 2010 Apr 8.

12. Dennler G, Scharber MC, Ameri Y, Denk P, Forberich K, Waldauf C, and Brabec CJ. Design rules for donors in bulk-heterojunction tandem solar cells-towards 15% energy-conversion efficiency. *Adv. Mater.* **2003**, *20(3)*, 579–583.

13. Gueymard C, Myers D, and Emery K. Proposed reference irradiance spectra for solar energy systems testing. *Solar Energy* **2003**, *73(6)*, 443–467.

14. Ito S, Nazeeruddin MK, Liska P, Comte P, Charvet R, Péchy P, Jirousek M, Kay A, Zakeeruddin SM, Grätzel G. Photovoltaic characterization of dye-sensitized solar cells: effect of device masking on conversion efficiency. *Prog. Photovoltaics* **2006**, *14*, 589–601.

15. Kushto GP and Lane PA. Organic photovoltaic cells using group 10 metallophthalocyanine electron donors. Unpublished data.

16. Mäckel H and Cuevas A. The spectral response of the open-circuit voltage: a new characterization tool for solar cells. *Solar Energy Mater. Solar Cells* **2004**, *81(2)*, 225–237.

17. Chamberlain GA. Organic solar cells—A review. *Solar Cells* **1983**, *8(1)*, 47–83.

18. Jurchescu OD, Baas J, and Palstra TTM. Effect of impurities on the mobility of single crystal pentacene. *Appl. Phys. Lett.* **2004**, *84*, 3061–3063.

19. Bao Z, Dodabalapur A, and Lovinger AJ. Soluble and processable regioregular poly(3-hexylthiophene) for thin film field-effect transistor applications with high mobility. *Appl. Phys. Lett.* **1996**, *69(26)*, 4108–4110.

20. Sirringhaus H, Brown PJ, Friend RH, Nielsen MM, Bechgaard K, Langeveld-Voss BMW, Spiering AJH, Janssen RAJ, Meijer EW, Herwig P, and de Leeuw DM. Two-dimensional charge transport in self-organized, high-mobility conjugated polymers. *Nature* **1999**, *401(6754)*, 685–688.

21. Sirringhaus H, Wilson RJ, Friend RH, Inbasekaran M, Wu W, Woo EP, Grell M, and Bradley DDC. Mobility enhancement in conjugated polymer field-effect transistors through chain alignment in a liquid-crystalline phase. *Appl. Phys. Lett.* **2000**, *77(3)*, 406–408.

22. Meier H. *Organic semiconductors*. Verlag Chemie, Weinheim **1974**, *372*, 429–459.

23. Harima Y, Yamashita K, and Suzuki H. Spectral sensitization in an organic *p-n* junction photovoltaic cell. *Appl. Phys. Lett.* **1984**, *45(10)*, 1144–1145.

24. Tang CW. Two-layer organic photovoltaic cell. *Appl. Phys. Lett.* **1986**, *48(2)*; 183–185.

25. Kroto HW, Heath JR, O'Brien SC, Curl RE, and Smalley RE. C_{60}—Buckminsterfullerene. *Nature* **1985**, *318(6042)*, 162–163.

26. Krätschmer W, Lamb LD, Fostiropocclos K, and Huffman DR. Solid C_{60}—A new form of carbon. *Nature* **1990**, *347(6291)*, 354–358.

27. Ohsawa Y and Saji T. Electrochemical detection of C_{60}^{6-} at low-temperature. *J. Chem. Soc. Chem. Comm.* **1992**, *10*, 781–782.

28. Priebe G, Pietzak B, and Könenberg R. Degtermination of transport parameters in fullerene films. *Appl. Phys. Lett.* **1997**, *71(15)*, 2160–2162.

29. Anthopoulos TS, Singh B, Marjanovic N, Sariciftci NS, Ramil AM, Sitter H, Colle M, and de Leeuw DM. High performance n-channel organic field-effect transistors and ring oscillators based on C_{60} fullerene films. *Appl. Phys. Lett.* **2006**, *89(21)*, 213504.

30. Hung RR and Grabowski JJ. A precise determination of the triplet energy of C_{60} by photoacoustlc calorimetry. *J. Phys. Chem.* **1991**, *95(16)*, 6073–6075.

31. Pettersson LAA, Roman LS, and Inganäs O. Modeling photocurrent action spectra of photovoltaic devices based on organic thin films. *J. Appl. Phys.* **1999**, *86(1)*, 487–496.

32. Kraabel B, Lee CH, McBranch D, Moses D, Sariciftci NS, and Heeger AJ. Ultrafast photoinduced electron-transfer in conducting polymer buckminsterfullerene composites. *Chem. Phys. Lett.* **1993**, *213(3–4)*, 389–394.

33. Smilowitz L, Sariciftci NS, Wu R, Gettinger C, Heeger AJ, and Wudl, F. Photoexcitation spectroscopy of conducting-polymer-C_{60} composites-photoinduced electron-transfer. *Phys. Rev. B* **1993**, *47(20)*, 13835–13842.

34. Murata K, Ito S, Takahasi K, and Hoffman BM. Long-lived excited state of C_{60} in C_{60}/phthalocyanine heterojunction solar cell. *Appl. Phys. Lett.* **1996**, *68(3)*, 427–429.

35. Pannemann Ch, Dyakonov V, Parisi J, Hild O, and Wöhrle D. Electrical characterisation of phthalocyanine-fullerene photovoltaic devices. *Synth. Met.* **2001**, *121(1–3)*, 1585–1586.

36. Peumans P, Bulovic V, and Forrest SR. Efficient photon harvesting at high optical intensities in ultrathin organic double-heterostructure photovoltaic diodes. *Appl. Phys. Lett.* **2000**, *76(19)*, 2650–2652.

37. Peumans P and Forrest SR. Small molecular weight organic thin-film photodetectors and solar cells. *J. Appl. Phys. Lett.* **2001**, *93(7)*, 3693–3723.

38. Xue J, Uchida S, Rand BP, and Forrest SR. Asymmetric tandem organic photovoltaic cells with hybrid planar-mixed molecular heterojunctions. *Appl. Phys. Lett.* **2004**, *85(23)*, 5757–5759.

39. Chen WB, Xiang HF, Xu ZX, Yan BP, Roy VAL, Che CM, and Lai PT. Improving efficiency of organic photovoltaic cells with pentacene-doped CuPc layer. *Appl. Phys. Lett.* **2007**, *91(19)*, 191109.

40. Placencia D, Wang WN, Shallcross RC, Nebesny KW, Brumbach M, and Armstrong NR. Organic photovoltaic cells based on solvent-annealed, textured titanyl phthalocyanine/C_{60} heterojunctions. *Adv. Func. Mater.* **2009**, *19(12)*, 1913–1921.

41. Pope M, Braams BJ, and Brenner HC. Diffusion of excitons in systems with non-planar geometry: theory. *Chem. Phys.* **2003**, *288(2–3)*, 105–112.

42. Theander M, Yartsev A, Zigmantas D, Sundström V, Mammo W, Andersson MR, and Inganäs O. Photoluminescence quenching at a polythiophene/C_{60} heterojunction. *Phys. Rev. B* **2000**, *61(19)*, 12957–12963.

43. Halls JJM, Pichler K, Friend RH, Moratti SC, and Holmes AB. Exciton diffusion and dissociation in a poly(p-phenylenevinylene)/C_{60} heterojunction photovoltaic cell. *Appl. Phys. Lett.* **1996**, *68(22)*, 3120–3122.

44. Stübinger T and Brütting W. Exciton diffusion and optical interference in organic donor–acceptor photovoltaic cells. *J. Appl. Phys.* **2001**, *90(7)*, 3632–3641.

45. Ioannidis A and Dodelet JP. Hole mobilities in trivalent metal phthalocyanine thin films. 1. Activated charge transport in time-of-flight measurements between 333 and 213 K for chloroaluminum phthalocyanine films with various amounts of disorder. *J. Phys. Chem. B* **1997**, *101(6)*, 891–900.

46. Hasobe T. Supramolecular nanoarchitectures for light energy conversion. *Phys. Chem. Chem. Phys.* **2010**, *12*, 44–57.

47. Gregg BA, Fox MA, and Bard AJJ. 2,3,7,8,12,13,17,18-Octakis (beta-hydroxyethyl)porphyrin (octaethanolporphyrin) and its liquid-crystalline derivatives—Synthesis and characterization. *J. Am. Chem. Soc.* **1989**, *111(8)*, 3024–3029.

48. Gregg BA, Fox MA, and Bard AJ. Photovoltaic effect in symmetrical cells of a liquid crystal porphyrin. *J. Phys. Chem.* **1990**, *94(4)*, 1586–1598.

49. Sergeyev S, Pisula W, and Geerts YH. Discotic liquid crystals: a new generation of organic semiconductors. *Chem. Soc. Rev.* **2007**, *36(12)*, 1902–1929.

50. Adam D, Schuhmacher P, Simmerer J, Haussling L, Siemensmeyer K, Etzbach KH, Ringsdorf H, and Haarer D. Fast photoconduction in the highly ordered columnar phase of a discotic liquid crystal. *Nature* **1994**, *371(6493)*, 141–143.

51. van de Craats AM, Warman JM, de Haas MP, Adam D, Simmerer J, Haarer D, and Schuhmacher P. The mobility of charge carriers in all four phases of the columnar discotic material hexakis(hexylthio)triphenylene: combined TOF and PR-TRMC results. *Adv. Mater.* **1996**, *8(10)*, 823–826.

52. Schmidt-Mende L, Fechtenkötter A, Müllen K, Moons E, Friend RH, and MacKenzie JD. Self-organized discotic liquid crystals for high-efficiency organic photovoltaics. *Science* **2001**, *293(5532)*, 1119–1122.

53. van de Craats AM, Warman JM, Fechtenkötter A, Brand JD, Harbison MA, and Müllen K. Record charge carrier mobility in a room-temperature discotic liquid-crystalline derivative of hexabenzocoronene. *Adv. Mater.* **1999**, *11(17)*, 1469–1472.

54. Schmidtke JP, Friend RH, Kastler M, and Müllen K. Control of morphology in efficient photovoltaic diodes from discotic liquid crystals. *J. Chem. Phys.* **2006**, *124(17)*, 174704.

55. Oukachmich M, Destruel P, Seguy I, Ablart G, Jolinat P, Archambeau S, Mabiala M, Fouet S, and Bock H. New organic discotic materials for photovoltaic conversion. *Solar Energy Mater. Solar Cells* **2005**, *85(4)*, 535–543.

56. Feng X, Liu M, Pisula W, Takase M, Li J, and Müllen K. Supramolecular organization and photovoltaics of triangle-shaped discotic graphenes with swallow-tailed alkyl substituents. *Adv. Mater.* **2008**, *20(14)*, 2684–2689.

57. Palermo V, Schwartz E, Finlayson CE, Liscio A, Otten MBJ, Trapani S, Müllen K, Beljonne D, Friend RH, Nolte RJM, Rowan AE, and Samor P. Macromolecular scaffolding: The relationship between nanoscale architecture and function in multichromophoric arrays for organic electronics. *Adv. Mater.* **2010**, *22(8)*, 81–88.

58. Tamayo AB, Tantiwiwat M, Walker B, and Nguyen TQ. Design, synthesis, and self-assembly of oligothiophene derivatives with a diketopyrrolopyrrole core. *J. Phys. Chem. C* **2008**, *112(39)*, 15543–15552.

59. Walker B, Tamayo AB, Dang XD, Zalar P, Seo JH, Garcia A, Tantiwiwat M, and Nguyen TQ. Nanoscale phase separation and high photovoltaic efficiency in solution-processed, small-molecule bulk heterojunction solar cells. *Adv. Func. Mater.* **2009**, *19(19)*, 3063–3069.

60. Hiramoto M, Fujiwara H, and Yokoyama M. *p-i-n* like behavior in three-layered organic solar cells having a co-deposited interlayer of pigments. *J. Appl. Phys.* **1992**, *72(8)*, 3781–3787.

61. Rostalski J and Meissner D. Monochromatic versus solar efficiencies of organic solar cells. *Solar Energy Mater. Solar Cells* **2000**, *61(1)*, 87–95.

62. Geens W, Aernouts T, Poortmans J, and Hadziioannou G. Organic co-evaporated films of a PPV-pentamer and C_{60}: model systems for donor/acceptor polymer blends. *Thin Solid Films* **2002**, *403–404*, 438–443.

63. Wang XH, Grell M, Lane PA, and Bradley DDC. Determination of the linear optical constants of poly(9,9-dioctylfluorene). *Synth. Met.* **2001**, *119(1–3)*, 535–536.

64. Pfeiffer M, Beyer A, Plönnigs B, Nollau A, Fritz T, Leo K, Schlettwein D, Hiller S, and Wöhrle D. Controlled *p*-doping of pigment layers by cosublimation: Basic mechanisms and implications for their use in organic photovoltaic cells. *Solar Energy Mater. Solar Cells* **2000**, *63(1)*, 83–99.

65. Maennig B, Drechsel J, Gebeyehu D, Simon P, Kozlowski F, Werner A, Li F, Grundmann S, Sonntag S, Koch M, Leo K, Pfeiffer M, Hoppe H, Meissner D, Sariciftci NS, Riedel I, Dyakonov V, and Parisi J. Organic p-i-n solar cells. *Appl. Phys. A* **2004**, *79(1)*, 1–14.

66. Hong ZR, Maennig B, Lessmann R, Pfeiffer M, Leo K, and Simon P. Improved efficiency of zinc phthalocyanine/C_{60} based photovoltaic cells via nanoscale interface modification. *Appl. Phys. Lett.* **2007**, *90(20)*, 203505.

67. Esenturk O, Melinger JS, Lane PA, and Heilweil EJ, Relative photon-to-carrier efficiencies of alternating nanolayers of zinc phthalocyanine and C_{60} films assessed by time-resolved terahertz spectroscopy. *J. Phys. Chem. C* **2009**, *113(43)*, 18842–18850.

68. Lane PA, Melinger JS, Esenturk O, and Heilweil EJ. Carrier dynamics of composite and layered films of zinc phthalocyanine and C_{60} measured by time-resolved terahertz spectroscopy. *SPIE Proc.* **2009**, *7416*, 7416Z1–7416Z8.

69. Hiramoto M, Suezaki M, and Yokoyama M. Effect of thin gold interstitial-layer on the photovoltaic properties of tandem organic solar-cell. *Chem. Lett.* **1990**, *3*, 327–330.

70. Yakimov A and Forrest SR. High photovoltage multiple-heterojunction organic solar cells incorporating interfacial metallic nanoclusters. *Appl. Phys. Lett.* **2002**, *80(9)*, 1667–1669.

71. Schwartz G, Maennig B, Uhrich C, Gnehr W, Sonntag S, Erfurth O, Wollrab E, Walzer K, and Pfieffer M. Efficient and long-term stable organic vacuum deposited tandem solar cells. *SPIE Proc.* **2009**, *7416*, 74160K.

72. Tsukamoto J, Ohigashi H, Matsumura K, and Takahashi A. A schottky-barrier type solar-cell using polyacetylene. *Jpn. J. Appl. Phys.* **1981**, *20(2)*, L127–L129.

73. Kanicki J and Fedorko P. Electrical and photovoltaic properties of trans-polyacetylene. *J. Phys. D Appl. Phys.* **1984**, *17(4)*, 805–817.

74. Glenis S, Horowitz G, Tourillon G, and Garnier F. Electrochemically grown polythiophene and poly(3-methylthiophene) organic photovoltaic cells. *Thin Solid Films* **1984**, *111(2)*, 93–103.

75. Koezuka H, Hyodo K, and MacDiarmid AG. Organic heterojunctions utilizing two conducting polymers: Poly(acetylene)/poly(N-methylpyrrole) junctions. *J. Appl. Phys.* **1985**, *58(3)*, 1279–1284.

76. Blom PWM, de Jong MJM, and Vleggaar JJM. Electron and hole transport in poly(p-phenylene vinylene) devices. *Appl. Phys. Lett.* **1996**, *68(23)*, 3308–3310.

77. Antoniadis H, Abkowitz MA, and Hsieh BR. Carrier deep-trapping mobility-lifetime products in poly(p-phenylenevinylene). *Appl. Phys. Lett.* **1994**, *65(16)*, 2030–2032.

78. Burn PL, Grice AW, Tajbakhsh A, Bradley DDC, and Thomas AC. Insoluble poly[2-(2′-ethylhexyloxy)-5-methoxy-1,4-phenylenevinylene] for use in multilayer light-emitting diodes. *Adv. Mater.* **1997**, *9*(15), 1171.

79. Greenham NC, Moratti SC, Bradley DDC, Friend RH, and Holmes AB. *Nature* **1993**, *365(6447)*, 628–630.

80. Wang YZ, Gebler DD, Lin LB, Blatchford JW, Jessen SW, and Wang HL, Epstein AJ. Alternating-current light-emitting devices based on conjugated polymers. *Appl. Phys. Lett.* **1996**, *68(7)*, 894–896.

81. Halls JJM, Walsh CA, Greenham NC, Marseglia EA, Friend RH, Moratti SC, and Holmes AB. Efficient photodiodes from interpenetrating polymer networks. *Nature* **1995**, *376(6540)*, 498–500.

82. Yu G and Heeger AJ. Charge separation and photovoltaic conversion in polymer composites with internal donor/acceptor heterojunctions. *J. Appl. Phys.* **1995**, *78(7)*, 4510–4515.

83. Moons E. Conjugated polymer blends: linking film morphology to performance of light emitting diodes and photodiodes. *J. Phys. Condens. Matter* **2002**, *14(47)*, 12235–12260.

84. Bates FS. Polymer-polymer phase-behavior. *Science* **1991**, *251(4996)*, 898–905.

85. Kimber RGE, Walker AB, Schröder-Turk GE, and Cleaver DJ. Bicontinuous minimal surface nanostructures for polymer blend solar cells, *Phys. Chem. Chem. Phys*. **2010**, *12(4)*, 844–851.

86. Tan ZA, Zhou EJ, Zhan XW, Wang X, Li YF, Barlow S, and Marder SR. Efficient all-polymer solar cells based on blend of tris(thienylenevinylene)-substituted poly-thiophene and poly[perylene diimide-alt-bis(dithienothiophene)]. *Appl. Phys. Lett*. **2008**, *93(7)*, 073309.

87. McNeill CR, Abrusci A, Zaumseil J, Wilson R, McKiernan MJ, Burroughes JH, Halls JJM, Greenham NC, and Friend RH. Dual electron donor/electron acceptor character of a conjugated polymer in efficient photovoltaic diodes. *Appl. Phys. Lett*. **2007**, *90(19)*, 193506.

88. Kietzke T, Horhold HH, and Neher D. Efficient polymer solar cells based on M3EH-PPV. *Chem. Mater*. **2005**, *17(26)*, 6532–6537.

89. Sariciftci NS, Smilowitz L, Heeger AJ, and Wudl F. Photoinduced electron-transfer from a conducting polymer to buckminsterfullerene. *Science* **1992**, *258(5087)*, 1474–1476.

90. Kraabel B, Lee CH, McBranch D, Moses D, Sariciftci NS, and Heeger AJ. Ultrafast photoinduced electron-transfer in conducting polymer buckminsterfullerene compos-ites. *Chem. Phys. Lett*. **1993**, *213(3–4)*, 389–394.

91. Smilowitz L, Sariciftci NS, Wu R, Gettinger C, Heeger AJ, and Wudl F. Pho-toexcitation spectroscopy of conducting-polymer-c(60) composites—photoinduced electron-transfer. *Phys. Rev. B* **1993**, *47(20)*, 13835–13842.

92. Morita S, Zakhidov AA, and Yoshino K. Doping effect of buckminsterfullerene in conducting polymer—change of absorption-spectrum and quenching of lumines-cence. *Solid State Comm*. **1992**, *82(4)*, 249–252.

93. Morita S, Kiyomatsu S, Yin XH, Zakhidov AA, Noguchi T, Ohnishi T, and Yoshino K. Doping effect of buckminsterfullerene in poly(2,5-dialkoxy-*p*-phenylenevinylene). *J. Appl. Phys*. **1992**, *74(4)*, 2860–2865.

94. Sariciftci NS, Braun D, Zhang C, Srdanov VI, Heeger AJ, Stucky G, and Wudl F. Semiconducting polymer-buckminsterfullerene heterojunctions: Diodes, photodi-odes, and photovoltaic cells. *Appl. Phys. Lett*. **1993**, *62(6)*, 585–587.

95. Yu G, Pakbaz K, and Heeger AJ. Semiconducting polymer diodes: Large size, low cost photodetectors with excellent visible-ultraviolet sensitivity. *Appl. Phys. Lett*. **1994**, *64(25)*, 3422–3424.

96. Ruoff RS, Tse DS, Malhotra R, and Lorents DC. Solubility of C_{60} in a variety of solvents. *J. Phys. Chem*. **1993**, *97(13)*, 3379–3383.

97. Taylor R and Walton DRM. The chemistry of fullerenes. *Nature* **1993**, *363(6431)*, 685–693.

98. Krätschmer W, Lamb L, Fostiropoulos K, and Huffman DR. Solid C_{60}: A new form of carbon. *Nature* **1990**, *347*, 354–358.

99. Wudl F. The chemical properties of buckminsterfullerene C_{60} and the birth and infancy of fulleroids. *Acc. Chem. Res*. **1992**, *25*, 157–161.

100. Hummelen JC, Knight BW, LePeq F, Wudl F, Yao J, and Wilkins CL. Preparation and characterization of fulleroid and methanofullerene derivatives. *J. Org. Chem*. **1994**, *60(3)*, 532–538.

101. Yu G, Gao J, Hummelen JC, Wudl F, and Heeger AJ. Polymer photovoltaic cells—enhanced efficiencies via a network of internal donor-acceptor heterojunctions. *Science* **1995**, *270(5243)*, 1789–1791.

102. Brabec CJ, Padinger F, Sariciftci NS, and Hummelen JS. Photovoltaic properties of conjugated polymer/methanofullerene composites embedded in a polystyrene matrix. *J. Appl. Phys.* **1999**, *85(9)*, 6866–6872.

103. Gebeyehu D, Brabec CJ, Padinger F, Fromherz T, Hummelen JC, Badt D, Schindler H, and Sariciftci NS. The interplay of efficiency and morphology in photovoltaic devices based on interpenetrating networks of conjugated polymers with fullerenes. *Synth. Met.* **2001**, *118(1–3)*, 1–9.

104. Shaheen SE, Brabec CJ, Sariciftci NS, Padinger F, Fromherz T, and Hummelen JC. 2.5% Efficient organic plastic solar cells. *Appl. Phys. Lett.* **2001**, *78(6)*, 841–843.

105. Brabec CJ, Shaheen SE, Winder C, Sariciftci NS, and Denk P. Effect of LiF metal electrodes on the performance of plastic solar cells. *Appl. Phys. Lett.* **2002**, *80(7)*, 1288–1290.

106. Schilinsky P, Waldauf C, and Brabec CJ. Recombination and loss analysis in polythiophene based bulk heterojunction photodetectors. *Appl. Phys. Lett.* **2002**, *81(20)*, 3885–3887.

107. Ma W, Yang C, Gong X, Lee K, and Heeger AJ. Thermally stable, efficient polymer solar cells with nanoscale control of the interpenetrating network morphology. *Adv. Func. Mater.* **2005**, *15(10)*, 1617–1622.

108. Li G, Shrotriya F, Huang JS, Tao Y, Moriarty T, Emergy K, and Yang Y. High-efficiency solution processible polymer photovoltaic cells by self-organization of polymer blends. *Nat. Mater.* **2005**, *4*, 864–868.

109. Kim Y, Cook S, Tuladhar SM, Choulis SA, Nelson J, Durrant JR, Bradley DDC, Giles M, Mcculloch I, Ha CK, and Ree M. A strong regioregularity effect in self-organizing conjugated polymer films and high-efficiency polythiophene:fullerene solar cells. *Nat. Mater.* **2006**, *5*, 197–203.

110. Chen TA, Wu X, and Rieke RD. Regiocontrolled synthesis of poly(3-alkylthiophenes) mediated by Rieke zinc: Their characterization and solid-state properties. *J. Am. Chem. Soc.* **1995**, *117(1)*, 233–244.

111. Chiu M, Jeng U, Su M, and Wei K. Morphologies of self-organizing regioregular conjugated polymer/fullerene aggregates in thin film solar cells. *Macromolecules* **2010**, *43*, 428–432.

112. Sunderberg M, Inganas O, Stafstrom S, Gustafsson G, and Sjogren B. Optical absorption of poly(3-alkylthiophenes) at low temperatures. *Solid State Commun.* **1989**, *71*, 435–439.

113. Shrotriya V, Yao Y, Li G, and Yang Y. Effect of self-organization in polymer/fullerene bulk heterojunctions on solar cell performance. *Appl. Phys. Lett.* **2006**, *89(6)*, 063505.

114. Melzer C, Koop EJ, Mihailetchi VD, and Blom PWM. Hole transport in poly(phenylenevinylene)/methanofullerene bulk-heterojunction solar cells. *Adv. Funct. Mater.* **2004**, *14*, 865–870.

115. Scharber MC, Mühlbacher D, Koppe M, Denk P, Waldauf C, Heeger J, and Brabec CJ. Design rules for donors in bulk-heterojunction solar cells-towards 10% energy-conversion efficiency. *Adv. Mater.* **2006**, *18(6)*, 789–794.

116. Zhu Z, Waller D, Gaudiana R, Morana M, Muhlbacher D, Scharber M, and Brabec CJ. Panchromatic conjugated polymers containing alternating donor/acceptor units for photovoltaic applications. *Macromolecules* **2007**, *40*(*6*), 1981–1986.

117. Mühlbacher D, Scharber M, Morana M, Zhu Z, Waller D, Gaudiana R, and Brabec C. High photovoltaic performance of a low-bandgap polymer. *Adv. Mater.* **2006**, *18*(*21*), 2884–2889.

118. Peet J, Heeger AJ, and Bazan GC. Plastic solar cells: Self-assembly of bulk heterojunction nanomaterials by spontaneous phase separation. *Acc. Chem. Res.* **2009**, *42*(*11*), 1700–1708.

119. Terrill RH, Postlethwaite TH, Chen CH, Poon CD, Terzis A, Chen A, Hutchison JE, Clark MR, Wignall G, Londono JD, Superfine R, Falvo M, Johnson CS, Samulski ET, and Murray RW. Monolayers in three dimensions: NMR, SAXS, thermal, and electron hopping studies of alkanethiol stabilized gold clusters. *J. Phys. Chem.* **1995**, *117*(*50*), 12537–12548.

120. Peet J, Kim JY, Coates NE, Ma WL, Moses D, Heeger AJ, and Bazan GC. Efficiency enhancement in low-bandgap polymer solar cells by processing with alkane dithiols. *Nat. Mater.* **2007**, *6*, 497–500.

121. Park SH, Roy A, Beaupré S, Cho S, Coates N, Moon JS, Moses D, Leclerc M, Lee K, and Heeger AJ. Bulk heterojunction solar cells with internal quantum efficiency approaching 100%. *Nat. Photonics* **2009**, *3*, 297–303.

122. Hoven CV, Dang XD, Coffin RC, Peet J, Nguyen TQ, and Bazan GC. Improved Performance of Polymer Bulk Heterojunction Solar Cells Through the Reduction of Phase Separation via Solvent Additives. *Adv. Mater.* **2010**, *22*, E63–E66.

123. Laird DW, Vaidya S, Li S, Mathai M, Woodworth B, Sheina E, Williams S, and Hammond T. Advances in Plexcore™ active layer technology systems for organic photovoltaics: Roof-top and accelerated lifetime analysis of high performance organic photovoltaic cells. *SPIE Proc.* **2008**, *6656*, 66560X.

124. Ross RR, Cardona CM, Guldi DM, Sankaranarayanan SG, Reese MO, Kopidakis N, Peet J, Walker B, Bazan GC, Van Keuren E, Holloway BC, and Drees M. Endohedral fullerenes for organic photovoltaic devices. *Nat. Mater.* **2009**, *8*, 208–212.

125. Ross RB, Cardona CM, Swain FB, Guldi DM, Sankaranarayanan SG, Van Keuren E, Holloway BC, and Drees M. Tuning conversion efficiency in metallo endohedral fullerene-based organic photovoltaic devices. *Adv. Func. Mater.* **2009**, *19*(*14*), 2332–2337.

126. Liang YY, Wu Y, Feng DQ, Tsai ST, Son HJ, Li G, and Yu LP. Development of new semiconducting polymers for high performance solar cells. *J. Am. Chem. Soc.* **2009**, *131*(*1*), 56–57.

127. Liang YY, Feng DQ, Wu Y, Tsai ST, Li G, Ray C, and Yu LP. Highly efficient solar cell polymers developed via fine-tuning of structural and electronic properties. *J. Am. Chem. Soc.* **2009**, *131*(*22*), 7792–7799.

128. Chen HY, Hou J, Zhang S, Liang YY, Yang G, Yang Y, Yu LP, Wu Y, and Li G. Polymer solar cells with enhanced open-circuit voltage and efficiency. *Nat. Photonics* **2009**, *3*, 649–653.

CHAPTER 9

Selective Molecular Assembly for Bottom-Up Fabrication of Organic Thin-Film Transistors

TAKEO MINARI[1,2], MASATAKA KANO[3], and KAZUHITO TSUKAGOSHI[1,4,5]

[1]MANA, NIMS, Tsukuba, Ibaraki, Japan; [2]RIKEN, Wako, Saitama, Japan; [3]Dai Nippon Printing Co., Ltd., Kashiwa, Chiba, Japan; [4]AIST, Tsukuba, Ibaraki, Japan; [5]CREST, JST, Kawaguchi, Saitama, Japan

9.1. INTRODUCTION

In a time of rapid advances and proliferation in electronics, and the growing movement toward a low-carbon society, the need is rapidly rising for people-friendly, eco-friendly electronic devices and production technologies. Next-generation devices that will use plastic electronics technology to achieve thinner size, lighter weight, and greater flexibility are expected to become key devices for a society of ubiquitous networks. The characteristics of organic molecules such as solubility and low-temperature processability are expected to enable device fabrication by relatively simple printing processes and bring substantially lower manufacturing costs. Organic field-effect transistors (OFETs), in particular, have reached the stage where their operating characteristics are fully comparable to those of amorphous silicon transistors and are now at a turning point in which development is contemplated or under way for many practical applications [1–8].

In contrast to vapor deposition processes, in which organic thin-film patterns are readily obtained with a metal mask, the spin-coating and casting processes that are normally used to obtain semiconductor films from solutions do not include an organic thin-film patterning capability. Without effective patterning, thin-film semiconductors pose substantial problems in terms of large leakage

Self-Organized Organic Semiconductors: From Materials to Device Applications, First Edition.
Edited by Quan Li.
© 2011 John Wiley & Sons, Inc. Published 2011 by John Wiley & Sons, Inc.

current and device cross talk. Various methods have been proposed for organic thin-film patterning in solution applications [9–14]. The bottom-up technique is particularly promising for this purpose, as a method of electronic device fabrication that utilizes the inherent functional properties of organic molecules for self-assembly and organization [15–24]. Depending on their structural configuration, functional groups, and electron state, they exhibit different assembly and organizational characteristics and can form into molecular groups that exhibit distinctive electronic functions. The establishment of a process for the self-organization of electronic devices through these molecular functions will provide the basis for an ecological, energy-saving method of mass production for electronic devices.

For self-organization of electronic devices, a method of crystal growth control to obtain selective growth at specific locations on the substrate is essential. In recent years, several methods for self-organized OFETs by selective organic crystal growth have been proposed [18–21, 25], including the methods of preprinting of regions with a high nucleation rate on the substrate [25] and molecular crystallization from electrode tips for direct deposition between source and drain electrodes [18].

This chapter describes the selective crystallization of organic semiconductors and its application to device fabrication, based on our proposed "surface-selective deposition" process for efficient self-formation of devices [19, 20]. In this process, functional organic molecules are patterned on the substrate, and their functional groups then interact with the semiconductor molecules to stimulate the spontaneous formation of a semiconductor active layer. Surface-selective deposition provides the ultimate in energy saving. Immediately after the formation of the simple molecular template on the substrate, the intrinsic action of the semiconductor molecular solution results in selective crystallization at specific regions and thus in the fabrication of a semiconductor active layer and an OFET device without the use of vacuum or ink-jet processes. This organic semiconductor layer patterning based on molecular self-assembly is expected to provide the following advantages for plastic electronics production:

(1) Large-area, high-throughput device fabrication: Device fabrication utilizing molecular self-assembly can provide energy-saving and high-throughput advantages that increase with production area and volume.

(2) Efficient utilization of organic semiconductor material: Surface-selective deposition is an on-demand device fabrication process that uses only the required material quantities at the required locations.

(3) Reduced plant and equipment requirements: Thin-film patterning can be achieved with UV irradiation equipment, with no need for vacuum-deposition equipment or ink-jet printers, and may thus become a core technology for fabrication of next-generation organic devices with low-cost printing.

9.2. FABRICATION OF OFET ARRAY BY SURFACE-SELECTIVE DEPOSITION

9.2.1. Principles of Surface-Selective Deposition

A prerequisite for selective crystal growth based on surface-selective deposition is the formation of modified surface regions mutually distinguished by their functional groups and thus by their effects on the organic semiconductor solution and its constituent molecules. Surface regions with high affinity for the solution and its constituent molecules, and thus "wettable" for that solution, tend to promote nucleation. Surface regions lacking interactivity with the newly coated semiconductor solution, in contrast, tend to repel the semiconductor, thus inhibiting crystallization in those regions. If these regions of different functional effect can be effectively formed into the desired pattern on the substrate, the resulting patterned surface will enable the nucleation and growth of a thin film of semiconductor crystals having the desired shape and location.

A conceptual drawing of selective crystal growth by surface-selective deposition is shown in Figure 9-1. Many different functional groups may be considered for formation of the functional surfaces. Here we describe surface modification by alkyl or fluoroalkyl groups to obtain lyophobic, or solution-repellent, surface regions and primarily by phenyl groups to obtain lyophilic, or solution-wettable, surface regions. These functional groups have been chosen in part because of their advantages in relation to process convenience and relative freedom from adverse effects on the electrical characteristics of the organic transistor devices after their formation. When a semiconductor solution is deposited on a surface patterned with these two types of groups, its differing interactions with the surface

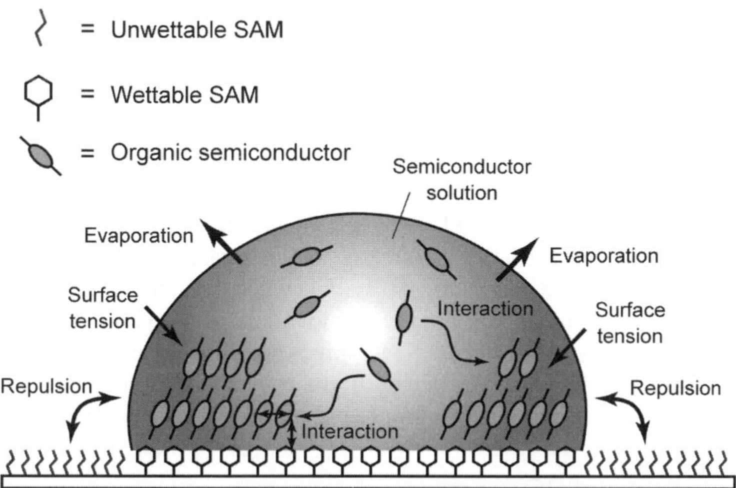

Figure 9-1   Conceptual drawing of surface-selective deposition.

regions enable selective crystallization of its organic semiconductor molecules in the phenyl regions.

This difference in wettability is believed to be attributable to a difference in charge distribution. An alkyl surface is electrically neutral and stable, and its charge distribution is unlikely to change even when coated with a polarized organic semiconductor solution as no stable energy state can be attained, and the solution will be repelled by the surface. On a phenyl-group modified surface, however, π-electrons are present in abundance, and there is greater flexibility in the charge distribution. With this type of surface, a dipole is readily induced when polarized semiconductor molecules or solvent approach the surface, and its solution wettability is increased, thereby stabilizing the contact between the surface and the solution. This type of surface, with its π-electrons, may also exhibit some degree of wettability by nonpolarized solutions, due to induction of electric dipole pairs in both the surface and the solution. As an alternative means of obtaining a surface with high solution wettability, its modification by highly polarized functional groups such as hydroxyls or aminos may also be considered, but such highly polarized surfaces are likely to act as charge traps in the fabricated transistor devices, and thus degrade the device's electrical characteristics. The use of a benzene ring functional group, which is in itself essentially nonpolar, is considered advantageous for this process.

9.2.2. Formation of Functional Molecule Template

Self-assembled monolayer (SAM) patterning may be conveniently employed to obtain a functional molecule template for the subsequent deposition from solution. The pattern may be formed either by using a polydimethylsiloxane (PDMS) stamp to selectively apply the SAM molecules to the target regions [26] or by first applying the SAM to the entire surface and then patterning it by vacuum ultraviolet (VUV, wavelength is 185 nm) irradiation [27]. We used the latter to first form a solution-repellent SAM pattern on the substrate, and then reacted a lyophilic SAM with the bare target regions to complete the SAM template. This process enables us to form the molecular template by SAM formation and VUV irradiation only. The process is shown schematically in Figure 9-2. A silicon substrate with a thermally oxidized silicon oxide layer is first cleaned, and then the surface is completely coated with hexamethyldisilazane (HMDS) to form a repellent SAM. Next, VUV irradiation is performed through a metal mask or by other means to expose target regions where organic crystal growth is desired. The repellent molecules are thus removed, leaving hydroxyl groups in the target regions. Finally, a lyophilic SAM composed of aromatic surface-modifying molecules is formed by reacting phenetyltrichlorosilane (PhTS) with the hydroxyl groups in the target regions, thus completing the molecular template.

9.2.3. Formation of OFET Array

The molecular template, formed as described above, is coated with an organic semiconductor solution to cause crystallization of the semiconductor molecules

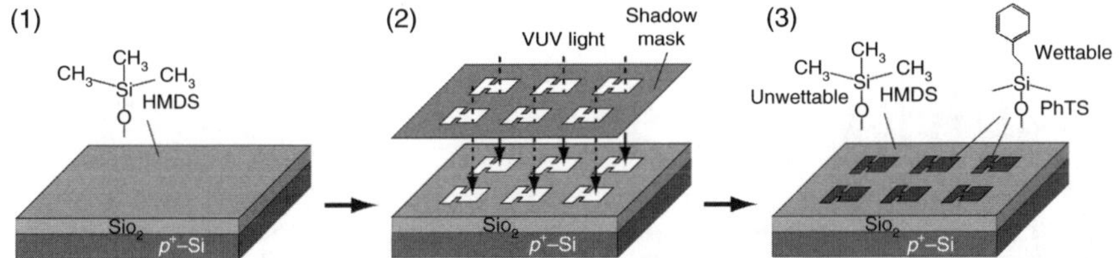

Figure 9-2 Schematic drawing of surface functional patterning with repellent (unwettable) SAM (HMDS) and lyophilic (wettable) SAM (PhTS) for formation of a template pattern of repellent HMDS regions and lyophilic PhTS on a silicon oxide substrate. (1) Application of repellent HMDS to silicon oxide surface. (2) VUV irradiation of target regions through shadow mask, and removal of HMDS from target regions. (3) Application of PhTS to surface, resulting in formation of aromatic lyophilic SAM in the target regions.

in the target regions. The results described here were obtained by using a 0.3 wt% toluene solution of dioctylquaterthiophene (8QT8) as the semiconductor solution. The semiconductor crystal growth in the target regions is driven by selective repulsion and affinity to different SAM surfaces. When the semiconductor solution is dropped onto the molecular template [Fig. 9-3(a)], the different surface energies of the surface-modifying functional groups result in selective adhesion of the solution to the aromatic surfaces only [Fig. 9-3(b)]. Evaporating the solution thus results in the formation of an organic semiconductor thin film that self-assembles into the desired configuration [Fig. 9-3(c)]. Source and drain electrodes can be formed after the deposition of organic film, resulting in the simultaneous formation of an OFET array over a wide area [Fig. 9-3(d)]. A single device is shown in Figure 9-3(e). The crystalline state of the organic thin film fabricated by surface-selective deposition depends on the type of solution used, the evaporation temperature, and other process conditions. Here, a toluene solution was used, and the semiconductor crystallization proceeded at atmospheric pressure and room temperature. As shown in Figure 9-3(f), this resulted in a thin film having an average crystal size of approximately 10 μm and a polycrystalline OFET having a 20-μm channel length.

The output and transfer characteristics of the self-organized OFETs are shown in Figure 9-4. As shown in Figure 9-4(a), the drain current of these OFETs increases nonlinearly at low drain voltages, presumably as an effect of contact resistance due to the relatively short length of the channels. As shown in Figure 9-4(b), an on/off ratio of 10^6 was obtained without hysteresis. The field-effect mobility (μ_{FET}) of this device is estimated to be 0.014 cm^2/V · s.

9.3. IMPROVEMENT OF SELF-ORGANIZED OFET PERFORMANCE WITH AROMATIC SAM

The carrier path of an OFET is known to be in the vicinity of the semiconductor/insulator interface, and efforts are frequently made to enhance OFET

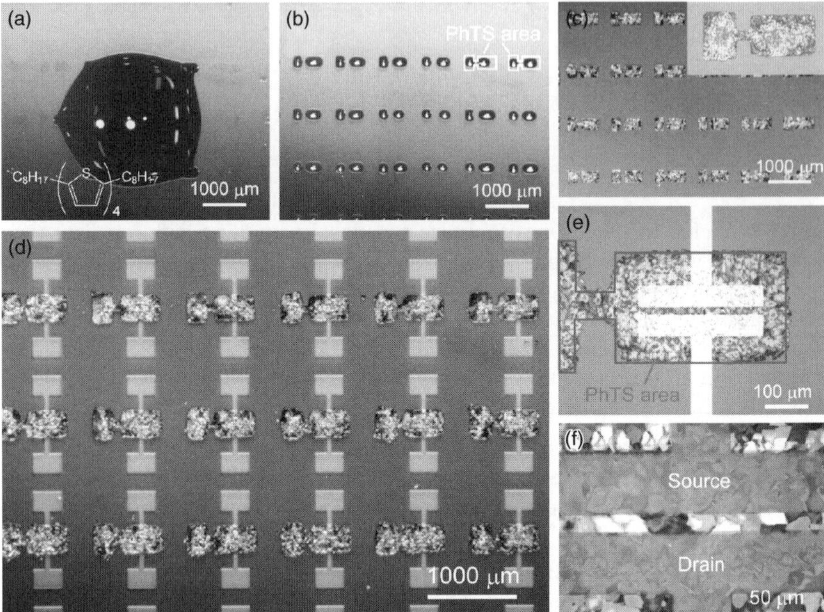

Figure 9-3 Polycrystalline OFET array formed by surface-selective deposition. (a) Organic semiconductor solution dropped onto substrate bearing surface functional pattern; inset shows the molecular structure of 8QT8. (b) Self-distribution of the semiconductor solution onto the lyophilic target-region surfaces. (c) Organic semiconductor crystallization and thin-film formation at target locations, by solvent evaporation; inset shows an individual pattern. (d) Simultaneous formation of the complete OFET array, with addition of top contact electrodes by vacuum evaporation. (e) Optical micrograph of single device. (f) Optical polarized micrograph of area surrounding the channel. (A full color version of this figure appears in the color plate section.)

characteristics by forming a SAM on the insulator surface and thus reduce the density of interface trap states [28–32]. In a case such as the present, in which the semiconductor layer of the OFET is formed by surface-selective deposition, lyophilic surface-modifying molecules will be present at this charge-accumulation interface. Accordingly, the material selected for the lyophilic surface-modifying molecules should not only be wettable to the semiconductor solution but should also contribute to the stability of the semiconductor/insulator interface and charge transport efficiency of the final device. In patterning by surface-selective deposition, various molecules can be used to obtain the lyophilic surface, so long as they include hydroxyl, amino, or some other type of polarized functional groups. To improve the charge transport efficiency of the device, however, aromatic functional groups are most desirable. The modification of semiconductor/insulator interfaces by an aromatic SAM has been reported to provide favorable device characteristics, even in an OFET fabricated by deposited pentacene [31].

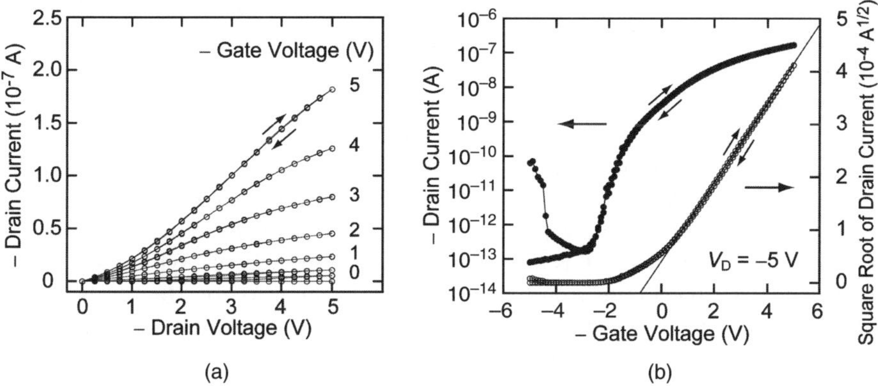

Figure 9-4   Output characteristics (a) and transfer characteristics (b) of a self-organized polycrystalline OFET fabricated by surface-selective deposition.

To confirm the advantage of using an aromatic SAM to obtain the lyophilic surface regions, we compared the bias stress of self-organized OFETs produced with and without the use of PhTS. The effect of bias stress in a p-type OFET is mainly a negative shift in threshold voltage (V_T) under continuous gate voltage application, and also an observable decrease in drain current with increasing stress time. Figure 9-5 shows the time change in drain current under gate-voltage stress in the two OFETs. Selective deposition can be obtained without the aromatic SAM (i.e., without SAM application after the removal of the HMDS by VUV

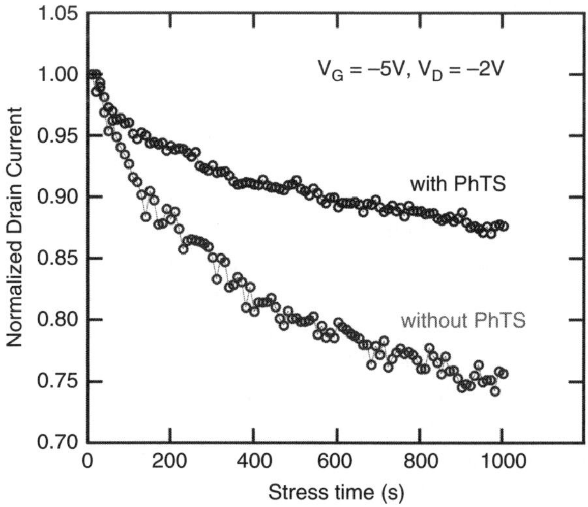

Figure 9-5   Decrease in drain current under bias stress in self-organized OFETs with and without an aromatic SAM; current values normalized to initial drain current. The use of PhTS in the lyophilic surface substantially reduces the current decrease.

irradiation) since the main end groups at the silicon oxide layer surface are presumably hydroxyls, and the surface is therefore wettable by the semiconductor solution. As shown by the results in Figure 9-5, however, the drop in drain current under bias stress is very rapid in the OFET obtained without PhTS but substantially moderated in that obtained with PhTS. Generally, the main cause of a shift in V_T in an OFET under bias stress is thought to be charge trapping at the semiconductor/insulator interface [33–46]. The improved stability resulting from the PhTS process may therefore be attributed to the replacement of the hydroxyl groups, which act as charge traps, with the PhTS phenyl groups. In this light, it is essential in electronic device self-formation to construct a molecular template that serves to enhance the performance of the resulting device, as well as providing appropriate lyophilic and lyophobic regions.

9.4. FORMATION OF SINGLE-CRYSTAL OFETs BY SURFACE-SELECTIVE DEPOSITION

With region-selective crystal growth by surface-selective deposition, organic single crystals can be grown directly between the source and drain electrodes of the OFET. Since local charge traps may occur at the grain boundaries of polycrystalline OFETs, OFETs with single-crystal, or single-grain, channels have been employed as a tool for investigation of the intrinsic conduction characteristics of OFETs [47–68]. The OFETs fabricated by rubrene single crystals have exhibited especially high mobility [54, 59, 60, 67]. The intrinsic charge transport mechanism has been discussed interms of the correlation between the crystal structure of single-crystal devices and their electrical characteristics [64, 65], and organic single crystals are also used as the channels in high-efficiency light emitting transistors [66].

Single-crystal OFETs are generally fabricated by manipulation of individual single crystals that have been grown by physical vapor transport or other methods [70], and then bonding them with static electricity to a substrate or electrode. The use of a single-crystal OFET as a practical electronic device is considered highly desirable because of the superior device characteristics. On the other hand, the technique of handling individual crystals is not suitable for practical applications, and the single crystals must in such cases be grown directly at specific locations. Surface-selective deposition is also applied to obtain the necessary growth control for such organic single crystals. The results of an experiment on direct formation of a single-crystal OFET by the solution process are shown in Figure 9-6. In this experiment, the organic semiconductor solution and the evaporation conditions were adjusted to obtain growth of the single crystal in the channel region of an OFET. Four bottom-contact electrodes (A to D) were formed on a silicon oxide layer of a substrate, and a SAM pattern was formed on the oxide surface. With a method similar to that described above, the surface area between the electrodes was modified with PhTS and the surface area outside the electrodes was modified with HMDS. The organic semiconductor 8QT8 solution was prepared in a concentration of 2.0 wt% by using tetrahydronaphthalene, which has a high boiling point, as solvent. When the semiconductor solution was dropped onto the substrate at 55°C, the solution adhered only to the wettable channel regions and the electrode surfaces. The temperature was maintained to control the solvent evaporation rate, and a semiconductor crystal formed on the channel regions and electrodes as shown in Figure 9-6(a). Here, although the organic crystal does

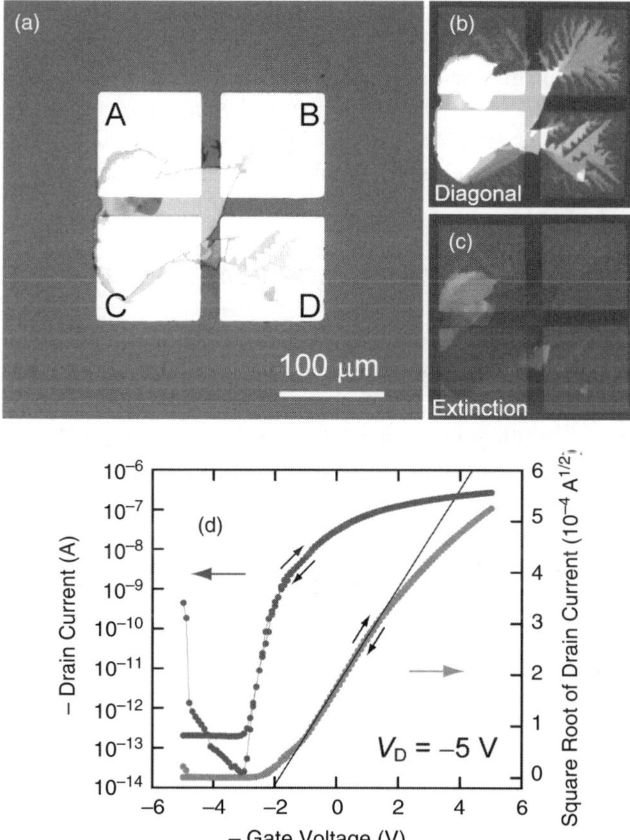

Figure 9-6 (a) Optical micrograph of organic single-crystal transistor directly self-organized by surface-selective deposition. (b), (c) Polarized optical micrographs of an organic single crystal transistor from (b) diagonal position and (c) extinction position (rotated 45°). (d) Current and voltage characteristics between electrodes A and B in a self-organized single-crystal OFET.

not completely cover the channel regions, single-crystal FETs have thus formed between electrodes A and B and between electrodes A and C. Polarized optical micrographs of this device are shown in Figure 9-6(b) and 9-6(c). Observations from the diagonal and extinction positions show that its growth on the electrodes is monocrystalline. The transfer characteristics measured between electrodes A and B are shown in Figure 9-6(d). The μ_{FET} of this single-crystal device is estimated to be 0.17 cm^2/V · s, which is at least one order of magnitude greater than that of a polycrystalline device. Additionally, an on/off ratio greater than 10^6 and a favorable subthreshold slope (S) of 0.19 V/dec were obtained. The single-crystal region between electrodes A and C had a μ_{FET} of 0.12 cm^2/V · s, indicating that the mobility of a single-crystal device is dependent upon crystallographic orientation. This electrical characteristic dependence on crystallographic orientation has been reported in rubrene and pentacene single-crystal FETs grown by physical vapor transport [67–69], but variations in device characteristics according to the crystallographic orientation are undesirable for practical applications, and a method for controlling the crystallographic orientation is needed.

9.5. FORMATION OF OFET ARRAY ON PLASTIC SUBSTRATE

The low process temperatures of organic semiconductors make them highly suitable for fabrication on lightweight, flexible plastic substrates [71–76]. Surface-selective deposition, moreover, with thin-film patterning obtained simply by UV irradiation and its application to the fabrication of electronic devices on plastic substrates will enable large-scale roll-to-roll production processes.

Here we describe the formation of an OFET array on a polyethylene naphthalate (PEN) substrate with a polyimide (PI) gate insulator patterned with lyophobic alkyl groups and lyophilic phenyl groups. Gate electrodes were first vacuum deposited on the PEN substrate, and a 420-nm PI layer was then applied as the gate insulator. A composite layer of tetramethoxysilane (TMS) and decyltrimethoxysilane (DTMS) was then applied to the PI insulation layer, to obtain efficient molecular template formation. This TMS-DTMS composite layer not only acts as a repellent layer because of the alkyl groups at its surface, but also functions as a base layer for the formation of phenyl group regions. The OFET channels were next formed by VUV irradiation of the TMS-DTMS layer to remove the alkyl groups from the OFET channel regions, and phenyltrichlorosilane (PTS) was finally applied to the channel regions, thus obtaining a phenyl SAM. A molecular template consisting of repellent and lyophilic regions was thus formed on the PI layer. Source and drain electrodes were then formed, and an organic semiconductor solution of dioctylbenzothienobenzothiophene (C8-BTBT) [77–79] in a chlorobenzene solution was dropped onto the patterned surface, to complete the OFET array by surface-selective deposition (Fig. 9-7).

Figure 9-8(b) shows an optical micrograph of the completed OFET array. As can be seen in Figure 9-8(c), the individual devices are fully isolated from each another by the organic thin-film patterning. Figure 9-8(d) shows a polarized

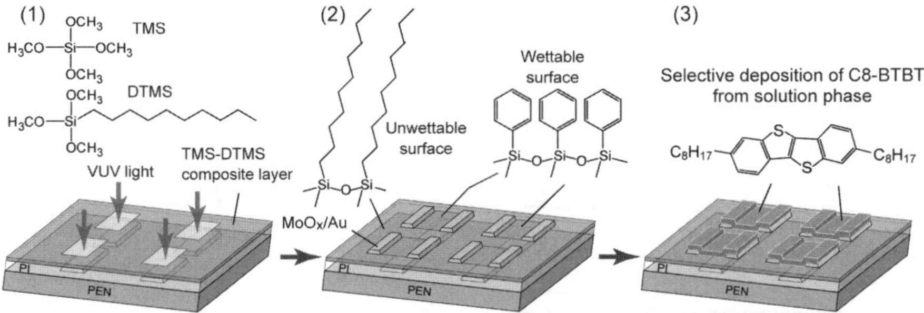

Figure 9-7 Schematic drawing of fabrication process for OFET array formed on plastic substrate by functional surface patterning. (1) TMS-DTMS composite layer formed on surface of PI insulator; surface repellent due to presence of DTMS alkyl chain, except in regions having alkyl groups removed by VUV irradiation. (2) PTS applied to surface, forming pattern of lyophilic SAM regions surrounded by repellent regions, and source and drain electrodes formed. (3) Solution of C8-BTBT is dropped onto the surface, for surface-selective deposition and thus crystallization in transistor channel regions only.

Figure 9-8 OFET array formed on a plastic substrate by surface-selective deposition. (a) Flexing plastic substrate and array. (b) Polarized optical micrograph of array. (c) Enlarged polarized optical micrograph of single device. (d) Polarized optical micrograph image of area around channel of the self-organized OFET. (A full color version of this figure appears in the color plate section.)

optical micrograph image of the channel area where the crystals have grown to a maximum size of approximately 100 μm. Typical device output and transfer characteristics are shown in Figure 9-9. The drain current exhibits a nonlinear increase at low drain voltage because of the relatively large contact resistance of the bottom contact electrodes. At high drain voltage, the drain current exhibits saturation characteristics, indicating ideal operation as a MOSFET. In terms of transfer characteristics, the current increases at an extremely steep rate in the subthreshold region, and there is no sign of hysteresis. The organic semiconductor

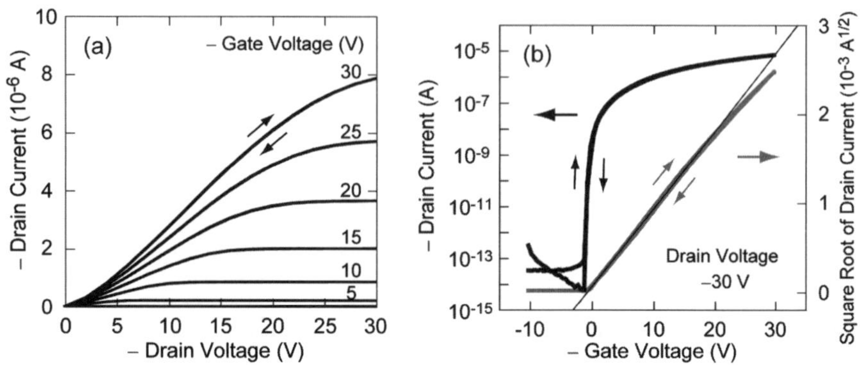

Figure 9-9 Output characteristics (a) and transfer characteristics (b) of OFET self-organized on plastic substrate by surface selective deposition.

layer pattern resulted in ideal switching characteristics, with an off current of approximately 10^{-14} A and an on/off ratio of 10^9.

9.6. EVALUATION OF VARIANCE IN CHARACTERISTICS OF SELF-ORGANIZED OFETs

A major practical problem related to OFETs is a tendency for large variation in device characteristics. In this light, electrical characteristics of 56 devices formed by surface-selective deposition were investigated for variation in key performance parameters. Figure 9-10 shows histograms of the μ_{FET}, V_T, and S of these devices. They are shown separately, as no correlation of note was found among the parameters. The variation in V_T and S was extremely small. The V_T was approximately 0 V for all devices, and all the data were within a range of ± 2 V. The S showed an average value of 0.18 V/dec, which is itself extremely small, and the standard deviation was 0.34. The variation in μ_{FET} was somewhat larger, however, with values ranging from 0.2 to 1.5 cm^2/V s.

As can be seen from the nonlinear rise in drain current shown in Figure 9-9(a), the self-organized OFETs exhibit an extremely large contact resistance at the bottom contact electrode and semiconductor interfaces. The rather large variation in mobility found for these OFETs may be attributable to large variation in their contact resistance. It may therefore be possible to reduce the variation in this characteristic, by further reducing the contact resistance.

9.7. INVERTER CIRCUIT CONFIGURED FROM SELF-ORGANIZED OFETs

The high level of device characteristics and homogeneity in V_T and S of self-organized OFETs indicates the feasibility of practical electronic circuit fabrication using these devices. Figure 9-11(a) shows an inverter circuit built with self-organized OFETs, and Figure 9-11(b) and 9-11(c) show its voltage transfer characteristics and inverter gain. Since the fabricated devices are both p-type FETs, the two devices shown in Figure 9-11(a) operate as load and drive transistors. The inverter exhibits steep switching characteristics even at a relatively low supply voltage (V_{DD}), and has a gain of 20 at a supply voltage of -15 V. These excellent characteristics are attributable to the outstanding subthreshold characteristics of the self-organized OFETs.

9.8. ALL-SOLUTION-PROCESSED ASSEMBLY OF OFET ARRAYS

We have also developed an all-solution fabrication process for OFET arrays, in which all components of devices spontaneously assemble with desired geometry. By formation of a surface molecular template as shown in the previous sections, complete OFET arrays have been patterned on a substrate, including gate and source/drain electrodes, active semiconductor layers, and other constituents in specified regions. Using a plastic substrate and a polymer insulating layer, we were thus able to produce a flexible OFET array by an all-solution process without using vacuum apparatuses.

In this process, we used a PEN sheet as the flexible plastic substrate. All processing was therefore performed at or below the PEN glass transition temperature of 150°C. Nanoparticle silver colloid (Nano-Ag) ink was used for the gate, source, and drain electrodes, and a soluble small molecule, C8-BTBT, was used for the semiconductor layers. We first performed the gate electrode patterning on the lyophobic PEN substrate surface. The gate assembly regions were exposed to VUV irradiation through a metal mask under ambient atmosphere, and the exposed surface regions were then made lyophilic by chemical surface modification. When the Nano-Ag ink was next applied to the substrate with an applicator, it thus selectively adhered only to these lyophilic regions. The substrate was then annealed at 150°C, thus forming the gate electrodes as shown in

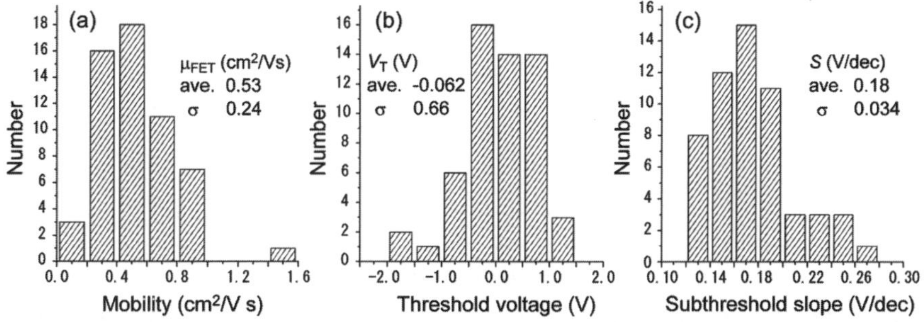

Figure 9-10 Histograms of variation in mobility (a), threshold voltage (b), and subthreshold slope (c) of OFETs self-organized on plastic substrate by surface-selective deposition.

Figure 9-12. A cross-linked polyvinylphenol (PVP) thin film of 505 nm thickness was applied as the gate insulator by spin-coating, together with a thin film of cardo epoxy polymer (approximately 10 nm thick) as the adhesion layer. Here poly(melamine-coformaldehyde) was used as a cross-linker of PVP [80], and the firing temperature for the insulator layer was held to 150°C or lower to avoid damage to the plastic substrate. The functional patterning of the gate insulator surface was performed in the following steps. First, to render the gate insulator surface lyophobic, a fluoroalkylsilane (FAS) SAM was formed by vapor-phase reaction. The FAS groups were then removed from the regions of the source and drain electrodes, and the surface thus rendered lyophilic, by exposure to VUV through a metal mask. In the same way as for the gate electrodes, the Nano-Ag ink was next applied, and thus selectively deposited only in these lyophilic regions. The assembly was then baked at 150°C, thus completing the formation of the source and drain electrodes (Fig. 9-12). Semiconductor layer assembly was next performed, using a C8-BTBT 2 wt% chlorobenzene solution. First, the channel regions were exposed to VUV irradiation, and after removal of the FAS groups, lyophilic phenyl-group regions were formed by vapor-phase reaction with phenyltrichrolosilane (PTS). The phenyl-group formation is indeed essential, to lower the level of trap formation between the organic semiconductor and

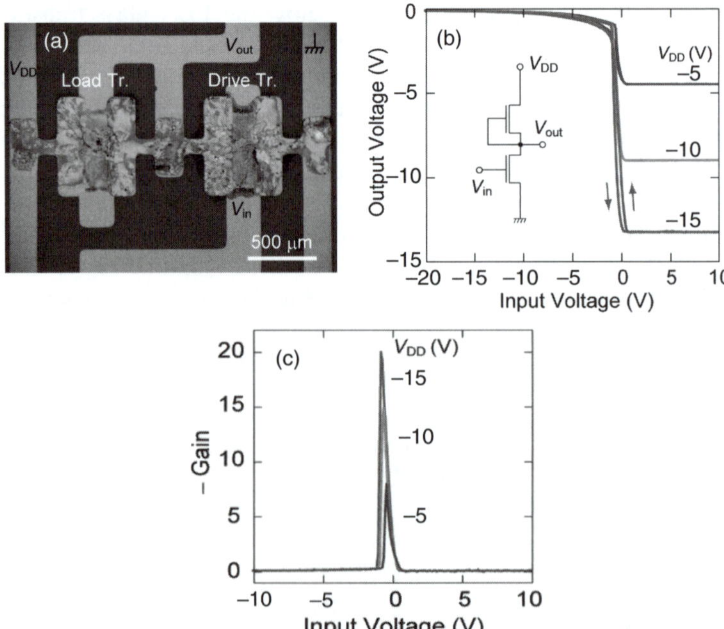

Figure 9-11   Inverter circuit self-organized on plastic substrate by surface-selective deposition. (a) Optical micrograph. (b) Voltage transfer characteristics (inset shows schematic circuit diagram.) (c) Gain of the inverter circuit with respect to supply voltage.

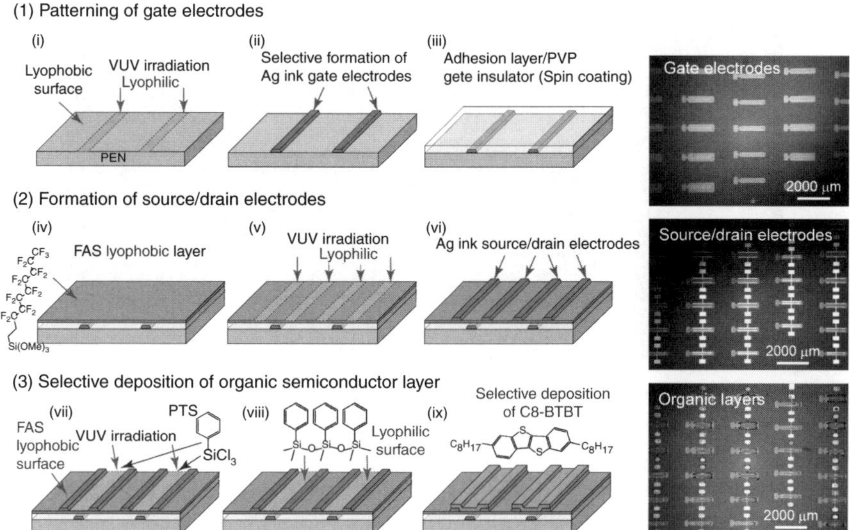

Figure 9-12 Schematic of OFET array fabrication on plastic substrate by all-solution process. (1) Surface-selective patterning and formation of gate electrodes on lyophobic PEN substrate by: (i) VUV irradiation of the gate electrode regions to render them lyophilic; (ii) selective deposition of nanoparticle silver on the lyophilic regions to form the gate electrodes; and (iii) formation of the gate insulator by spin coating. (2) Formation of source and drain electrodes by: (iv) surface coating by fluoroalkylsilane to form lyophobic SAM; (v) VUV irradiation of the source and drain electrode regions to render them lyophilic, and (vi) selective deposition of nanoparticle silver to form the source and drain electrodes. (3) Formation of organic semiconductor layers by: (vii) VUV irradiation of the channel regions followed by coating with PTS, a lyophilic SAM, to render OFET channel regions lyophilic; (viii) thus the pattern of lyophilic channel regions surrounded by lyophobic regions was obtained; and (ix) C8-BTBT solution drop-casting on the surface to selectively crystallize molecules in the transistor channel regions only, as a process of selective deposition. (A full color version of this figure appears in the color plate section.)

the insulator and thus greatly improve the bias stress characteristics of the final device. In the final step, the C8-BTBT solution was drop-cast on the surface-patterned substrate, thus selectively coating the channel regions with the organic semiconductor layer and completing the OFET array (Fig. 9-12).

With a plastic substrate and a polymer insulator, the result is a fully flexible OFET array, as shown in the inset in Figure 9-13(a). The self-assembled OFET array obtained by the all-solution process is shown in the optical micrograph in Figure 9-13(a). The total patterning of the electrodes and the semiconductor layers results in complete separation between the devices, which is essential for reduction of gate leakage current and interdevice cross talk but is difficult to achieve in other organic semiconductor solution processes. A polarized optical micrograph of a single device is shown in Figure 9-13(b).

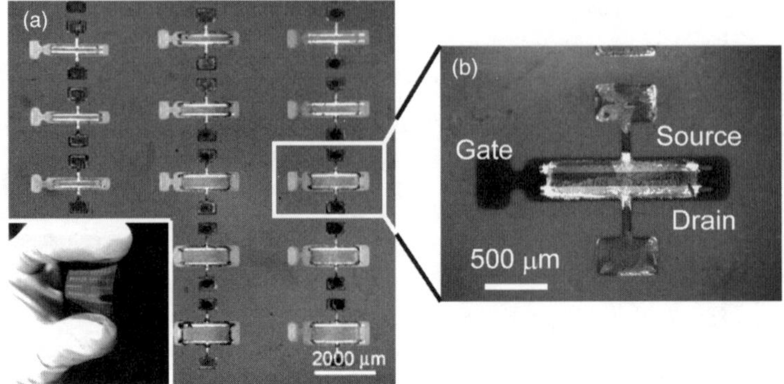

Figure 9-13 OFET array assembled on plastic substrate by surface-selective deposition in the all-solution process. (a) Optical micrograph of OFET array obtained by the total self-assembly process. Inset shows the assembled OFET array on the plastic substrate. (b) Enlarged polarized optical micrograph of a single device.

The output and transfer characteristics of the self-assembled OFET are shown in Figure 9-14(a) and (b). The nonlinear rise in drain current in the low-drain-voltage region [Fig. 9-14(a)] is attributable to the high contact resistance of the Nano-Ag electrodes. Since a C8-BTBT film has deep valence band (VB) level ($E_V = 5.7$ eV), a high injection barrier may exist at the metal/organic interface because of energy mismatch between the Fermi level of Ag electrodes and the VB of organic semiconductor. It would be necessary to use a material with a larger work function for the electrodes or to form a charge injection layer on the electrode surface for further improvement in charge injection. In the high-drain-voltage regions, on the other hand, the drain current shows saturation characteristics, which confirms ideal MOSFET operation. As shown in Figure 9-14(b), hysteresis-free transfer characteristics were achieved with good

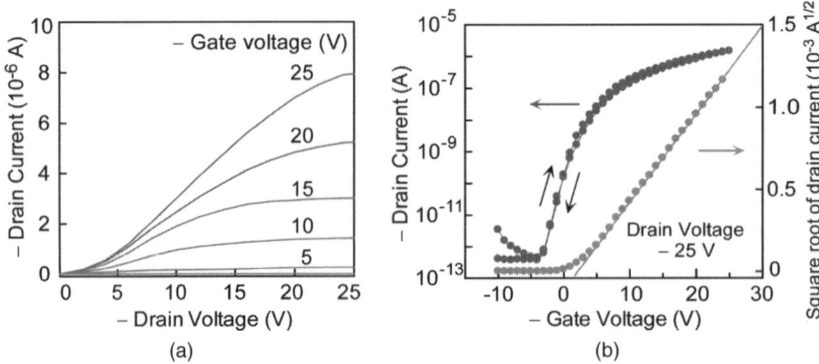

Figure 9-14 Typical output (a) and transfer (b) characteristics of OFETs self-assembled on plastic substrate by the all-solution process.

subthreshold properties. The gate leak current is substantially reduced to the order of 10^{-11} A by the complete patterning of the electrodes and the semiconductor layers (not shown), and thus an on-off ratio of 10^6 was obtained. The typical μ_{FET} of self-assembled OFETs was estimated to be 0.016 cm^2/V · s. For further heightening of device characteristics, it will be necessary to further lower the contact resistance.

9.9. CONCLUSION

In this chapter we have described region-selective organic crystal growth with functional surface patterning using a SAM and its application to the formation of a number of devices. Surface-selective deposition can be performed on practically any base material on which a molecular template can be formed, and its use with plastic substrates and polymer insulators as base materials will enable the production of flexible electronic devices by a solution deposition. With its capability for self-organized formation of organic electronic devices by a solution process on a flexible base material, it has the potential to become the basic technology for development and implementation of roll-to-roll device fabrication.

With the rapid advances occurring in information technology, the need is also rising rapidly for development of new electronic devices that are both human- and eco-friendly and their production technology. The simple, convenient processes described in this chapter, consisting essentially of ultraviolet irradiation and solution coating, are expected to dramatically change electronics production, which currently requires processes that consume large amounts of energy and have a high environmental impact, through the fabrication of electronic devices by molecular interactions for self-assembly, and to lead to further future reductions in energy consumption.

ACKNOWLEDGMENTS

The authors would like to thank Prof. Takimiya (Hiroshima University) and Dr. Maeda (Dai Nippon Printing Co.) for their generous cooperation. We would also like to thank Dr. T. Miyadera (National Institute of Advanced Industrial Science and Technology) and Prof. S. D. Wang (Soochow University) for fruitful discussions. We are grateful to Dr. Ikeda and Mr. Kanoh (Nippon Kayaku Co.) for providing the C8-BTBT. This study was supported in part by Grants-In-Aid for Scientific Research (Grant Nos. 17069004, 21241038, and 21750197) from the Ministry of Education, Culture, Sport, Science, and Technology of Japan.

REFERENCES

1. Horowitz, G. Organic field-effect transistors. *Adv. Mater.* **1998**, *10*, 365.

2. Dimitrakopoulos, C. D. and Malenfant, P. R. L. Organic thin film transistors for large area electronics. *Adv. Mater.*, **2002**, *14*, 99.

3. Forrest, S. R. The path to ubiquitous and low-cost organic electronic appliances on plastic. *Nature* **2004**, *428*, 911.

4. Braga, D. and Horowitz, G. High-performance organic field-effect transistors. *Adv. Mater.* **2009**, *21*, 1473.

5. Gelinck, G. H., Huitema, H. E. A.,. Veenendaal, E., Cantatore, E., Schrijnemakers, L., Putten, J. B. P. H., Geuns, T. C. T., Beenhakkers, M., Giesbers, J. B., and Huisman, B.-H. Flexible active-matrix displays and shift registers based on solution-processed organic transistors. *Nat. Mater.* **2004**, *3*, 106.

6. Eder, F., Klauk, H., Halik, M., Zschieschang, U., Schmid, G., and Dehm, C. Organic electronics on paper. *Appl. Phys. Lett.* **2004**, *84*, 2673.

7. Yagi, I., Hirai, N., Noda, M., Imaoka, A., Miyamoto, Y., Yoneya, N., Nomoto, K., Kasahara, J., Yumoto, A., and Urabe, T. A full-color, top-emission AM-OLED display driven by OTFTs. *SID Symposium Digest* **2007**, *38*, 1753.

8. Sekitani, T., Takamiya, M., Noguchi, Y., Nakano, S., Kato, Y., Sakurai, T., and Someya, T. A large-area wireless power-transmission sheet using printed organic transistors and plastic MEMS switches. *Nat. Mater.* **2007**, *6*, 413.

9. Bao, Z., Rogers, J. A., and Katz, H. E. Printable organic and polymeric semiconducting materials and devices. *J. Mater. Chem.* **1999**, *9*, 1895.

10. Pardo, D. A., Jabbour, G. E., and Peyghambarian, N. Application of screen printing in the fabrication of organic light-emitting devices. *Adv. Mater.* **2000**, *12*, 1249.

11. Kim, C. Burrows, P. E., and Forrest, S. R. Micropatterning of organic electronic devices by cold-welding. *Science* **2000**, *288*, 831.

12. Chabinyc, M. L., Salleo, A., Wu, Y., Liu, P., Ong, B. S., Heeney, M., and McCulloch, I. Lamination method for the study of interfaces in polymeric thin film transistors. *J. Am. Chem. Soc.* **2004**, *126*, 13928.

13. Dickey, K. C., Subramanian, S., Anthony, J. E., Han, Li-H., Chen, S., and Loo, Y.-L. Large-area patterning of a solution-processable organic semiconductor to reduce parasitic leakage and off currents in thin-film transistors. *Appl. Phys. Lett.* **2007**, *90*, 244103.

14. Liu, S., Becerril, H. A., LeMieux, M. C., Wang, W. M., Oh, J. H., and Bao, Z. Direct patterning of organic-thin-film-transistor arrays via a dry-taping approach. *Adv. Mater.* **2009**, *21*, 1266.

15. Chabinyc, M. L., Wong, W. S., Salleo, A., Paul, K. E., and Street, R. A. Organic polymeric thin-film transistors fabricated by selective dewetting. *Appl. Phys. Lett.* **2002**, *81*, 4260.

16. Choi, H. Y., Kim, S. H., and Jang, J. Self-organized organic thin-film transistors on plastic. *Adv. Mater.* **2004**, *16*, 732.

17. Liu, S., Wang, W. M., Mannsfeld, S. C. B., Locklin, J., Erk, P., Gomez, M., Richter, F., and Bao, Z. Solution-assisted assembly of organic semiconducting single crystals on surfaces with patterned wettability. *Langmuir* **2007**, *23*, 7428.

18. Gundlach, D. J., Royer, J. E., Park, S. K., Subramanian, S., Jurchescu, O. D., Hamadani, B. H., Moad, A. J., Kline, R. J., Teague, L. C., Kirillov, O., Richter, C. A., Kushmerick, J. G., Richter, L. J., Parkin, S. R., Jackson, T. N., and Anthony, J. E. Contact-induced crystallinity for high-performance soluble acene-based transistors and circuits. *Nat. Mater.* **2008**, *7*, 216.

19. Minari, T., Kano, M., Miyadera, T., Wang, S. D., Aoyagi, Y., Seto, M., Nemoto, T., Isoda, S., and Tsukagoshi, K. Selective organization of solution-processed organic field-effect transistors. *Appl. Phys. Lett.* **2008**, *92*, 173301.

20. Minari, T., Kano, M., Miyadera, T., Wang, S. D., Aoyagi, Y., and Tsukagoshi, K. Surface selective deposition of molecular semiconductors for solution-based integration of organic field-effect transistors. *Appl. Phys. Lett.* **2009**, *94*, 093307.

21. Mannsfeld, S. C. B., Sharei, A., Liu, S., Roberts, M. E., McCulloch, I., Heeney, M., Bao, Z. Highly efficient patterning of organic single-crystal transistors from the solution phase. *Adv. Mater.* **2008**, *20*, 4044.

22. Kim, S. H., Choi, D., Chung, D. S., Yang, C., Jang, J., Park, C. E., Park, S. H. K. High-performance solution-processed triisopropylsilylethynyl pentacene transistors and inverters fabricated by using the selective self-organization technique. *Appl. Phys. Lett.* **2008**, *93*, 113306.

23. Park, S. K., Mourey, D. A., Subramanian, S., Anthony, J. E., and Jackson, T. N. Non-relief-pattern lithography patterning of solution processed organic semiconductors. *Adv. Mater.* **2008**, *20*, 4145.

24. Chang, K. J., Yang, F. Y., Liu, C. C., Hsu, M. Y., Liao, T. C., Cheng, H. C. Self-patterning of high-performance thin film transistors. *Org. Electronics* **2009**, *10*, 815.

25. Briseno, A. L., Mannsfeld, S. C. B., Ling, M. M., Liu, S., Tseng, R. J., Reese, C., Roberts, M. E., Yang, Y., Wudl, F., and Bao, Z. (2006). Patterning organic single-crystal transistor arrays. *Nature* **2006**, *444*, 913.

26. Aizenberg, J., Black, A. J., and Whitesides, G. M. Control of crystal nucleation by patterned self-assembled monolayers. *Nature* **1999**, *398*, 495.

27. Sugimura, H., Ushiyama, K., Hozumi, A., and Takai, O. Micropatterning of alkyl- and fluoroalkylsilane self-assembled monolayers using vacuum ultraviolet light. *Langmuir* **2000**, *16*, 885.

28. Chua, L.-L., Zaumseil, J., Chang, J.-F., Ou, E. C. W., Ho, P. K. H., Sirringhaus, H., and Friend, R. H. General observation of n-type field-effect behaviour in organic semiconductors. *Nature* **2005**, *434*, 194.

29. Yagi, I., Tsukagoshi, K., and Aoyagi, Y. Modification of the electric conduction at the pentacene/SiO_2 interface by surface termination of SiO_2. *Appl. Phys. Lett.* **2005**, *86*, 103502.

30. Goldmann, C., Krellner, C., Pernstich, K. P., Haas, S., Gundlach, D. J., and Batlogg, B. Determination of the interface trap density of rubrene single-crystal field-effect transistors and comparison to the bulk trap density. *J. Appl. Phys.* **2006**, *99*, 034507.

31. Kumaki, D., Yahiro, M., Inoue, Y., and Tokito, S. Air stable, high performance pentacene thin-film transistor fabricated on SiO_2 gate insulator treated with β-phenethyltrichlorosilane. *Appl. Phys. Lett.* **2007**, *90*, 133511.

32. Mathijssen, S. G. J., Kemerink, M., Sharma, A., Cölle, M., Bobbert, P. A., Janssen, R. A. J., and Leeuw, D. M. (2008). Charge trapping at the dielectric of organic transistors visualized in real time and space. *Adv. Mater.* **2008**, *20*, 975.

33. Schoonveld, W. A., Oostinga, J. B., Vrijmoeth, J., and Klapwijk, T. M. Charge trapping instabilities of sexithiophene thin film transistors. *Synth. Met.* **1999**, *101*, 608.

34. Matters, M., Leeuw, D. M., Herwig, P. T., and Brown, A. R. Bias-stress induced instability of organic thin film transistors. *Synth. Met.* **1999**, *102*, 998.

35. Zilker, S. J., Detcheverry, C., Cantatore, E., and Leeuw, D. M. (2001). Bias stress in organic thin-film transistors and logic gates. *Appl. Phys. Lett.* **2001**, *79*, 1124.

36. Salleo, A. and Street, R. A. Light-induced bias stress reversal in polyfluorene thin-film transistors. *J. Appl. Phys.* **2003**, *94*, 471.

37. Gomes, H. L., Stallinga, P., Dinelli, F., Murgia, M., Biscarini, F., Leeuw, D. M., Muck, T., Geurts, J., Molenkamp, L. W., and Wagner, V. Bias-induced threshold voltages shifts in thin-film organic transistors. *Appl. Phys. Lett.* **2004**, *84*, 3184.

38. Kagan, C. R., Afzail, A., and Graham, T. O. Operational and environmental stability of pentacene thin-film transistors. *Appl. Phys. Lett.* **2005**, *86*, 193505.

39. Salleo, A., Endicott, F., and Street, R. A. Reversible and irreversible trapping at room temperature in poly(thiophene) thin-film transistors. *Appl. Phys. Lett.* **2005**, *86*, 263505.

40. Sekitani, T., Iba, S., Kato, Y., Noguchi, Y., Someya, T., and Sakurai, T. Suppression of DC bias stress-induced degradation of organic field-effect transistors using postannealing effects. *Appl. Phys. Lett.* **2005**, *87*, 073505.

41. Jung, T., Dodabalapur, A., Wenz, R., and Mohapatra, S. Moisture induced surface polarization in a poly(4-vinyl phenol) dielectric in an organic thin-film transistor. *Appl. Phys. Lett.* **2005**, *87*, 182109.

42. Goldmann, C., Gundlach, D. J., and Batlogg, B. Evidence of water-related discrete trap state formation in pentacene single-crystal field-effect transistors. *Appl. Phys. Lett.* **2006**, *88*, 063501.

43. Hwang, D. K., Lee, K., Kim, J. H., Im, S., Park, Ji H., and Kim, E. Comparative studies on the stability of polymer versus SiO_2 gate dielectrics for pentacene thin-film transistors. *Appl. Phys. Lett.* **2006**, *89*, 093507.

44. Kalb, W. L., Mathis, T., Haas, S., Stassen, A. F., and Batlogg, B. Organic small molecule field-effect transistors with Cytop gate dielectric: Eliminating gate bias stress effects. *Appl. Phys. Lett.* **2007**, *90*, 092104.

45. Miyadera, T., Wang, S. D., Minari, T., Tsukagoshi, K., and Aoyagi, Y. Charge trapping induced current instability in pentacene thin film transistors: Trapping barrier and effect of surface treatment. *Appl. Phys. Lett.* **2008**, *93*, 033304.

46. Sirringhaus, H. Reliability of organic field-effect transistors. *Adv. Mater.* **2009**, *21*, 3859.

47. Horowitz, G., Garnier, F., Yassar, A., Hajlaoui, R., and Kouki, F. Field-effect transistor made with a sexithiophene single crystal. *Adv. Mater.* **1996**, *8*, 52.

48. Schoonveld, W. A., Vrijmoeth, J., and Klapwijk, T. M. Intrinsic charge transport properties of an organic single crystal determined using a multiterminal thin-film transistor. *Appl. Phys. Lett.* **1998**, *73*, 3884.

49. Schoonveld, W. A., Wildeman, J., Fichou, D., Bobbert, P. A., Wees, B. J., and Klapwijk, T. M. Coulomb-blockade transport in single-crystal organic thin-film transistors. *Nature* **2008**, *404*, 977.

50. Chwang, A. B., and Frisbie, C. D. Field effect transport measurements on single grains of sexithiophene: Role of the contacts. *J. Phys. Chem. B*. **2000**, *104*, 12202.

51. Ichikawa, M., Yanagi, H., Shimizu, Y., Hotta, S., Suganuma, N., Koyama, T., and Taniguchi, Y. Organic field-effect transistors made of epitaxially grown crystals of a thiophene/phenylene co-oligomer. *Adv. Mater.* **2002**, *14*, 1272.

52. Podzorov, V., Pudalov, V. M., and Gershenson, M. E. Field-effect transistors on rubrene single crystals with parylene gate insulator. *Appl. Phys. Lett.* **2003**, *82*, 1739.

53. Takeya, J., Goldmann, C., Haas, S., Pernstich, K. P., Ketterer, B., and Batlogg, B. Field-induced charge transport at the surface of pentacene single crystals: A method to study charge dynamics of two-dimensional electron systems in organic crystals. *J. Appl. Phys*. **2003**, *94*, 5800.

54. Podzorov, V., Sysoev, S. E., Loginova, E., Pudalov, V. M., and Gershenson, M. E. Single-crystal organic field effect transistors with the hole mobility 8cm^2/ Vs. *Appl. Phys. Lett*. **2003**, *83*, 3504.

55. Boer, R. W. I., Klapwijk, T. M., and Morpurgo, A. F. Field-effect transistors on tetracene single crystals. *Appl. Phys. Lett*. **2003**, *83*, 4345.

56. Newman, C. R., Chesterfield, R. J., Merlo, J. A., and Frisbie, C. D. Transport properties of single-crystal tetracene field-effect transistors with silicon dioxide gate dielectric. *Appl. Phys. Lett*. **2004**, *85*, 422.

57. Minari, T., Nemoto, T., and Isoda, S. Fabrication and characterization of single-grain organic field-effect transistor of pentacene. *J. Appl. Phys*. **2004**, *96*, 769.

58. Minari, T., Nemoto, T., and Isoda, S. Temperature and electric-field dependence of the mobility of a single-grain pentacene field-effect transistor. *J. Appl. Phys*. **2007**, *99*, 034506.

59. Hulea, I. N., Fratini, S., Xie, H., Mulder, C. L., Iossad, N. N., Rastelli, G., Ciuchi, S., and Morpurgo, A. F. Tunable Fröhlich polarons in organic single-crystal transistors. *Nat. Mater*. **2006**, *5*, 982.

60. Takeya, J., Yamagishi, M., Tominari, Y., Hirahara, R., Nakazawa, Y., Nishikawa, T., Kawase, T., Shimoda, T., and Ogawa, S. Very high-mobility organic single-crystal transistors with in-crystal conduction channels. *Appl. Phys. Lett*. **2007**, *90*, 102120.

61. Minari, T., Miyadera, T., Tsukagoshi, K., Hamano, T., Aoyagi, Y., Yasuda, R., Nomoto, K., Nemoto, T., and Isoda, S. Scaling effect on the operation stability of short-channel organic single-crystal transistors. *Appl. Phys. Lett*. **2007**, *91*, 063506.

62. Calhoun, M. F., Sanchez, J., Olaya, D., Gershenson, M. E., and Podzorov, V. (**2007**). Electronic functionalization of the surface of organic semiconductors with self-assembled monolayers. *Nat. Mater*. **2007**, *7*, 84

63. Kawasugi, Y., Yamamoto, H. M., Hosoda, M., Tajima, N., Fukunaga, T., Tsukagoshi, K., and Kato, R. Strain-induced superconductor/insulator transition and field effect in a thin single crystal of molecular conductor. *Appl. Phys. Lett*. **2008**, *92*, 243508.

64. Minari, T., Seto, M., Nemoto, T., Isoda, S., Tsukagoshi, K., and Aoyagi, Y. Molecular-packing-enhanced charge transport in organic field-effect transistors based on semi-conducting porphyrin crystals. *Appl. Phys. Lett*. **2007**, *91*, 123501.

65. Reese, C., Roberts, M. E., Parkin, S. R., and Bao, Z. Tuning crystalline solid-state order and charge transport via building-block modification of oligothiophenes. *Adv. Mater*. **2009**, *21*, 3678.

66. Takenobu, T., Bisri, S. Z., Takahashi, T., Yahiro, M., Adachi, C., and Iwasa, Y. High current density in light-emitting transistors of organic single crystals. *Phys. Rev. Lett*. **2008**, *100*, 066601.

67. Sundar, V. C., Zaumseil, J., Podzorov, V., Menard, E., Willett, R. L., Someya, T., Gershenson, M. E., and Rogers, J. A. Elastomeric transistor stamps: Reversible probing of charge transport in organic crystals. *Science* **2004**, *303*, 1644.

68. Lee, J. Y., Roth, S., and Park, Y. W. Anisotropic field effect mobility in single crystal pentacene. *Appl. Phys. Lett*. **2006**, *88*, 252106.

69. Reese, C. and Bao, Z. High-resolution measurement of the anisotropy of charge transport in single crystals. *Adv. Mater.* **2007**, *19*, 4535.

70. Kloc, Ch., Simpkins, P. G., Siegrist, T., and Laudise, R. A. Physical vapor growth of centimeter-sized crystals of α-hexathiophene. *J. Cryst. Growth* **1997**, *182*, 416.

71. Gelinck, G. H., Huitema, H. E. A., Veenendaal, E., Cantatore, E., Schrijnemakers, L., Putten, J. B. P. H., Geuns, T. C. T., Beenhakkers, M., Giesbers, J. B., Huisman, B.-H., Meijer, E. J., Benito, E. M., Touwslager, F. J., Marsman, A. W., Rens B. J. E., and Leeuw, D. M. Flexible active-matrix displays and shift registers based on solution-processed organic transistors. *Nat. Mater.* **2004**, *3*, 106.

72. Zhou, L., Wanga, A., Wu, S.-C., Sun, J., Park, S., and Jackson, T. N. All-organic active matrix flexible display. *Appl. Phys. Lett.* **2006**, *88*, 083502.

73. Sheraw, C. D., Zhou, L., Huang, J. R., Gundlach, D. J., Jackson, T. N., Kane, M. G., Hill, I. G., Hammond, M. S., Campi, J., Greening, B. K., Francl, J., and West, J. Organic thin-film transistor-driven polymer-dispersed liquid crystal displays on flexible polymeric substrates. *Appl. Phys. Lett.* **2002**, *80*, 1088.

74. Amundson, K., Ewing, J., Kazlas, P., McCarthy, R., Albert, J. D., Zehner, R., Drzaic, P., Rogers, J., Bao, Z., and Baldwin, K. Flexible, active-matrix display constructed using a microencapsulated electrophoretic material and an organic-semiconductor-based backplane. *SID Symposium Digest* **2001**, *32*, 160.

75. Lieshout, P. J. G., Huitema, H. E. A., Veenendaal, E., Schrijnemakers, L. R. R., Gelinck, G. H., Touwslager, F. J., and Cantatore, E. System-on-plastic with organic electronics: A flexible QVGA display and integrated drivers. *SID Symposium Digest* **2004**, *35*, 1290.

76. Maeda, H., Matsuoka, M., Nagae, M., Honda, H., and Kobayashi, H. Active-matrix backplane with printed organic TFTs for QR-LPD. *SID Symposium Digest* **2007**, *38*, 1749.

77. Ebata, H., Izawa, T., Miyazaki, E., Takimiya, K., Ikeda, M., Kuwabara, H., and Yui, T. Highly soluble [1]benzothieno[3,2-b]benzothiophene (BTBT) derivatives for high-performance, solution-processed organic field-effect transistors. *J. Am. Chem. Soc.* 2007, *129*, 15732.

78. Izawa, T., Miyazaki, E., and Takimiya, K. Molecular ordering of high-performance soluble molecular semiconductors and re-evaluation of their field-effect transistor characteristics. *Adv. Mater.* **2008**, *20*, 3388.

79. Kano, M., Minari, T., and Tsukagoshi, K. Improvement of subthreshold current transport by contact interface modification in *p*-type organic field-effect transistors. *Appl. Phys. Lett.* **2009**, *94*, 143304.

80. Klauk, H., Halik, M., Zschieschang, U., Schmid, G., and Radlik, W. High-mobility polymer gate dielectric pentacene thin film transistors. *J. Appl. Phys.* **2002**, *92*, 5259.

Self-Organized Organic Semiconductors: From Materials to Device Applications, First Edition.
Edited by Quan Li.
© 2011 John Wiley & Sons, Inc. Published 2011 by John Wiley & Sons, Inc.